PERFORMANCE MEASUREMENT AND MANAGEMENT CONTROL: GLOBAL ISSUES

STUDIES IN MANAGERIAL AND FINANCIAL ACCOUNTING

Series Editor: Marc J. Epstein

Recent Volumes:

Volume 6:	Throughout Modeling: Financial Information Used by Decision Makers
Volume 7:	Applications of Fuzzy Sets and the Theory of Evidence to Accounting II
Volume 8:	Corporate Governance, Accountability, and Pressures to Perform: An International Study
Volume 9:	The January Effect and Other Seasonal Anomalies: A Common Theoretical Framework
Volume 10:	Organizational Change and Development in Management Control Systems: Process Innovation for Internal Auditing and Management Accounting
Volume 11:	US Individual Federal Income Taxation: Historical, Contemporary and Prospective Policy Issues
Volume 12:	Performance Measurement and Management Control: A Compendium of Research
Volume 13:	Information Asymmetry: A Unifying Concept for Financial and Managerial Accounting Theories
Volume 14:	Performance Measurement and Management Control: Superior Organization Performance
Volume 15:	A Comparative Study of Professional Accountants' Judgements
Volume 16:	Performance Measurement and Management Control: Improving Organizations and Society
Volume 17:	Non-Financial Performance Measurement and Management Practices in Manufacturing Firms: A Comparative International Analysis
Volume 18:	Performance Measurement and Management Control: Measuring and Rewarding Performance
Volume 19:	Managerial Attitudes Toward a Stakeholder Prominence within a Southeast Asia Context
Volume 20:	Performance Measurement and Management Control: Innovative Concepts and Practices
Volume 21:	Reputation Building, Website Disclosure and the Case of Intellectual Capital
Volume 22:	Achieving Global Convergence of Financial Reporting Standards: Implications from the South Pacific Region
Volume 23:	Globalization and Contextual Factors in Accounting: The Case of Germany
Volume 24:	An Organizational Learning Approach to Process Innovations: The Extent and Scope of Diffusion and Adoption in Management Accounting Systems

STUDIES IN MANAGERIAL AND FINANCIAL
ACCOUNTING VOLUME 25

PERFORMANCE MEASUREMENT AND MANAGEMENT CONTROL: GLOBAL ISSUES

EDITED BY

ANTONIO DAVILA
University of Navarra, Barcelona

MARC J. EPSTEIN
Rice University, TX, USA

JEAN-FRANÇOIS MANZONI
INSEAD Global Leadership Centre, Singapore

Emerald

United Kingdom – North America – Japan
India – Malaysia – China

Emerald Group Publishing Limited
Howard House, Wagon Lane, Bingley BD16 1WA, UK

First edition 2012

British Library Cataloguing in Publication Data
A catalogue record for this book is available from the British Library

ISBN: 978-1-78052-910-3
ISSN: 1479-3512 (Series)

CONTENTS

LIST OF CONTRIBUTORS

Maria Concepcion Alvarez-Dardet	Department of Accounting and Finance, Pablo Olavide University, Spain
Maria Ceu-Alves	NECE, University of Beira Interior, Covilhä, Portugal
Pedro Araujo	Department of Accounting and Finance, University of Cádiz, Spain
Selena Aureli	Faculty of Economics, Urbino University, Italy
Regino Banegas	Department of Business Administration, University of Castilla-La Mancha, Spain
Adriano Leal Bruni	Federal University of Bahia (UFBA), Brazil
Franco Cescon	Department of Economics and Statistics, University of Udine, Italy
Antonio Davila	IESE Business School, University of Navarra, Barcelona, Spain
Marc J. Epstein	Jones Graduate School of Business, Rice University, Houston, TX, USA
Juliano Almeida De Faria	Federal University of Bahia (UFBA), Brazil
Raquel Florez	Department of Accounting and Finance, Pablo Olavide University, Spain
Luís Francisco	Management School, Polytechnic Institute of Tomar, Portugal
Pierre Guernaccini	Public Research Centre Henri Tudor, Luxembourg

Jane Q. He	Department of Accounting and Corporate Governance Faculty of Business and Economics, Macquarie University, Sydney, Australia
Thierry Hirtz	Public Research Centre Henri Tudor, Luxembourg
Andreas Hoffjan	Chair of Management Accounting and Control, TU Dortmund University, Germany
Jennifer Kunz	Department of Accounting, Goethe University Frankfurt am Main, Germany
Maik Lachmann	Junior Professorship of Management Accounting and Control, TU Dortmund University, Germany
Chong M. Lau	Department of Accounting and Corporate Governance, Faculty of Business and Economics, Macquarie University, Sydney, Australia
Antonio Leotta	Faculty of Economics, University of Catania, Italy
Stefan Linder	Department of Accounting & Management Control, ESSEC Business School, Paris, France
Montserrat Manzaneque	Department of Business Administration, University of Castilla-La Mancha, Spain
Jean-François Manzoni	INSEAD Global Leadership Centre, Singapore
Elena Merino	Department of Business Administration, University of Castilla-La Mancha, Spain
Andrei Mineev	Bodø Graduate School of Business, University of Nordland, Norway

Gilca Garcia De Oliveira	Federal University of Bahia (UFBA), Brazil
Emilia Pentti	School of Economics, Department of Accounting, Aalto University, Finland
Jan O. Piontkowski	Chair of Management Accounting and Control, TU Dortmund University, Germany
Juan M. Ramon	Department of Accounting and Finance, Pablo Olavide University, Spain
Daniela Ruggeri	Faculty of Economics, University of Catania, Italy
Federica Salvatori	Department of Business Administration, University of Modena and Reggio Emilia, Italy
Jose M. Sanchez	Department of Business Organization, University of Cádiz, Spain
Márcio Santos Sampaio	Federal University of Bahia (UFBA), Brazil
Lukas D. Schuchardt	Chair of Management Accounting and Control, TU Dortmund University, Germany
Rodrigo Silva De Souza	Federal University of Bahia (UFBA), Brazil
Sônia Maria Da Silva Gomes	Federal University of Bahia (UFBA), Brazil
Mari Tuomaala	School of Engineering, Department of Energy Technology, Aalto University, Finland
Maria Velez	Department of Accounting and Finance, University of Cádiz, Spain
Tuija Virtanen	School of Economics, Department of Accounting, Aalto University, Finland

PREFACE

This book collects representative papers from the sixth conference on Performance Measurement and Management Control held in Nice, in September 2011. The conference has been running every two years since 2001 with increasing participation each edition. The number of participants as well as the quality of the papers highlight the relevance of this topic in the evolving world of organizational management.

The papers in this book capture the diversity of the conference. Performance measurement and management control is a rich field with multiple topics being investigated all the way from financial measures of performance to behavioral implications of management control systems. This variety of interests requires a variety of research methods. The papers in the conference also reflected diverse research methods, some of them relying heavily on quantitative data while others relied on various types of qualitative methods. Some of them were focused on this year's theme of Performance Measurement and Management Control: Global Issues, while others were on other aspects of the topic. Finally, the conference brings together researchers and practitioners from around the world. This fact reinforces the universal character of the topics presented.

This book is a sample of the topics, research methods, and geographical diversity. The selection process for participating at the conference and a second one to be part of this book has led to a high quality, interesting, and diverse set of readings. The various chapters address important topics for management from a carefully crafted research design. Their reading is interesting and informative.

The first part of the book summarizes the plenary sessions. Marc Epstein provided an engaging talk comparing governance across the globe. He emphasized how governance means different things in different countries and how the associated structures respond to the particular idiosyncrasies of each region. Each approach to governance has its own advantages and disadvantages that Marc carefully detailed. Jean-François Manzoni addressed the intersection between ethics, behavior, and control mechanisms. He highlighted why people may fail to do the right thing and how organizations can use their culture and systems to encourage an ethical behavior. Finally, Antonio Davila's chapter provides an overview of the different topics in the conference highlighting the dynamism of the field and the quality of the research evidence. Performance measurement and management control as reflected in the conference and this book is a dynamic and rich field.

The fourteen selected paper cover a variety of topics. Part II presents interesting papers covering topics around performance measurement in traditional fields such as manufacturing but also applied to emerging fields such as sustainability and public sector management. Part III collects interesting papers in interorganizational settings, a topic that has attracted a lot of attention in the research community as companies use more complex structures around networks of partners to deliver their value to the market. Part IV also addresses a topic that is gathering force in the research community. The papers in this part study how risk management is structured in various organizations. The recent events across the world have highlighted deficiencies in the way we interpreted and managed risk. These papers are a step in better solving this challenge. Part V presents papers around the topic of management control systems as motivational tools. This topic is a traditional one almost since the inception of control as a field of inquiry, yet these papers present novel perspectives on the topic. Finally, Part VI is devoted to important new trends. The papers in this section look at issues of governance, public sector, and nonprofit organizations. All these topics are gaining relevance both in research and in practice. On the latter, society is demanding from public sector and nonprofit organizations to increase the efficiency of their operations; this transformation is only possible through the adaptation (and development) of management tools to the requirements of these organizations.

A conference of this magnitude is only possible through the collaboration of many people. Since the first edition, the *Performance Measurement and Management Control*, EIASM through Graziella Michelante have been a central piece to make the conference happen. Graziella and her team not only execute the operations of the conference flawlessly, but do it with a positive attitude that all participants appreciate. Eric Cauvin has been cochair of the conference for several editions now and has provided valuable input on the research side. His help is also appreciated. CIMA has also been an important partner through these years; both through their financial support and the opportunities that it brings to participants to push forward their research agenda. The support of Margaret de Sosa at Rice University has been critical to the success of both the conference and this volume. We all thank their contribution. We hope that the reader will find in this book information relevant to their research and their practice. As a summary of the themes of the conference, it provides insights into where knowledge on performance measurement and management control is going. These are central aspects to the management of organizations; the progress reflected in this book reinforces the dynamism and relevance of this field going forward.

Antonio Davila
Marc J. Epstein
Jean-François Manzoni
Editors

PART I
PERFORMANCE MEASUREMENT AND MANAGEMENT CONTROL: GLOBAL ISSUES

GOVERNING GLOBALLY: CONVERGENCE, DIFFERENTIATION, OR BRIDGING

Marc J. Epstein

ABSTRACT

Purpose – *As corporations and capital markets become more global, it is increasingly important to understand the differences in corporate governance practices.*

Approach – *This chapter provides a framework for the implementation of corporate governance that can be used globally for study and adaptation. It also describes three corporate governance systems (Anglo-American, Communitarian, and Emerging Markets) and provides an analysis and comparison of how the framework for corporate governance is applied differently, and how success should be evaluated differently, in these three systems. Lastly, it considers the possibility of convergence toward a global system of corporate governance.*

Practical implications – *There is significant heterogeneity in corporate governance worldwide but there are universal aspects, such as roles, responsibilities, inputs, and processes, which result in effective corporate governance. Understanding the similarities and differences enables researchers and managers to work with multiple systems in different*

Performance Measurement and Management Control: Global Issues
Studies in Managerial and Financial Accounting, Volume 25, 3–39

ISSN: 1479-3512/doi:10.1108/S1479-3512(2012)0000025004

countries where corporations and stakeholders have varying objectives, structures, and internal and external determinants.

Value of chapter – *This chapter presents a comparison of the three systems that is critical for further study of global practices. Additionally, the internal and external determinants that impact the varying corporate governance systems are analyzed to more carefully consider the performance measures that account for differences in objectives, motivations, and performance.*

INTRODUCTION

To evaluate corporate governance practices globally and to understand how and when these practices lead to success is of interest to managers and researchers alike. It is also critical for the analysis of organizational performance and for any proposal for the redesign and improvement of corporate government systems.

Over the last 10 years, I (along with Marie-Josee Roy) proposed a model for corporate governance performance that consists of four inputs and six processes that we determined are the critical elements (or keys) for corporate governance success. After a decade, this model is still used by both managers and researchers. Further, the model has proven to be generalizable, applying globally to both for profit and nonprofit organizations. However, the application is significantly different in various countries and under different corporate governance systems. This raises interesting and significant research questions such as which corporate governance determinants and elements are more critical for success under different contexts, company and national policies, and practices. Understanding these differences is critical for researchers and managers for either the study or implementation of corporate governance systems globally.

Further, some have suggested that countries will (and should) move toward convergence of global governance practices. However, some disagree that such convergence could occur in either law or in practice. While there may be some convergence in function, there is most likely to be bridging based on the common practice of good governance.

This chapter provides a framework for the implementation of corporate governance that can be used globally for study and adaptation. It also describes three corporate governance systems (Anglo-American, Communitarian, and Emerging Markets) and provides an analysis and comparison of how the framework for corporate governance is applied differently, and should be measured differently, in these three systems. Lastly, it considers the possibility of convergence toward a global system of corporate governance.

CORPORATE GOVERNANCE CHALLENGES ARE GLOBAL

In January 2009, just six months after Satyam and its politically influential chairman, Ramalinga Raju, were honored by the World Council for Corporate Governance, the company and its senior leadership were the subject of the largest corporate fraud investigation in India's history (Behan, 2009). Shortly after New Year's Day 2009, Raju admitted to fabricating 70 billion rupees ($1.5 billion) of Satyam's assets and 95% of the previous year's revenue. Despite its record for good governance, being listed on the New York Stock Exchange (NYSE), and hiring auditors from a reputable accounting firm, this large, respected outsourcing firm saw its market value fall over 80% in a 24-hour period after its chairman's confession (Kripalani, 2009; Thakurta, 2009). Days after the scandal was announced, the existing board of Satyam was disbanded and new directors were appointed by the government. Long before Raju's confession, shareholders had expressed dissatisfaction with his leadership. Many have charged that the board of directors and its members failed to meet its most basic responsibilities.

Senior executives are not the only perpetrators of corporate fraud. A scandal at Germany's largest car maker, Volkswagen (VW), erupted in 2005 when it was discovered that managers and labor representatives had received improper benefits from the company and its suppliers. A quid pro quo agreement had developed between managers and labor that centered around the important role that union representatives play on German boards. In return, the local government, which was the controlling owner of the corporation, was satisfied with the ability to guarantee regional job security. Despite its control of VW, the state only owned 18% of the corporation (Bloomberg Businessweek, 2005). However, it seems that the other shareholders may not have been receiving equal benefits and that the board seems to have been unwilling to intervene. The VW board was caught in a series of conflicts of interest, and found itself unable or unwilling to protect its minority owners from actions that did not maximize the returns to all.

Not all corporate governance failures involve financial fraud. In 2010, BP, one of the world's largest oil and gas companies, saw its market value fall by over 50% in less than 60 days after an explosion at a deep water drilling rig off the U.S. Gulf Coast. The incident killed 11 employees and resulted in the largest marine oil spill in history. The government investigation of the explosion revealed many causes and guilty parties. The fact that management failed to adequately address known risks prior to the disaster is considered an important factor. Many suggest that the board failed to put the right processes in place to ensure that reasonable safety measures were taken to prevent such a disaster (Casselman & Gold, 2010). The oil spill cost BP billions of dollars in compensation for victims and cleanup costs (Bergin, 2010; Power & Casselman, 2011).

These stories occur far too often in corporations all over the world. Enron, Worldcom, and Tyco in the United States and Parmalat in Italy are significant corporate governance failures that have been extensively discussed. Sometimes these failures are fraud and sometimes the result of poor oversight. However, they have caused us to reexamine the structures, systems, and processes of corporate governance.

Corporate governance is the system by which corporations are directed, controlled, and made accountable to shareholders and other stakeholders (Demirag, 1998). While managers, employees, and owners are often involved in corporate governance, corporate boards are ultimately responsible for ensuring good corporate governance (Behan, Ducharme, & Hossack, 2006; Epstein & Roy, 2002). Corporate governance failures such as those outlined above occur in a wide variety of companies and industries, and in different countries around the world. Despite increasing global attention to corporate governance, failures are likely to continue. There are significant challenges in corporate governance, as these cases demonstrate. However, the principles presented here provide a foundation and guidance for board and corporate practice globally. They are implemented in different ways throughout the world, leading to tremendous diversity between and within countries. An examination of these principles and global practices provides guidance as to improving both board and corporate performance, and understanding cross-national differences in corporate governance.

A MODEL FOR IMPROVING BOARD AND CORPORATE PERFORMANCE

Corporations across the globe all aim to improve their performance, though their measurement of success is often defined differently. To improve board

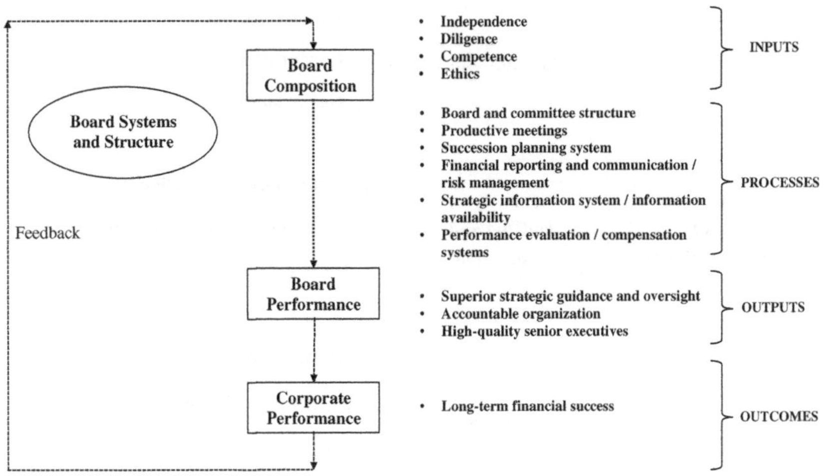

Fig. 1. Corporate Board Performance Model. *Source*: Adapted from Epstein and Roy (2002).

and corporate performance, a clear understanding of the roles and responsibilities of corporate boards is necessary. Though these roles are interpreted differently in different settings, successful implementation requires an understanding of the causal chain that leads to success. The Corporate Board Performance Model (Fig. 1) describes the inputs, processes, outputs, and outcomes that are critical for board and corporate success (Epstein & Roy, 2002, 2003).

ROLES AND RESPONSIBILITIES

Boards have three primary roles and responsibilities (Epstein & Roy, 2010):

(1) ***Accountability***: The board of directors must implement systems to (a) hold corporate managers and executives accountable to the shareholders, (b) ensure that pertinent financial and nonfinancial information is disclosed to both internal and external stakeholders, (c) ensure that the corporation acts ethically and transparently, and (d) create and review corporate policies related to internal controls, risk management, disclosures, compliance, codes of ethics, and board structure.
(2) ***Senior-level staffing and evaluation***: The board of directors has the responsibility to ensure that effective management is in place. In

Fig. 2. Board Role and Responsibilities. *Source*: Adapted from Epstein and Roy (2007).

addition, the board evaluates senior managers, determines appropriate compensation, and creates succession plans for the directors and management team.

(3) ***Strategic oversight*:** The executive team is responsible for developing and implementing corporate strategy. The board, however, has an important role in providing input, reviewing and approving corporate strategy, and providing oversight for major corporate decisions and strategy implementation.

To fulfill its corporate governance responsibilities, the board must actively engage in these roles and responsibilities (Wong, 2011). These are interpreted differently in different countries. However, there is general agreement on the three roles and responsibilities discussed here (Epstein & Roy, 2007, 2010; Pierce, 2009) and outlined and summarized in Fig. 2.

INPUTS AND PROCESSES

Board composition is an important input into corporate performance (Behan et al., 2006; Levrau & Van den Berghe, 2009). Boards must determine the appropriate qualifications and criteria for selecting board members, such as board member independence, competence, diligence, and ethics (Epstein & Roy, 2010). Independence helps prevent conflicts of interest. Additionally, many countries have given independent directors exclusive power in nominating new directors, determining compensation for senior executives, and in overseeing the financial auditing process (New York Stock Exchange, 2003; Organization for Economic Cooperation and Development, 2004; Wong, 2011). Diligence is required, since board members must put in a significant amount of time and preparation in order to participate effectively (Epstein & Roy, 2002; Pozen, 2010a, 2010b). Competence ensures that the board of directors has the skills and knowledge, like financial literacy and relevant experience, to understand specific company issues and challenges (Epstein & Roy, 2006). Ethics are required to prevent fraud and ensure that the corporation acts according to legal and community standards (Epstein & Roy, 2006; Wong, 2011). Now more than ever, a reputation for integrity and honesty is an important quality for board members (Bagley, 2006).

These inputs interact through a series of systems and structures that lead to superior board performance. These include important board structural issues, such as board organization, leadership, size and committee structures; board functional issues, such as productive meetings; and, board systems, such as performance evaluation, compensation, and financial reporting (Epstein & Roy, 2004a). Boards also need to develop an effective strategy to empower board members with the relevant, clear, and complete information necessary to make timely and effective decisions (Epstein & Roy, 2010). These board processes allow the board members to transform their individual and collective competencies into board and corporate performance and are essential to providing boards with relevant and reliable information.

OUTPUTS AND OUTCOMES

In evaluating performance, determining board and executive compensation, and reporting firm financial performance, the board must actively aggregate a diverse information set and analyze it thoroughly. Providing superior strategic guidance, holding the organization and managers accountable, and evaluating senior executives to ensure quality performance are also important outputs for which the board is responsible. With this strategy and the inputs and processes described above, boards can achieve superior performance as well as the desired outcome of long-term corporate financial success.

To effectively fulfill their roles and responsibilities, boards transform composition (inputs) through structure, function, and systems (processes) to achieve better board performance (outputs) and better corporate performance (outcomes). Boards looking to improve outputs and outcomes must have a clear definition of both board and corporate performance. In order to use the model for improved corporate performance, corporate boards must have a working understanding of the desired output and ultimate outcome in the particular corporate governance system in which they wish to operate.

The effects of board performance on corporate performance have proven difficult to demonstrate. Many studies examining this relationship have shown inconclusive results, though some have shown a positive relationship (Dalton, Daily, Ellstrand, & Johnson, 1998; De Jong, 1997; Larcker, Richardson, & Tuna, 2007; Levrau & Van den Berghe, 2009). This chapter began with several descriptions of how poor corporate governance can destroy firm value. Improving board performance can ultimately help prevent the kinds of corporate governance failures that can prove disastrous to many companies. In this way, better inputs, processes, and outcomes lead to better corporate performance.

GLOBAL CORPORATE GOVERNANCE SYSTEMS

The Corporate Board Performance Model described above portrays how the inputs and processes of corporate boards can improve board and corporate performance. This model and its elements are surprisingly applicable globally. However, implementation of the model differs globally as a result of country differences like culture, history, economic development, and legal systems. Differences in definitions of corporate performance also lead to differences in existing corporate practices. Improving board inputs requires careful attention to the legal, geographical, cultural, and historical origins of current board inputs and processes.

International differences in corporate governance have led to various classifications of corporate governance systems (Choi, Raman, Usoltseva, & Lee, 1999; Denis & McConnell, 2003, 2005; Garcia-Meca & Sanchez-Ballesta, 2010; Kakabadse, Kakabadse, & Yavuz, 2009; La Porta, Lopez-de-Silanes, Shleifer, & Vishny, 1998, 1999). Country characteristics, such as culture, norms, legal systems, and regulations, impact governance practices (Aggarwal, Erel, Stulz, & Williamson, 2009; Doidge, Andrew Karolyi, & Stulz, 2007; Larcker & Tayan, 2011). Economic and financial development of the country also affects governance (Doidge et al., 2007). The following classifications and their elements are not descriptive of any one country. Additionally, the regulations and governance practice in each country within these classifications do differ. But, the corporate governance systems can generally be described in the following three classifications:

1. Anglo-American system
2. Communitarian system
3. Emerging Market system

The countries in each of these three systems define corporate and board performance differently and have evolved a variety of board inputs and processes to increase the desired board and corporate performance. General characteristics of each system are described in Table 1. Though there are differences in structure and systems for historical, geographical, financial, and other reasons, there are management control and performance measurement and management principles that are relevant and critical globally to improve corporate governance. This chapter provides these principles and their application throughout the world and a framework to understand their application.

Anglo-American Corporate Governance System

The United States, the United Kingdom, Australia, New Zealand, Canada, and South Africa have developed many similar corporate governance mechanisms, and a similar understanding of successful corporate performance due to similarities in legal structures and cultures. The Anglo-American system of governance evolved in countries to which the United Kingdom exported its common law legal system during its colonial period in the seventeenth to nineteenth centuries (La Porta et al., 1998). The common law legal system was originally developed to protect private landholders from nobility (Matoussi & Jardak, 2009). Currently, the common law system that exists in many of the countries under this system sets a broad legal precedent that protects shareholders and prevents the managers of the firm from acting against shareholder interests (Shleifer & Vishny, 1997).

In the Anglo-American system, corporate performance is generally defined in terms of the maximization of shareholder wealth and focuses on shareholder primacy (Bainbridge, 2002; Balachandran, Dossi, Van der Stede, 2010; Larcker & Tayan, 2011). Companies in this system have diverse

Table 1. General Characteristics of Global Corporate Governance Systems.

	Anglo-American	Communitarian	Emerging market
Examples of countries	United States, United Kingdom, Australia, Canada, South Africa	Japan, Germany, Belgium, Scandinavia	China, Eastern Europe, India, Brazil, Mexico, Russia
General characteristics	• Shareholder-centric • Market centered • Unitary board structure • Boards primarily composed of nonexecutive directors (and independent directors) • Common law legal system • High levels of disclosure and more rules on disclosure • Large pay incentives for managers, including pay for performance	• Stakeholder-centric • Bank centered • Two-tier board structure (supervisory and management) • Labor, founding family, and bank are common members – interlocking common • Civil law legal system • Moderate levels of disclosure • Pay incentives moderate	• Stakeholder-centric • Government/ Family centered • Board structure varies • Few independent board members • Legal systems relatively weak • Low levels of disclosure • Pay incentives smaller

Sources: Brunetti and Cescon (1998), Choi et al. (1999), Cottrell (2009), Denis and McConnell (2005), Doi (1998), Epstein and Roy (2002), Fackler (2008), Glendon et al. (1994), Husted and Serrano (2002), International Finance Corporation (2008), Jensen and Meckling (1976), N. Kakabadse, Kakabadse et al. (2009), La Porta et al. (1998), Larcker and Tayan (2011), McFarlan et al. (2009), Merryman (1969), OICU-IOSCO (2007a, 2007b), Pierce (2009), A. M. Reed (2002), Schmidt (1998), Shleifer and Vishny (1997), Sjoberg (2009), and Stout (2011).

ownership structures and exist primarily to generate returns for their owners (Jensen & Meckling, 1976; Matoussi & Jardak, 2009). Additionally, directors are usually elected by shareholders and can be replaced if shareholders are not satisfied with results (Bagley, 2006).

Since well-developed financial markets heavily influence corporate governance practices, the Anglo-American system is considered market based (Kaplan, 1997). Financial firms control a majority of equity, and a large share of corporate ownership is traded frequently on large, public exchanges such as the NYSE and London Stock Exchange (LSE) (Demirag,

1998; Kakabadse, Tsvetkov, Uldatsov, & Kakabadse, 2009). These exchanges provide guidance and may define limits of board inputs and processes as they often set their own requirements for board structure, independence, meetings, and performance evaluations (Kakabadse, Tsvetkov et al., 2009; NYSE, 2003; Pierce, 2009). While some countries have chosen to implement reforms on a "comply or explain" basis, others have made them outright requirements (Kakabadse, Tsvetkov et al., 2009; Larcker & Tayan, 2011; NYSE, 2003; Pierce, 2009).

Corporate boards under this system are primarily composed of nonexecutive directors who represent the interests of large shareholders. Additionally, these boards also have independent directors with no financial interest in the future performance of the firm (Choi et al., 1999; Pierce, 2009). These independent directors play a key oversight role on the board, and often guide decisions on director compensation, monitor the audit process, and oversee the director nomination process. Critics of the Anglo-American system have argued that boards are frequently too large (with as many as 20 members) and that directors have often failed to dedicate enough time to their roles and responsibilities (Epstein & Roy, 2007; Pozen, 2010a, 2010b).

Boards are typically led by a chairman who may also serve as the chief executive officer (CEO) of the company. However, this practice is becoming less common. Canada, for example, has only 30% of CEOs serving also as chairman of the board. Even in the United States, where there is still a great deal of chairman–CEO overlap, only 16.5% of new CEOs appointed in 2009 were also chairman of the board (Favaro, Karlsson, & Neilson, 2010; PRNewswire, 2010).

Anglo-American boards are characterized by committees that perform many of the board's functions. This will typically include an audit committee (oversees the financial auditing process), a nomination committee (oversees the nomination of new directors), and a compensation committee (determines executive and director compensation). These committees often have requirements as to who can serve on them, and what percentage of the committee members must be independent directors (Pierce, 2009).

In the Anglo-American system, boards are now usually expected to perform rigorous oversight and evaluations of performance, of senior managers and the corporation (Behan et al., 2006; Epstein & Roy, 2005, 2007; Pierce, 2009). It is common to structure executive compensation so that a large percentage of that executive compensation is tied to the financial performance of the corporation (Bryan, Nash, & Patel, 2006). Furthermore, the Anglo-American system emphasizes transparency and disclosure on a company's audited financial status, board composition, and corporate strategy (Larcker & Tayan, 2011).

Communitarian Corporate Governance System

In much of continental Europe and in Japan, a Communitarian corporate governance system has evolved. This system has its origins in the Romano-Germanic civil law codes developed over a millennium ago (Glendon, Gordon, & Osakwe, 1994; Merryman, 1969). The civil law legal system uses statutes and codes as a means of regulating performance (Matoussi & Jardak, 2009). Countries with civil law legal codes have adopted fewer laws to protect shareholder rights than countries with common law legal codes, and have evolved different corporate governance regimes as a result. Corporate governance under this system relies heavily on banks for financing instead of capital markets.

In Communitarian corporate governance systems, the goals of corporate performance are defined differently than in the Anglo-American system. Whereas the Anglo-American governance system focuses on maximizing return to shareholders, the Communitarian system provides for a stakeholder-centric approach to governance (Balachandran et al., 2010; Larcker & Tayan, 2011). These stakeholders often include lending partners, members of supply chains, customers, employees, and members of society (Cadbury, 2006; Choi et al., 1999). These stakeholders all have a role in the long-term success of the company and thus are considered in decisions and actions affecting the company (Cadbury, 2006). Instead of relying on strong legal protections for investors, corporate governance is embedded more in these multidimensional relationships, so the Communitarian system is considered relationship based (Aguilera & Jackson, 2003; Kaplan, 1997). Corporations in Communitarian countries typically rely heavily on trust in these relationships, in addition to contracts, to conduct important business transactions (Choi et al., 1999). Monitoring and disciplinary action are often carried out by these relationships with banks and between corporations (Kaplan, 1997).

Due to the relationships between customers, employees, managers, and owners in the Communitarian system, there is frequent interaction between the parties. This allows them to wield strong control over corporations and hold them accountable. In Germany and Japan, for example, it is common for firms and banks to cross-invest in their suppliers and customers, creating conglomerates of corporations (Larcker & Tayan, 2011). As a result, many large banks in these countries and other European nations own significant portions of their borrowers' equity, and many suppliers own significant portions of their customers' equity (Cadbury, 2006). These controlling companies play a much more active role in deciding corporate governance standards than they do in the Anglo-American model.

Critics of the Communitarian system have argued that it fails to protect minority shareholders and lacks representation for outside investors (Dahya, Dimitrov, & McConnell, 2008). However, some countries in the Communitarian system are implementing codes that strengthen shareholders' oversight of corporate boards, pay practices, and accounting and auditing policies (Iwata, 2008).

Financial markets in the Communitarian system do not typically hold as much power as in Anglo-American countries. Instead, corporations rely on banks for financing. Often, representatives from these banks sit on the board of directors for a company that they finance (Balasubramanian, Barua, Bhagavatula, & George., 2011; Larcker & Tayan, 2011). As a result, board inputs and processes are often structured to balance the competing interests of all company stakeholders (Choi et al., 1999; A. Kakabadse et al., 2009; La Porta et al., 1998).

There are typically two separate boards in the Communitarian system. The executive board is responsible for managerial decisions for the corporation, such as strategy, marketing, and product development. The supervisory board is comprised primarily of outside directors elected by shareholders and oversees the management board, has authority over the CEO, and makes decisions regarding capital investments. The supervisory board in Germany is required to have at least half of its members elected by employees (Larcker & Tayan, 2011; Schmidt, 1998). Similarly, in Japan, employees play an active role in the supervisory board and help ensure that the corporation has employees' interests in mind. However, supervisory boards also have representatives from founding families or national banks (Larcker & Tayan, 2011).

Emerging Market Corporate Governance System

Emerging Market countries, such as China, Brazil, India, Mexico, Russia, and much of Eastern Europe, still have relatively young legal systems that have yet to fully develop the legal framework necessary to classify their corporate governance systems into the other categories (Choi et al., 1999). In China, for example, the first corporations were founded centuries after countries with more defined corporate governance systems began their corporate history (Goetzmann & Koll, 2005; La Porta et al., 1998). Additionally, countries in the Emerging Market system differ in legal traditions, language, and culture (Black, de Carvalho, & Gorga, 2011). The result is a significant uncertainty and dynamism with regards to how corporate governance will likely evolve in these countries (Choi et al., 1999; D. Reed, 2002; Wright, Buck, & Filatotchev, 2005).

Despite these uncertainties, there are still recognizable patterns. Due to a lack of defined legal systems, shareholders typically lack credible legal protections (Shleifer & Vishny, 1997). Boards are therefore expected to vigorously monitor and protect shareholder interests. Additionally, low levels of disclosure and a lack of transparency make it difficult for shareholders to exercise informed control over the company (Larcker & Tayan, 2011).

Corporate governance in Emerging Markets tends to be stakeholder-centric and centered around family or government relationships. Large corporate shareholders dominate boards and are well known worldwide. For example, about 30 Korean family groups that are organized as chaebols (business conglomerates) control almost 40% of that country's economy (Schmidt, 1998). In India, company insiders and their families own approximately 45% of the equity value of all Indian companies (Larcker & Tayan, 2011).

In addition to these large corporate groups, there is a large amount of state participation in the economy (Balasubramanian et al., 2011). In some countries this participation is a by-product of postcolonial governments, while in others it is an orchestrated national policy aimed at achieving specific economic goals (Ahunwan, 2002; A. M. Reed, 2002). Even in countries where the government has privatized its corporate holdings, the government often retains significant control over these corporations. In China, for example, the government still highly regulates equity markets and denies many companies the ability to trade their shares on exchanges for political reasons (Larcker & Tayan, 2011; McFarlan, Xu, & Manty, 2009). The government also maintains effective control of many corporations through mandatory board requirements and multi-vote shares (McFarlan et al., 2009; Rajagopalan & Zhang, 2008).

Board structure in the Emerging Market system varies between countries and even among corporations in the same country; some corporations have adopted the two-tier board structure of Communitarian countries, while others utilize the unitary board structure of the Anglo-American system (Kakabadse, Kakabadse, & Yang, 2009; International Organization of Securities Commissions (OICU-IOSCO), 2007b). Countries in this system have also used different methods for protecting the interests of minority stakeholders besides the majority owners of these corporations, such as establishing requirements for nonexecutive directors (A. Kakabadse et al., 2009; McFarlan et al., 2009; A. M. Reed, 2002). Despite these attempts, many of these requirements have been largely circumvented (Kakabadse, Tsvetkov et al., 2009; McFarlan et al., 2009; Mitchell, 2008). Due to cross-country differences in these Emerging Market countries, the major corporate governance challenges may require different regulations to manage them appropriately (Balasubramanian, Black, & Khanna, 2010).

APPLICATION OF MODEL GLOBALLY

The applications of the various model components to the three global corporate governance systems are summarized in Tables 2 and 3. The diversity of these practices, both across and within these systems, as well as in board roles and responsibilities, is explored in the sections below.

Table 2. Internal Determinants of Corporate Governance Performance.

Corp. Governance Mechanism	Anglo-American	Communitarian	Emerging Markets
Board composition, systems, and structure			
Board composition	• Balance between internal and independent directors	• Few independent directors	• Independent directors rare
Board structure	• Unitary	• Two tier	• Varies
Board accountability	• Shareholders	• Stakeholders (including entire value chain and employees)	• Stakeholders and government or founding family
Executive compensation	• Large performance pay	• Moderate performance pay	• Little performance pay

Sources: Brunetti and Cescon (1998), Cottrell (2009), Denis and McConnell (2003, 2005), Doi (1998), Exley and Hudson (2006), Hermalin and Weisbach (2003), Husted and Serrano (2002), A. Kakabadse et al. (2009), N. Kakabadse, Kakabadse et al. (2009), N. Kakabadse, Yang et al. (2009), Larcker and Tayan (2011), McFarlan et al. (2009), OICU-IOSCO (2007a, 2007b), Sjoberg (2009), and Stout (2011).

Table 3. External Determinants of Corporate Governance Performance.

Corp. Governance Mechanism	Anglo-American	Communitarian	Emerging Markets
Markets			
• Financial markets	• Strong and active	• Weak and not commonly used	• Volatile
• Investment purpose	• Short-term return	• Long-term return	• Policy and political goals
• Methods of finance	• Financial markets	• Bank credit and retained earnings	• Private and state-owned banks
Legal systems			
• Legal history	• Common law system	• Civil law system	• Combined systems that are still rapidly evolving
• Transactional methods	• Contracts	• Relationship-based transactions	• Relationship-based transactions
Ownership and control structures			
• Ownership structure	• Diverse individual and institutional ownership	• Concentrated family and corporate ownership	• Concentrated family, corporate, and government ownership
• Minority shareholder protections	• Strong	• Moderate	• Weak
• Dominant control mechanisms	• Voting and board representation	• Cross-holding, pyramidal groups, lending relationships	• Internal and external mechanisms

Sources: Aguilera and Jackson (2003), Ahunwan (2002), Bagley (2006), Black et al. (2011), Cadbury (2006), Choi et al. (1999), Dahya et al. (2008), Glendon et al. (1994), Jensen and Meckling (1976), Kakabadse, Tsvetkov et al. (2009), Kaplan (1997), La Porta et al. (1998), Larcker and Tayan (2011), McFarlan et al. (2009), Merryman (1969), Mitchell (2008), A. M. Reed (2002), Schmidt (1998), and Shleifer and Vishny (1997).

As discussed above, boards of directors are usually expected to fulfill three separate but related roles and responsibilities (Epstein & Roy, 2010): accountability, senior-level staffing and evaluation, and strategic oversight.

The application of these roles and responsibilities varies significantly in the three corporate governance systems.

Accountability

Worldwide, there is general agreement that boards should focus on accountability (OECD, 2004). While across the world the board is seen as a vehicle of accountability, the primary stakeholders that the board ultimately attempts to serve differs widely among systems. In the United States and other Anglo-American countries, it has traditionally been understood that boards of directors are responsible for holding managers accountable to shareholders (Hermalin & Weisbach, 2003; Stout, 2011). In the Communitarian system, the responsibilities of the board are often less prescribed by law. In many Communitarian countries, it is common for the corporation, management, and board to be accountable to other stakeholders, such as employees, suppliers, and customers (Denis & McConnell, 2003, 2005). In the Emerging Market system, boards are typically accountable to the controlling shareholders in the corporation. Sometimes, as is often the case in China, this shareholder is the state (McFarlan et al., 2009). Other times, as in South Korea and Mexico, this shareholder is a family or corporate conglomerate (Fackler, 2008; Husted & Serrano, 2002).

Senior-Level Staffing and Evaluation

Evaluating management, determining compensation, and making staffing changes are important roles for the board in all three governance systems (OECD, 2004). The ability to evaluate the management team, and adjust compensation accordingly, allows the board to ensure a return for shareholders. The extent to which this role is carried out differs in each governance system. In the Anglo-American system, many corporations give extremely large performance pay packages to managers in exchange for relentless attention to financial returns (Sjoberg, 2009). In the Communitarian and the Emerging Market systems, pay incentives and the role of the board in determining pay are considerably smaller (Brunetti & Cescon, 1998; Sjoberg, 2009).

Strategic Oversight

In all three governance systems, strategic oversight is an important role for the board (OECD, 2004). In the Anglo-American system, managers focus on strategy formulation and the board plays a more passive role in approving important strategic moves (Epstein & Roy, 2002). In the Communitarian system, the board is more assertive in its strategic oversight role (Schmidt, 1998). And in the Emerging Markets system, many board members view the formulation and oversight of strategy as the primary role of the board (N. Kakabadse, Kakabadse et al., 2009).

In a unitary board structure, the board of directors is responsible for developing strategy as well as overseeing its execution. In the two-tier board structure, supervision is separate from execution and therefore the management board oversees strategy (Cadbury, 2006).

These three roles are ultimately the mechanisms that ensure executives fulfill shareholder expectations regarding corporate performance. While these specific roles and responsibilities manifest themselves differently in different settings, fulfilling these roles is a crucial aspect of improved corporate performance (Epstein & Roy, 2010).

INTERNAL DETERMINANTS OF CORPORATE BOARD PERFORMANCE

Board performance and corporate performance are significantly impacted by internal and external factors. As described in Table 2, these internal determinants differ substantially between the three global governance systems.

These factors are mostly under the control of corporate boards and their effectiveness in the management of these factors affects board success. These determinants are

1. *board composition* including size, number of independent members, skills and competencies, etc.;
2. *board systems and structure* including board organizations and committees, information flow, operating systems, etc.; and
3. *performance evaluation and compensation systems* including the role in senior-level staffing, succession planning, compensation systems, etc.

Board Composition

The board's effectiveness in successfully performing their roles is determined by the board's composition, systems, and processes. Board composition choices determine board competencies, skills, and power structures, and are an important aspect of board performance (Epstein & Roy, 2002).

The OECD (Organization for Economic Cooperation and Development) Principles for Corporate Governance (2004) and other international guidelines address issues surrounding board composition (OICU-IOSCO, 2007a). These guidelines suggest that the board should make use of nonexecutive directors for certain board functions and that board composition should be disclosed publicly and transparently. The European Union also suggests that boards should balance the number of executive and nonexecutive directors, and that certain committees should be primarily made up of nonexecutive and independent directors (International Finance Corporation, 2008).

While the definitions and differences vary from country to country, nonexecutive directors are typically defined as those who do not hold executive-level positions at the corporation. The nonexecutive directors are primarily responsible for overseeing the executive directors and designing and implementing succession plans (Exley & Hudson, 2006). Independent directors should not work for the corporation or any other company that has a business relationship with the corporation and should also not benefit from the financial performance of the corporation, except through their compensation as directors (Garcia-Meca & Sanchez-Ballesta, 2010).

The increasing focus on nonexecutive and independent directors worldwide is driven by arguments that these corporate outsiders improve the outcomes associated with key board of directors' decisions (OICU-IOSCO, 2007a). Boards with more independent directors have been shown to outperform other firms (Hermalin & Weisbach, 2003). Board independence has also shown to be significantly related to firm value, including in developed countries and in countries with poor investor protection and a controlling shareholder (Aggarwal et al., 2009; Dahya et al., 2008). This has been found to be especially true in countries with weak legal protection for minority shareholders, such as those under a Communitarian system (Dahya et al., 2008).

Many corporate governance experts, international guidelines, and national corporate laws advocate for increased board member independence and increased numbers of nonexecutive directors on boards (Epstein & Roy, 2002; International Finance Corporation, 2008; OECD, 2004). However, a board should not be any larger than is necessary to compose an appropriate mix of knowledge and experience (Cadbury, 2006).

In each corporate governance system, the role of nonexecutive and independent directors is markedly different. In the Anglo-American system, independent directors are common among listed firms and are involved in many of the board's core activities. There are often stringent requirements for board independence (Cottrell, 2009). In addition, these board members have a significant amount of power. In Canada, for example, 70% of companies have independent directors serving as chairman of the board (Behan et al., 2006).

In the Communitarian system of corporate governance, independent directors are less common but still hold a considerable amount of power. National codes and companies typically do not stress the independence of board directors as in the Anglo-American system. For example, most board members for corporations in Japan, Italy, and much of continental Europe are internal in nature (Brunetti & Cescon, 1998; Doi, 1998). While there are often strenuous regulations for director independence, the proportion of independent directors is much smaller (Garcia-Meca & Sanchez-Ballesta, 2010). This is largely a result of co-determination laws that mandate high levels of employee participation on the supervisory board in order to balance the interests of shareholders and employees (Larcker & Tayan, 2011). However, many of these countries are increasingly stressing requirements for diversity in board composition, such as quotas on the percentage of females in the boardroom (Fouche & Treanor, 2006; Oxford Analytica, 2010; Patel, 2010).

The Emerging Market system features many of the regulatory requirements of the Anglo-American system with regard to independent directors. However, truly independent directors are uncommon and wield little power relative to other controlling interests on the board. In China, for example, listed companies are required to have two independent directors, and stock exchanges recommend that one-third of directors be independent (N. Kakabadse, Kakabadse et al., 2009; Kakabadse, Yang, & Sanders, 2009; McFarlan et al., 2009). Despite these requirements and recommendations, typically only 5% of Chinese directors show signs of true independence, and of these, many are still strongly influenced by the government (N. Kakabadse, Yang, et al., 2009). In Mexico, independent directors are also rare and likely strongly influenced by the controlling family or corporation's presence on the board (Husted & Serrano, 2002).

Board Systems and Structure

We have already discussed the primary differences in board roles, responsibilities, and composition across governance systems. The mechanisms used to translate these components into decisions that critically affect board and corporate performance also vary. These mechanisms, known generally as board systems and structure, are an important determinant of corporate performance as well.

There are strong theoretical arguments that board systems and structure, such as leadership structures, size, and decision-making processes, affect corporate performance since they impact accountability, transparency, and trust between board members. However, the results of empirical studies to examine the effects of various board systems and structures are mixed, despite strong qualitative evidence that board systems and structure are critical determinants of board performance and corporate performance (Levrau & Van den Berghe, 2009). Many corporate governance guidelines stress the role of board systems and structure in determining corporate performance as a result since these systems and structures map board composition into board performance (Epstein & Roy, 2010).

Boards of directors in different governance systems vary first in overall structure (OICU-IOSCO, 2007a). In the Anglo-American system, there is typically a unitary board that combines executive directors and nonexecutive directors, some of which typically are required to be independent of the firm (Cottrell, 2009; Exley & Hudson, 2006). Boards are organized into committees that perform much of the board's functions. These committees usually have requirements as to who can serve on them, and what percentage of the committee members must be independent directors (Pierce, 2009).

In the Communitarian system, the executive directors and nonexecutive directors are divided into two separate entities. The executive board, composed primarily of senior-level management, is mainly responsible for the strategic direction of the firm. The supervisory board, composed of company insiders, such as employees, and company outsiders, such as independent accountants, is primarily responsible for overseeing executive management, accounting, senior-level staffing, and evaluation (A. Kakabadse et al., 2009).

In the Emerging Market system, the board structure can vary by country and firm (OICU-IOSCO, 2007b). There is typically less adherence to standards of independence for supervisory board members. Additionally, countries in the Emerging Market system often allow the largest shareholders, usually the state or family groups, to control the boards (Husted & Serrano, 2002; A. Kakabadse et al., 2009; N. Kakabadse, Kakabadse et al., 2009; N. Kakabadse, Yang et al., 2009).

Performance Evaluation and Compensation Systems

Performance evaluation and compensation systems differ between governance systems as well (Balachandran et al., 2010). In the Anglo-American system, boards are expected to perform rigorous evaluations (Behan et al., 2006; Epstein & Roy, 2007; Pierce, 2009). It is common to structure executive pay so that a large percentage of that pay is tied to the financial performance of the corporation (Bryan et al., 2006; Denis & McConnell, 2003, 2005). Evaluation procedures have only begun to become part of the Communitarian system, and most companies in the Emerging Markets system lack explicit evaluation mechanisms (Denis & McConnell, 2003, 2005; Pierce, 2009). In the Communitarian system and Emerging Markets system, performance pay is significantly less common and accounts for a much smaller percentage of overall executive compensation than in Anglo-American countries (Balachandran et al., 2010).

There are other board processes that help determine the ways that board composition is transformed into board and corporate performance that do not vary systematically between corporate governance systems. For example, productive meetings are an important board process and occur in many corporations, and in many different settings, across the world. Furthermore, some board processes such as risk management procedures and information systems can vary widely among corporations, regardless of corporate governance systems (Wong, 2011).

EXTERNAL DETERMINANTS OF CORPORATE BOARD PERFORMANCE

There are also external factors that affect how boards of directors can perform their roles and responsibilities. The following are three external factors that determine key components of corporate governance:

1. *Markets* determine the parameters for changes in ownership and choices of capital structure.
2. *Legal systems* determine protections for shareholders and define the parameters of economic transactions.
3. *Ownership and control structures* determine who is in control of corporate decisions.

In the Anglo-American, Communitarian, and Emerging Market corporate governance systems, these mechanisms are used to ensure that corporations act according to national expectations regarding the governance of corporations. To improve performance through the mechanisms they can control, boards must also better understand the external determinants of structure, processes, and performance (Table 3). However, it is unclear which external determinant has the greatest impact on corporate board performance, as studies and arguments have varying opinions (Easterbrook, 1997; Matoussi & Jardak, 2009).

Markets

The existence of markets for ownership and control of corporations is an important aspect of corporate governance. Markets often act as a disciplining mechanism for corporations (Larcker & Tayan, 2011). Recent studies report that access to effective markets and high market valuation of a company is positively related to good corporate governance practices (Cheung, Jiang, Limpaphayom, & Lu, 2008, 2010; Doidge et al., 2007). Further, if the board mechanisms for governance fail, the potential for drastic changes like takeovers increase. This process keeps management accountable for maintaining the firm value to prevent an outside party from vying for control (Denis & McConnell, 2003, 2005). Takeovers are most common in the Anglo-American system, particularly the United States (Denis & McConnell, 2003, 2005; La Porta et al., 1998). They are uncommon in the Communitarian system and difficult to execute in the Emerging Market system (Denis & McConnell, 2003, 2005; Doi, 1998). The existence of a controlling shareholder in the Communitarian and Emerging Market systems often makes the possibility of a takeover irrelevant in terms of governance practices (Bebchuk & Hamdani, 2009).

Beyond the market for corporate control, there are markets for corporate ownership (stock markets) that determine the return available to stockholders in a corporation. Companies in countries where efficient capital markets exist often have more diverse ownership (Easterbrook, 1997). In the Anglo-American system, these markets are quite active. Over 6,000 domestic companies are listed publicly in the United States, and over 1,000 are listed in the United Kingdom. (A. Kakabadse et al., 2009). Only a relatively small number of firms are listed in Communitarian countries. In Germany, for example, there are less than half as many listed companies as in the United Kingdom (A. Kakabadse et al., 2009). As a result, the market capitalization and volume traded on debt and equity markets in these countries is small (Choi et al., 1999, La Porta et al., 1998). In Emerging Markets, stock exchanges are developing rapidly. There is a great deal of variation across these countries, however, as their economies are dynamic and growing. China, for example, has become a "dominant force" in the Initial Public Offering (IPO) market in the last two years with its large exchange in Hong Kong (Cowan, 2011). Yet, in Mexico, only 3% of daily market turnover is a result of stock activity. Since these markets are developing so rapidly, it is difficult to predict their future size, scope, or effect on corporate governance.

Markets play another important role in corporate governance. There are typically strict listing requirements that govern corporate policies for all companies that list on a stock exchange. In the Anglo-American system, these regulations have become a dominant mechanism for implementing corporate governance reforms. The NYSE, Toronto Stock Exchange (TSX), LSE, and others have adopted mandatory listing requirements that include more stringent corporate governance measures for publically traded companies (Behan et al., 2006; McFarlan et al., 2009). While Anglo-American countries such as Canada and the United Kingdom were the first to implement corporate governance reform through their stock exchanges, many Communitarian and Emerging Market stock exchanges have followed suit more recently (N. Kakabadse, Tsvetkov et al., 2009). The system of "comply and explain" corporate governance, where companies must disclose and explain the extent to which they follow guidelines, that was established by the TSX and LSW has also been implemented in Mexico and Turkey (Husted & Serrano, 2002; N. Kakabadse, Tsvetkov et al., 2009).

As markets become increasingly global, corporate governance reforms have begun to spread through these market institutions. It is becoming increasingly common for companies in any of the corporate governance systems to adopt aspects of another system in order to list on a particular stock exchange. Studies show that firms in Emerging Markets are valued higher in markets if they are perceived to have better governance by the standards of that market (Klapper & Love, 2004; Morey, Gottesman, Baker, & Godridge, 2009). For this reason, access to global capital markets is often an incentive for firms to adopt better governance practices (Doidge et al., 2007). In Russia, for example, some companies adopt Anglo-American standards of governance by recruiting more independent directors in order to list up to 12% of their shares on an American stock exchange. For many, this serves as a signal that these companies are more trustworthy and may win a company more joint ventures, international acquisitions, or partnerships

(Oseichuk, Kakabadse, & Kakabadse, 2009). Conversely, the stringent listing requirements in the United States impose costs on corporations that many are unwilling to pay. This is causing some multinational companies, such as Daimler and Deutsche Bank, to reconsider their listings in the United States. While some companies are improving their governance to list on well-respected exchanges, others are willing to forgo the cachet of a large exchange in order to avoid the costs of listing (Fuhrmans & Stevens, 2010).

Legal Systems

Legal systems have been found to affect other aspects of corporate governance, including investor protection for minority shareholders, ownership structure, and financial markets (Larcker & Tayan, 2011; Matoussi & Jardak, 2009). Through the evolution of different legal traditions, countries evolved different legal rights for shareholders in corporations. Legal systems, in addition to the culture and political environment, are an important aspect of how investors are protected (Matoussi & Jardak, 2009).

Many have divided the world into two wide legal traditions: the English common law system, prominent in the United Kingdom and its former colonies, and the Romano-Germanic civil law system, prominent in Germany, France, Italy, and their former colonies (La Porta et al., 1998; Matoussi & Jardak, 2009). Whereas the civil law system relies heavily on the codification of laws from abstract principles, the common law system puts much of the burden on judges to interpret laws (Glendon et al., 1994; Merryman, 1969). Additionally, countries that have a common law legal system provide more rights to shareholders than countries with a legal system based on civil law (Larcker & Tayan, 2011).

The common law tradition has led to the development of the stylized Anglo-American corporate governance system. Since English common law countries evolved much stronger legal rights for shareholders than civil law countries, this system exhibits the best legal protection for investors (Matoussi & Jardak, 2009). Additionally, these laws tend to be better enforced in common law countries.

The civil law system has led to the development of the stylized Communitarian corporate governance system. Some researchers suggest that in the absence of legal protections for shareholders, countries develop other mechanisms to ensure that capital is allocated efficiently (Morck & Steier, 2005). Civil law countries have developed a wide variety of these mechanisms, such as concentrated ownership and control, mandatory dividends, and limited equity markets. As a result, civil law countries have been able to develop active corporate sectors despite weak legal protections for the owners of equity capital.

Countries in the Emerging Markets system have only recently begun to develop the legal framework necessary to support the development of a corporate sector (Choi et al., 1999; La Porta et al., 1998; Millar, Eldomiaty, Choi, & Hilton, 2005). Many have even fewer shareholder protections than in civil law countries and have evolved a severely concentrated ownership structure to compensate for these weak protections. In addition, the state often plays a much more active role as a corporate owner since many corporations attempt to engage in rapid industrialization and pro-growth economic strategies (A. M. Reed, 2002).

Ownership and Control Structures

Concentrated ownership and minority control mechanisms can be viewed as corporate governance mechanisms that evolve in the absence of basic shareholder protections. Additionally, ownership structures impact many governance practices. For example, board practices that protect minority shareholders can be irrelevant and possibly damaging in organizations with a controlling shareholder. Board systems must adapt to address the different issues and problems that shareholders face under different ownership structures (Bebchuk & Hamdani, 2009). With fewer powers guaranteed by legal systems, it is more efficient for owners of corporations to concentrate their ownership in order to ensure a fair return on their investment and check the power of managers (La Porta et al., 1998; Shleifer & Vishny, 1997).

As a result, the most concentrated ownership structures are usually in Emerging Market economies. One of the most glaring examples of this concentration is South Africa, where four business groups control over 56% of the country's large corporations. In Hong Kong, a small group of business tycoons control most large companies through concentrated shareholdings. In China, the state still owns majority holdings in most of the large firms (Mitchell, 2008). In Mexico, a small group of families also controls much of the corporate sector (Husted & Serrano, 2002).

There is considerable ownership concentration in Communitarian countries as well. In Germany and Japan, for example, large banks and corporate groups can control large portfolios of companies. In fact, there are very few corporations that are considered widely held by a diverse set of shareholders in Communitarian countries (La Porta et al., 1999). In many of these countries, families and governments do not always hold majority stakes in the companies they control. They use a variety of control tactics to ensure that even a fairly small equity stake in a company can effectively give the largest shareholder power over that company. These tactics include cross-shareholding, where two or more companies controlled by the same family group or government invest in each other, and pyramidal shareholding, where a company owns a sizable minority stake in a holding company that owns smaller stakes in more companies (Bebchuck, Kraakman, & Triantis, 1999). The intercorporate network formed often utilizes its control and power over the board and management to steer the company in the direction that they prefer (Bebchuk & Hamdani, 2009).

In Anglo-American countries, like the United States and United Kingdom, financial institutions, including pension funds, own more than 60% of all equity capital (Demirag, 1998). Their ownership is quite dispersed, however. Anglo-American countries are typically characterized as the only countries in the world that live up to the Berle–Means ideal of widely held corporations (La Porta et al., 1999; Berle & Means, 1932). In this system, countries lack a controlling shareholder and therefore officers and directors have more control over decisions. Boards are responsible for ensuring that these decisions are in the best interest of the shareholders. Governance practices in these countries must address the possibility of management acting opportunistically in a manner that negatively affects shareholders (Bebchuk & Hamdani, 2009).

Closely related to ownership and control of corporations is the existence and enforcement of minority shareholder rights. These rights deal with voting privileges, corporate meetings, and dividends. Typically, Emerging Market economies have the weakest rights and Anglo-American economies have the strongest rights, with Communitarian countries falling somewhere in between. The exception to this rule is the mandatory dividend, which typically is only found in Communitarian countries and often substitutes for other shareholder rights (La Porta et al., 1998). While these rights are also sometimes found in Emerging Markets, they are much less frequently enforced (Choi et al., 1999).

In Emerging Market economies, the state is frequently the controlling owner and can use its political power to derive wide control powers over corporations (Choi et al., 1999; La Porta et al., 1999; McFarlan et al., 2009). In China, for example, rules surrounding the boards of directors of major corporations and financial markets are structured to give the state control over corporations (Goetzmann & Koll, 2005; N. Kakabadse, Kakabadse et al., 2009; McFarlan et al., 2009). In some Emerging Market countries, such as India, Brazil, and Russia, the state has begun to turn control of corporations over to the private sector. However, many governments continue to wield significant power due to their ability to control the legal framework and market environment (Rajagopalan & Zhang, 2008; A. M. Reed, 2002; N. Kakabadse, Kakabadse et al., 2009).

MEASURING PERFORMANCE OF CORPORATE GOVERNANCE SYSTEMS

As previously mentioned, the three systems of corporate governance developed in response to various influences in their country, including historical, cultural, legal, and social. For this reason, it is difficult to compare the three systems and determine a "best" system of corporate governance without an acknowledgement of these differing influences and objectives. As studies of other governance-rating methodologies show, using a single metric or system to analyze corporate governance worldwide is likely to produce inaccurate results (Bebchuk & Hamdani, 2009). Instead, companies in each system should be evaluated based on principles and performance measures specific to the characteristics and goals of that system. There are advantages and disadvantages to the various elements in each of the three corporate governance systems. Thus, for example, corporate governance systems that are primarily shareholder-centric may choose a narrower set of performance metrics than systems that are more broadly stakeholder-centric.

INTERNATIONAL DIVERSITY AND IMPROVING CORPORATE GOVERNANCE

Markets, legal systems, and ownership and control structures have strong impacts on board composition and processes. In the face of certain market-imposed restrictions, legal stipulations, or strongly concentrated ownership structures, it may be particularly difficult to make changes to board composition and processes to drive improved corporate performance. In these situations, it is critically important to understand the diversity of board composition and processes as well as the determinants of board composition and processes in order to identify the most effective levers of governance reform. The role of markets and the benefits of listing on a major stock exchange, including a better international reputation and access to more potential partnerships, can be leveraged to impose governance reforms (Oseichuk et al., 2009). Similarly, helping implement better legal protections for shareholders may help free the board of strong ownership concentration and allow the board to implement governance reforms (La Porta et al., 1998).

Even where the board of directors operates with enough power to make changes to corporate governance mechanisms, the cross-national differences outlined here inform their choices. To drive improved performance, boards can make changes to their composition, systems, and structure. The inputs and processes outlined in the Corporate Board Performance Model can be utilized to improve performance as well. Composing boards of directors with increased independence, diligence, competence, and ethics can help boards better perform their duties (Epstein & Roy, 2004b). These four inputs are aided by a variety of board processes that can also be used to improve performance:

- Board structure
- Productive meetings
- Succession planning system
- Financial reporting and risk management
- Strategic information system
- Performance evaluation and compensation systems

These six processes and the four inputs are 10 keys by which boards can improve their performance along their three high-level roles and responsibilities: providing superior strategic guidance, a more accountable organization, and high-quality senior executives.

RECENT DEVELOPMENTS AND CONVERGENCE IN GLOBAL CORPORATE GOVERNANCE

While most visible in Emerging Market economies, corporate governance continues to evolve worldwide both within and among governance systems (Cheung et al., 2010). The OECD Principles of Corporate Governance (2004) that were approved by all member countries demonstrate an increasingly important trend: companies all over the world are trying to solve similar corporate governance problems and are increasingly agreeing on common solutions (Gregory, 2006). Some shareholders and investors have even called for stronger worldwide corporate governance laws (Iwata, 2008). However, the discussion of corporate governance convergence continues.

There is evidence to suggest both that corporate governance mechanisms are converging worldwide and that corporate governance systems remain as different as ever (Denis & McConnell, 2003, 2005). One expert argues that any convergence that has occurred is mostly superficial and that governance will continue to be influenced by local characteristics because of the varying patterns of economic development worldwide and the different industries (Iwata, 2008; Sify, 2008). Another prominent writer argues that in the past 7 to 10 years, there has been convergence toward a common model – a hybrid between the Communitarian system and the Anglo-American (Sify, 2008). In an analysis of 37 countries, the authors suggest that convergence is occurring globally at the country level in legal systems. They also suggest, however, that the convergence is among groups of economically dependent countries and that corporate governance between these groups remains divergent. They find little convergence in corporate practices, supporting the conclusion that governance may be converging in principle, but that these principles have not yet systematically affected corporate behavior (Khanna, Kogan, & Palepu, 2006).

Differences in culture, politics, legal systems, values, business financing arrangements, and ownership structure impact corporate governance and explain most of the variation between countries in governance practice (Gregory, 2006; Matoussi & Jardak, 2009; Sify, 2008). Therefore, though some convergence of these systems is occurring, significant diversity in corporate governance will likely exist for some time. Although globalization has led to some standardization, corporate governance practices remain diverse due to the various local forces affecting practices (Balachandran et al., 2011; Larcker & Tayan, 2011). Various studies show that the advantages of each system of corporate governance depend on firm and country characteristics, and that governance is not "one-size fits all" (Balasubramanian et al., 2010).

While the OECD has focused on ways to implement these convergent solutions through legal systems, others have argued that convergence in corporate governance through legal frameworks may not be a likely outcome (Coffee, 1999; La Porta, Lopez-de-Silanes, Shleifer, & Vishny, 2000). These researchers focus on the concept of "functional convergence," by which investors and boards of directors enact better corporate governance through listing or taking voluntary actions (such as listing on foreign stock exchanges). They suggest that this convergence trend may be significantly more likely than international legal system convergence. Additionally, it has been suggested that flexibility, either by comply-or-explain or by the ability to choose from different governance levels, be considered in convergence since beneficial corporate governance practices vary across countries (Black et al., 2011). Future convergence may be influenced by regulations or requirements established by international institutional investors or by global capital markets (Cadbury, 2006; Sify, 2008). However, convergence as a result of these influences would be a convergence of standards and function, not structures (Cadbury, 2006).

As corporations, capital markets, and corporate governance are becoming more global, there is likely to be more bridging in theory and practice than either differentiation or convergence. It is becoming increasingly common for companies in any of the corporate governance systems to adopt aspects of another system in order to list on a particular stock exchange. The net effect of these changes remains to be seen.

However, other large-scale studies and experiments with corporate governance systems are needed to provide further evidence on the comparative effectiveness of different systems. There may also be a need to develop and implement further methodologies for assessing governance in companies with different systems and processes, such as ownership and shareholder structure, instead of trying to produce global governance standards (Bebchuk & Hamdani, 2009).

CONCLUSION

There is significant heterogeneity in corporate governance practice worldwide. Corporations in the Anglo-American system operate differently than in the Communitarian system or the Emerging Market system. Yet, despite these differences, there are principles that help define superior corporate practice globally. Boards have universal roles and responsibilities. In order to fulfill them effectively, they must leverage inputs and processes to generate the desired outputs and outcomes.

As corporations, capital markets, and corporate governance are becoming more global, there is likely to be an increase in bridging of theory instead of differentiation or convergence in governance systems. It is important to understand each system in order to be able to work with multiple systems where corporations and stakeholders have different objectives in different countries. The Corporate Board Performance Model provides a model to evaluate each system and the different mechanisms each system uses to implement these principles. There exists a significant need for further research to evaluate governance based on the differentiation and accommodation of the various systems as well as on the implementation of the model and the effectiveness of relevant management control and performance measurement mechanisms.

REFERENCES

Aggarwal, R., Erel, I., Stulz, R., & Williamson, R. (2009). Differences in governance practices between US and foreign firms: Measurement, causes, and consequences. *Review of Financial Studies*, *23*(3), 3131–3169.

Aguilera, R. V., & Jackson, G. (2003). The cross-national diversity of corporate governance: Dimensions and determinants. *Academy of Management Review*, *28*, 447–465.

Ahunwan, B. (2002). Corporate governance in Nigeria. *Journal of Business Ethics*, *37*, 269–287.

Bagley, C. E. (2006). Shareholder primacy is a choice, not a legal mandate. In M. J. Epstein & K. O. Hanson (Eds.), *The accountable corporation: Corporate governance* (Vol. 1). Westport, CT: Praeger Publishers.

Bainbridge, S. M. (2002). Director v. shareholder primacy in the convergence debate. *The Transnational Lawyer*, *16*, 45–62.

Balachandran, K. R., Dossi, A., & Van der Stede, W. A. (2010). Corporate governance research in "The Rest of the World". *Journal of Accounting Auditing & Finance*, *25*, 523–530.

Balasubramanian, B., Barua, S. K., Bhagavatula, S., & George, R. (2011). *Board interlocks and their impact on corporate governance: The Indian experience. Coping with corporate cholesterol.* Bangalore, India: Centre for Corporate Governance and Citizenship (CCGC) and Indian Institute of Management Bangalore (IIMB).

Balasubramanian, B., Black, B. S., & Khanna, V. (2010). The relation between firm-level corporate governance and market value: A case study of India. *Emerging Markets Review*, *11*, 319–340.

Bebchuk, L. A., & Hamdani, A. (2009). The elusive quest for global governance standards. *University of Pennsylvania Law Review*, *157*, 1263–1317.

Bebchuck, L. A., Kraakman, R., & Triantis, G. (1999). *Stock pyramids, cross-ownership, and dual class equity: The creation and agency costs of separating control from cash flow rights.* Harvard Law School John M. Olin Center for Law, Economics, and Business Discussion Paper Series, Paper No. 249.

Behan, B. A. (2009). Governance lessons from India's Satyam. *Bloomberg Business Week*, January 16. Available at http://www.businessweek.com/managing/content/jan2009/ca20090116_465633.htm

Behan, B. A., Ducharme, J., & Hossack, R. D. (2006). Special issues in corporate governance in Canada. In D. A. Nadler, B. A. Behan & M. B. Nadler (Eds.), *Building better boards: A blueprint for effective governance*. Mercer Delta Consulting.

Bergin, T. (2010). Cost of oil spill could exceed $14 billion. *Reuters*, May 2. Available at http://www.reuters.com/article/2010/05/02/us-bp-oilspill-costs-idUSTRE6412H820100502

Berle, A., & Means, G. (1932). *The modern corporation and private property*. London, UK: MacMillan.

Black, B. S., de Carvalho, A. G., & Gorga, E. (2011). *What matters and for which firms for corporate governance in emerging markets? Evidence from Brazil (and other BRIK countries)*. ECGI-Finance Working Paper No. 306/2011. Retrieved from http://ssrn.com/abstract=1832404. Accessed on February 9, 2012.

Bloomberg Businessweek (2005). The Real Scandal at Volkswagen. *Bloomberg Businessweek*, July 18.

Brunetti, G., & Cescon, F. (1998). Capital markets and corporate governance in Italian companies: A study of short term pressures. In I. S. Demirag (Ed.), *Corporate governance, accountability, and pressures to perform: An international study*. Hartford, CT: JAI Press.

Bryan, S. H., Nash, R. C., & Patel, A. (2006, March). *The structure of executive compensation: International evidence from 1996-2004*. Retrieved from http://ssrn.com/abstract=891207

Cadbury, A. (2006). The rise of corporate governance. In M. J. Epstein & K. O. Hanson (Eds.), *The accountable corporation: Corporate governance* (Vol. 1). Westport, CT: Praeger Publishers.

Casselman, B., & Gold, R. (2010). BP decisions set stage for disaster. *The Wall Street Journal*, May 27. Available at http://online.wsj.com/article/SB10001424052748704026204575266560930780190.html

Cheung, Y., Jiang, P., Limpaphayom, P., & Lu, T. (2008). Does corporate governance matter in China? *China Economic Review*, *19*(3), 460–479.

Cheung, Y.-L., Jiang, P., Limpaphayom, P., & Lu, T. (2010). Corporate governance in China: A step forward. *European Financial Management*, *16*(1), 94–123.

Choi, C. J., Raman, M., Usoltseva, O., & Lee, S. S. (1999). Political embeddedness in the new triad: Implications for emerging economies. *Management International Review*, *39*, 257–275.

Coffee, J. C. (1999). The future as history: The prospects for global convergence in corporate governance and its implications. *Northwestern University Law Review*, *93*, 641–708.

Cottrell, J. (2009). Governance down-under: An overview of corporate governance in Australia. In A. Kakabadse & N. Kakabadse (Eds.), *Global boards: One desire, many realities*. Basingstoke, Hampshire, UK: Palgrave Macmillan.

Cowan, L. (2011). IPOs, the recovery: Starring China. *The Wall Street Journal*, January 3. Available at http://online.wsj.com/article/SB10001424052748704610904576032170947326628.html

Dahya, J., Dimitrov, O., & McConnell, J. J. (2008). Dominant shareholders, corporate boards, and corporate value: A cross-country analysis. *Journal of Financial Economics*, *87*(1), 73–100.

Dalton, D. R., Daily, C. M., Ellstrand, A. E., & Johnson, J. L. (1998). Meta-analytic reviews of board composition, leadership structure, and financial performance. *Strategic Management Journal*, *19*, 269–290.

De Jong, H. W. (1997). The governance structure and performance of large European corporations. *Journal of Management and Governance*, *1*, 5–27.

Demirag, I. S. (1998). Financial markets, corporate governance, and short-term pressures in the United Kingdom. In I. S. Demirag (Ed.), *Corporate governance, accountability, and pressures to perform: An international study*. Hartford, CT: JAI Press.

Denis, D. K., & McConnell, J. J. (2003). International corporate governance. *Journal of Financial and Quantitative Analysis*, *38*(1).

Denis, D. K., & McConnell, J. J. (2005). International corporate governance. In K. Keasey, S. Thompson & M. Wright (Eds.), *Corporate governance: Accountability, enterprise, and international comparisons*. Chichester, West Sussex, UK: Wiley.

Doi, N. (1998). Corporate governance, corporate control, and R&D in Japanese manufacturing. In I. S. Demirag (Ed.), *Corporate governance, accountability, and pressures to perform: An international study*. Hartford, CT: JAI Press.

Doidge, C., Andrew Karolyi, G., & Stulz, R. M. (2007). Why do countries matter so much for corporate governance? *Journal of Financial Economics*, *86*(1), 1–39.

Easterbrook, F. H. (1997). International corporate differences: Markets or law? *Journal of Applied Corporate Finance*, *9*(4), 23–30.

Epstein, M. J., & Roy, M. J. (2002). *Measuring and improving the performance of corporate boards*. Mississauga, ON: The Society of Management Accountants of Canada (CMA Canada).

Epstein, M. J., & Roy, M. J. (2003). Designing a board evaluation system for corporate directors. *NACD-Directors Monthly*, December, pp. 13–17.

Epstein, M. J., & Roy, M. J. (2004a). How does your board rate? *Strategic Finance, February*, 24–31.

Epstein, M. J., & Roy, M. J. (2004b). Improving the performance of corporate boards: Identifying and measuring the key drivers of success. *Journal of General Management*, *29*(3), 1–23.

Epstein, M. J., & Roy, M. J. (2005). Evaluating and monitoring CEO performance: Evidence from U.S. compensation committee reports. *Corporate Governance*, *5*(4), 75–87.

Epstein, M. J., & Roy, M. J. (2006). Measuring the effectiveness of corporate boards and directors. In M. J. Epstein & K. O. Hanson (Eds.), *The accountable corporation: Corporate governance* (Vol. 1). Westport, CT: Praeger Publishers.

Epstein, M. J., & Roy, M. J. (2007). *Strategic management of information for boards*. Mississauga, ON: The Society of Management Accountants of Canada, (CMA Canada).

Epstein, M. J., & Roy, M. J. (2010). Corporate governance is changing: Are you a leader or a laggard? *Strategic Finance, October*, 31–37.

Exley, M., & Hudson, B. (2006). U.K. corporate governance. In D. A. Nadler, B. A. Behan & M. B. Nadler (Eds.), *Building better boards: A blueprint for effective governance*. Mercer Delta Consulting.

Fackler, M. (2008). South Korea faces question of corporate control. *The New York Times*, April 24. Available at http://www.nytimes.com/2008/04/24/business/worldbusiness/24samsung.html

Favaro, K., Karlsson, P. O., & Neilson, G. L. (2010). CEO succession 2000–2009: A decade of convergence and compression. *Strategy + Business, Summer*(59), 77–89.

Fouche, G., & Treanor, J. (2006). In Norway, a woman's place is in the boardroom. *The Guardian*, January 9. Available at http://www.guardian.co.uk/money/2006/jan/09/business.workandcareers

Fuhrmans, V., & Stevens, L. (2010). Symbolic shift: Why Daimler is delisting. *The Wall Street Journal*, May 18. Available at http://www.efinancialnews.com/story/2010-05-18/daimler-delist-nyse

Garcia-Meca, E., & Sanchez-Ballesta, J. P. (2010). The association of board independence and ownership concentration with voluntary disclosure: A meta-analysis. *European Accounting Review*, *19*, 603–627.

Glendon, M. A., Gordon, M. W., & Osakwe, C. (1994). *Comparative legal traditions: Text, materials, and cases on the civil and common law traditions with special references to French, German, and English*. St. Paul, MN: West Publishing Co.

Goetzmann, W., & Koll, E. (2005). The history of corporate ownership in China: State patronage, company legislation, and the issue of control. In R. K. Morck (Ed.), *A history of corporate governance around the world: Family business groups to professional managers*. Chicago, IL: University of Chicago Press.

Gregory, H. J. (2006). International corporate governance: A gradual if incomplete convergence. In M. J. Epstein & K. O. Hanson (Eds.), *The accountable corporation: Corporate governance* (Vol. 1). Westport, CT: Praeger Publishers.

Hermalin, B. E., & Weisbach, M. S. (2003). Boards of directors as an endogenously determined institution: A survey of the economic literature. *Economic Policy Review – Federal Reserve Bank of New York*, *9*, 7–26.

Husted, B. W., & Serrano, C. (2002). Corporate governance in Mexico. *Journal of Business Ethics*, *37*, 337–348.

International Finance Corporation. (2008). *The E.U. approach to corporate governance: Essentials and recent developments*. Washington, DC: IFC.

Iwata, E. (2008). Corporate governance gets more transparent worldwide. *USA Today*, February 18. Available at http://www.usatoday.com/money/companies/management/2008-02-17-corporate governance_N.htm

Jensen, M., & Meckling, W. (1976). Theory of the firm: Managerial behavior, agency costs, and ownership structure. *Journal of Financial Economics*, *3*, 305–360.

Kakabadse, A., Kakabadse, N., & Yavuz, O. (2009). Turkish chairmen: Contrasting the art of dialogue against the discipline for governance. In A. Kakabadse & N. Kakabadse (Eds.), *Global boards: One desire, many realities*. Basingstoke, Hampshire, UK: Palgrave Macmillan.

Kakabadse, N., Kakabadse, A., & Yang, C. (2009). The Chinese boardroom: Roles, dynamics, and relationships. In A. Kakabadse & N. Kakabadse (Eds.), *Global boards: One desire, many realities*. Basingstoke, Hampshire, UK: Palgrave Macmillan.

Kakabadse, N., Tsvetkov, O., Uldatsov, V., & Kakabadse, A. (2009). Boards of steel: The role and contribution of the non-executive director. In A. Kakabadse & N. Kakabadse (Eds.), *Global boards: One desire, many realities*. Basingstoke, Hampshire, UK: Palgrave Macmillan.

Kakabadse, N., Yang, H., & Sanders, R. (2009). The recruitment and evaluation of boards of directors in China. In A. Kakabadse & N. Kakabadse (Eds.), *Global boards: One desire, many realities*. Basingstoke, Hampshire, UK: Palgrave Macmillan.

Kaplan, S. N. (1997). Corporate governance and corporate performance: A comparison of Germany, Japan, and the US. *Journal of Applied Corporate Finance*, *9*(4), 86–93.

Khanna, T., Kogan, J., & Palepu, K. (2006). Globalization and similarities in corporate governance: A cross-country analysis. *Review of Economics and Statistics*, *88*(1), 69–90.

Klapper, L. F., & Love, I. (2004). Corporate governance, investor protection, and performance in emerging markets. *Journal of Corporate Finance*, *10*(5), 703–728.

Kripalani, M. (2009). India's Madoff? Satyam scandal rocks outsourcing industry. *Bloomberg Businessweek*, January 7. Available at http://www.businessweek.com/globalbiz/content/jan2009/gb2009017_807784.htm

La Porta, R., Lopez-de-Silanes, F., Shleifer, A., & Vishny, R. W. (1998). Law and finance. *The Journal of Political Economy*, *106*(6), 1113–1155.

La Porta, R., Lopez-de-Silanes, F., Shleifer, A., & Vishny, R. W. (1999). A corporate ownership around the world. *Journal of Finance*, *54*, 471–517.

La Porta, R., Lopez-de-Silanes, F., Shleifer, A., & Vishny, R. W. (2000). Investor protection and corporate governance. *Journal of Financial Economics*, *58*, 3–27.

Larcker, D. F., Richardson, S. A., & Tuna, I. (2007). Corporate governance, accounting outcomes, and organizational performance. *The Accounting Review*, *82*(4), 963–1008.

Larcker, D. F., & Tayan, B. (2011). *Corporate governance matters*. Upper Saddle, NJ: FT Press.

Levrau, A., & Van den Berghe, L. (2009). Identifying key determinants of effective boards of directors. In A. Kakabadse & N. Kakabadse (Eds.), *Global boards: One desire, many realities*. Basingstoke, Hampshire, UK: Palgrave Macmillan.

Matoussi, H., & Jardak, M. K. (2009). *International corporate governance and finance: legal, cultural and political explanations*. The Economic Research Forum Working Papers, 503. Retrieved from http://www.erf.org.eg/CMS/uploads/pdf/1226392646_15thAC_INST_Hamadi_Matoussi.pdf. Accessed on February 9, 2012.

McFarlan, F. W., Xu, J., & Manty, T.Y. (2009). *Corporate governance in China: Current practice, key problems*. Harvard Business School, Case 9-309-058.

Merryman, J. H. (1969). *The civil law tradition: An introduction to the legal systems of Western Europe and Latin America*. Stanford University Press.

Millar, C., Eldomiaty, T., Choi, C. J., & Hilton, B. (2005). Corporate governance and institutional transparency in emerging markets. *Journal of Business Ethics*, *59*, 163–174.

Mitchell, T. (2008). Chinese board structures 'lead to confusion'. *The Financial Times*, April 3. Available at http://www.ft.com/cms/s/0/6a80bc66-0117-11dd-a0c5-000077b07658.html#axzz1qd6SuN3R

Morck, R. K., & Steier, L. (2005). The global history of corporate governance. In R. K. Morck (Ed.), *A history of corporate governance around the world: Family business groups to professional managers*. Chicago, IL: University of Chicago Press.

Morey, M., Gottesman, A., Baker, E., & Godridge, B. (2009). Does better corporate governance result in higher valuations in emerging markets? Another examination using a new data set. *Journal of Banking & Finance*, *33*(2), 254–262.

New York Stock Exchange (2003). Final NYSE Corporate Governance Rules. Available at http://www.nyse.com/pdfs/finalcorpgovrules.pdf

OICU-IOSCO (2007a, March). *Board independence of listed companies*. Final Report from the Technical Committee of the International Organization of Securities Commissions, in consultation with the OECD.

OICU-IOSCO (2007b, December). *Corporate governance practices in emerging markets*. Report from the Emerging Markets Committee of the International Organization of Securities Commissions.

Organization for Economic Cooperation and Development (OECD). (2004). Principles of Corporate Governance. Available at http://www/oecd.org/dataoecd/32/18/31557724.pdf

Oseichuk, E., Kakabadse, N., & Kakabadse, A. (2009). Workings of the board: Case of the international joint venture "Arman," Kazakhstan. In A. Kakabadse & N. Kakabadse (Eds.), *Global boards: One desire, many realities* (pp. 107–139). Basingstoke, Hampshire, UK: Palgrave Macmillan.

Oxford Analytica (2010). *Europe may impose quotas for women*. *Forbes.com* (August 24). Available at http://www.forbes.com/2010/08/23/european-union-women-quotas-business-oxford.html

Patel, T. (2010). French women storm the corporate boardroom. *Bloomberg Businessweek*, June 10. Available at http://www.businessweek.com/magazine/content/10_25/b4183015410606.htm

Pierce, C. (2009). International board evaluation: Trends and practices. In N. Kakabadse & A. Kakabadse (Eds.), *Global boards: One desire, many realities*. Basingstoke, Hampshire, UK: Palgrave Macmillan.

Power, S., & Casselman, B. (2011). White House probe blames BP, industry in Gulf blast. *The Wall Street Journal*, January 6. Available at http://online.wsj.com/article/SB10001424052748704405704576064122843672118.html

Pozen, R. C. (2010a, December 5). *The case for professional boards*. The Harvard Law School Forum on Corporate Governance and Financial Regulation.

Pozen, R. C. (2010b). The big idea: The case for professional boards. *Harvard Business Review*, *88*(120), 50–58.

PRNewswire. (2010). Few boards splitting the chairman and CEO roles among highest market cap U.S. companies. Media Release, PRNewswire, November 1. Retrieved from http://www.prnewswire.com/news-releases/few-boards-splitting-the-chairman-and-ceo-roles-among-highest-market-cap-us-companies-106429973.html

Rajagopalan, N., & Zhang, Y. (2008). Corporate governance reforms in China and India: Challenges and opportunities. *Business Horizons*, *51*, 55–64.

Reed, A. M. (2002). Corporate governance reforms in India. *Journal of Business Ethics*, *37*, 249–268.

Reed, D. (2002). Corporate governance reforms in developing countries. *Journal of Business Ethics*, *37*, 337–348.

Schmidt, R. (1998). Financial markets, corporate governance, and short term pressures in Germany. In I. S. Demirag (Ed.), *Corporate governance, accountability, and pressures to perform: An international study*. Hartford, CT: JAI Press.

Shleifer, A., & Vishny, R. W. (1997). A survey of corporate governance. *Journal of Finance*, *52*, 737–783.

Sify (2008, February 5). *Is global model of corporate governance likely or desirable?* Retrieved from http://www.sify.com/finance/is-global-model-of-corporate-governance-likely-or-desirable-news-fixed-income-jegm8Dgifgf.html. Accessed on February 23, 2008.

Sjoberg, O. (2009). Corporate governance and earnings inequality in the OECD countries 1979–2000. *European Sociological Review*, *25*(5), 519–533.

Stout, L. (2011, February18). *New thinking on shareholder primacy*. Law-Econ Research Paper No. 11-04, University of California, Los Angeles, CA. Retrieved from http://ssrn.com/abstract=1763944. Accessed on February 9, 2012.

Thakurta, P. G. (2009, January 9). *Satyam scam questions corporate governance*. Inter Press Service. Retrieved from http://ipsnews.net/news.asp?idnews=45347. Accessed on February 9, 2012.

Wong, S.C.Y. (2011). *Elevating board performance: The significance of director mindset, operating context, and other behavioral and functional considerations*. Law & Economics Research Paper No. 11-12, Northwestern University School of Law, Chicago, IL. Retrieved from http://ssrn.com/abstract=1832234. Accessed on February 9, 2012.

Wright, M., Buck, T., & Filatotchev, I. (2005). Corporate governance in transition economies. In K. Keasey, S. Thompson & M. Wright (Eds.), *Corporate governance: Accountability, enterprise, and international comparisons*. Chichester, West Sussex, UK: Wiley.

BUILDING AND NURTURING A HIGH-PERFORMANCE–HIGH-INTEGRITY CORPORATE CULTURE

Jean-François Manzoni

ABSTRACT

Purpose – *This chapter reviews the challenges faced by top management teams as they strive to create corporate cultures that combine a high performance with a strong sense of integrity.*

Methodology/approach – *The chapter integrates diverse theories from organizational research and cognitive psychology, as well as published accounts of ethical breakdowns, to shed new light on the barriers to corporate integrity.*

Findings – *The chapter distinguishes between two major types of ethical breakdowns.*

Conscious transgressions, where the individuals know what they should or should not do, but choose nonetheless to follow the unethical path, a decision that they then need to rationalize and which often places them on a slippery slope.

Performance Measurement and Management Control: Global Issues
Studies in Managerial and Financial Accounting, Volume 25, 41–63

ISSN: 1479-3512/doi:10.1108/S1479-3512(2012)0000025005

Unconscious transgressions, where the individuals do not even realize that they are making an inappropriate decision, as they fall prey to ethical fading or to other cognitive biases.

Practical implications – *The chapter proposes that top management plays a key role in establishing a climate where employees can speak up, emphasizing the importance of all stakeholders, and investing in training to increase awareness of the cognitive biases that support transgressions.*

Social implications – *The chapter recommends that management educators must alert students more forcefully to the personal and organizational repercussions of "minor" ethical transgressions; increase student awareness of key cognitive concepts, including ethical fading and other mental biases; and highlight the possible dysfunctions of intuitive remedies, like incentives or rules and regulations.*

Originality/value of chapter – *The chapter provides a clearer analysis of the causes of ethical breakdowns, allowing for more effective prevention.*

Modern organizations – for profit or not – face increasingly high demands for performance. In the case of for-profit organizations, these demands come from or are exacerbated by increasingly global financial markets, where capital can flow relatively easily to greater returns elsewhere. Not-for-profit organizations also face increasing demands for effectiveness and efficiency, from financially strapped governments, demanding tax-payers/citizens, and/or from donors hoping to obtain more impact from their contributions.

The search for the secrets of high organizational performance has hence been intense for decades and has yielded a number of well-known books along the way, some of which have stood the test of time better than others.

The search for high integrity in organizations is probably a slightly more recent phenomenon but has clearly intensified over the last few years. In part, this intensification is due to increasing expectations from society and customers, which have often led to increased regulation (including much higher fines) and lower tolerance for behavior perceived to be unethical or otherwise inappropriate. Transgressions are also harder to hide, as employees increasingly have (regulation-supported) incentives and means to speak up. And once identified (thanks to employee support or not), transgressions are likely to be disseminated much more quickly and pervasively than before – via the power of the Internet, social media, and even traditional media – thus having a much greater impact on employee reputation than before.

Senior leaders hence look for ways to boost both "performance" and "integrity" and, ideally, to embed them into the "culture" of the organization so their effect remains high beyond the existing leaders' tenure and focus.

ORGANIZATIONAL CULTURE

It is now generally accepted that organizations that enjoy lasting success do so in part because they have developed a "strong" and "positive organizational culture."

> While there will always be some counter-example, evidence provided by management thinkers like Kotter and Heskett (1992), Collins and Porras (1994) and O'Reilly and Pfeffer (2000) strongly suggests this is indeed the case.

> Many well-known organizations certainly also seem to believe so: General Electric, for example, explained in its 2008 Annual Report – and still features on their web site in January 2012 – that "at GE, we consider our culture to be among our innovations. Over decades our leaders have built GE's culture into what it is today – a place for creating and bringing big ideas to life. Today, that culture is the unifying force for our many business units around the world." (General Electric, 2012)

> Similarly, Schein (2008, p. 362), a life-long student of organizational cultures, believes that, "It can be argued that the only thing of real importance that leaders do is to create and manage culture, that the unique talent of leaders is their ability to understand and work with culture," and that "what distinguishes leadership from management (…) is that leaders create and change culture."

But what is organizational culture, exactly? Discussing the culture of an organization with its employees and managers is often a frustrating exercise in part because while we all understand the word "culture," we often understand it differently. Try this exercise: ask a group of executives what they understand by "culture." You will likely get a long list of words that, while related, mean very different things!

At its most basic, culture is said to be "the way we do things around here." More sophisticated definitions include the following:

> A pattern of shared basic assumptions that the group learned as it solved its problems (…), that has worked well enough to be considered valid and, therefore, to be taught to new members as the correct way to perceive, think and feel in relation to those problems. (Schein, 1992, p. 9)

> The specific collection of values and norms that are shared by people and groups in an organization and that control the way they interact with each other and with stakeholders outside the organization. (Hill & Jones, 2001, p. 68)

> The set of shared mental assumptions that guide interpretation and action in organizations by defining appropriate behavior for various situations. (Ravasi & Schultz, 2006, p. 437)

My own favorite definition is that of Goffee and Jones (1998, p. 15): "Culture comes down to a common way of thinking, which drives a common way of acting." This definition captures the fact that culture is about the way people *think*, translates into the way people *behave*, and that culture refers to a pattern of behavior that is reasonably *pervasive* throughout the organization (which can mean across countries, functions, departments, hierarchical levels, and across time).

Changing an organization's culture hence requires modifying the "common way of thinking" of its members. As a result, many organizations have over the years launched "cultural change programs," where the goal was to modify employee attitudes. Such efforts tend to fail because they miss a counterintuitive but long-known aspect of human functioning: Human beings tend to act their way into new attitudes much more than they think their way into new behaviors.

This is not a recent discovery: In his *Nicomachean Ethics*, written almost 2,500 years ago, Aristotle explained that "we do not act rightly because we have virtue or excellence, but we rather have those because we have acted rightly." He explained that human beings "acquire a particular quality by constantly acting a particular way .... Moral excellence comes about as a result of habit. You become just by performing just actions, temperate by performing temperate actions, brave by performing brave actions."

Reshaping the culture of an organization hence requires *reshaping the behavior of enough people, for long enough* for them to *internalize the new behavior*, i.e., for the new behavior to become "a habit."

I discussed in previous EIASM workshops (Manzoni, 2006, 2008) how reshaping employee behavior requires the alignment of the signals sent by all the managerial levers to which employees are exposed. There are several ways to describe such levers. Pascale and Athos (1981) provided a model emphasizing seven S's (Strategy, Structure, Systems, Skills, Staff, Style, and Shared values). Galbraith (1995) proposed the "Star Model," featuring strategy, structure, process, rewards, and people. I tend to use a slightly different framework, presented in Fig. 1.

These levers must send consistent enough signals, and must do so for a long enough period for employees to internalize the new behavior. How long is long

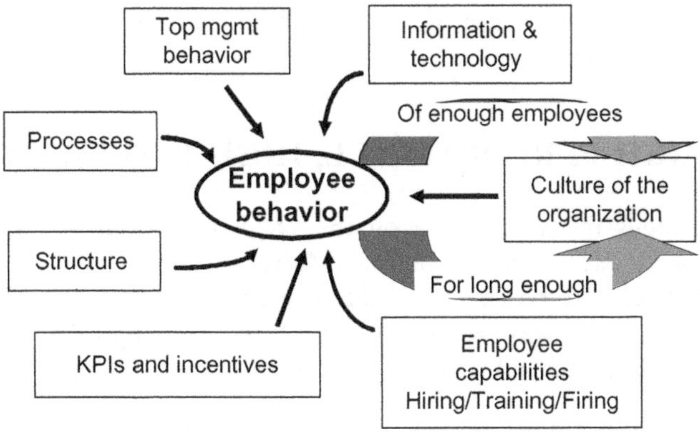

Fig. 1. Reshaping an Organizational Culture.

enough depends on a number of endogenous and exogenous factors, including how large the required change is, how deeply anchored are the previous behaviors, how individually powerful and collectively aligned are the forces applied to the employees, how many individuals need to be affected, how geographically dispersed these individuals are, and how easy it is to monitor behavior and enforce the changes. But it typically takes years rather than months or weeks for people to (a) learn effectively to produce new behaviors that are significantly different from their initial programming and (b) perfect this learning to a high enough degree for the new behavior to become a habit.

Tesco – the British-based but increasingly international retailer – provides a fantastic illustration of this alignment process. As discussed in Barsoux and Manzoni (2008), Tesco deploys a series of organizational practices and mechanisms that are often individually smart but rarely earth-shattering; most organizations do most of what Tesco does most of the time, but very few do all of it all the time over several years. Tesco has enjoyed exceptional success over the last 15 years and has created a remarkable corporate culture characterized by high employee engagement, customer-focused innovation, and excellence in execution because its managerial levers are

- incredibly *aligned*, with the organization's strategy and with one another;
- applied with *great intensity* (i.e., people do not just "do things" to "tick the box" and satisfy the process requirements; they really invest time and energy into the activity);

- *consistent over time*, driven by a management team that has been together for over a decade.

HIGH PERFORMANCE, HIGH INTEGRITY

I will not focus too much here on the "high performance" side of this equation, as I have done so before (Manzoni, 2002, 2004, 2006). High performance tends to be an elusive characteristic that rests on a large set of factors that are not always equally present. Still, companies that outperform their peers over nontrivial periods of time tend to

- propose products and services that are highly valued by customers and consumers (i.e., the products and services are perceived to provide great value for their price);
- have effective and efficient enough operational and managerial processes to capture some of the value created for customers;
- display a strong performance orientation, typically characterized by the setting of demanding objectives and their pursuit with great focus, energy, and intensity.

The "high integrity" side of the equation is no simpler. High integrity starts with a basic *compliance* dimension, whereby the organization's (direct and indirect) employees respect the laws and regulations that apply to their activities. That's of course a necessary condition, but it is by no means a sufficient one. High integrity requires more than respecting laws and regulations. It implies that the organization's direct and indirect employees will do "the right thing" for the organization's various constituencies, including customers, employees, suppliers, communities, and society in general.

The definition of the "right thing" is by no means clear-cut nor is it necessarily consistent over time and across countries and industries. Still, I think it is fair to say that as a general rule and across the world, society has over the last few years become increasingly demanding vis-à-vis organizations, in terms of their physical and social impact on communities, the appropriateness of certain business practices, or the distribution of wealth between shareholders, managers, and employees.

Different organizations (operating in the same industry and country) may also argue that "the right thing to do" is different for them than for their competitors. To some extent, expected behavior indeed depends on the organization's brand proposition and the reputation it wants to create for itself. (For example, Tesco has over the years worked hard to be perceived on the side of consumers and more recently on the side of communities. Banks have more rarely taken such a stand, but now that Standard Chartered Bank has forcefully communicated it's new "Here for Good" tag line, it is likely to be held to a higher standard than some of its competitors that have not made such a promise.)

The factors underlying organizational performance are probably too numerous and feature too much lag and measurement error for us ever to get an uncontroversial answer to the question: What is the long-term relationship between high performance and high integrity? There is a certain amount of anecdotal evidence suggesting that over the long run this relationship is generally positive. Studies such as Kotter and Heskett (1992) or Collins and Porras (1994) go in that direction too.

It is increasingly clear that breakdowns in integrity can be extremely costly. Financial service organizations such as UBS and Société Générale have also suffered considerable losses because of employee misconduct. The disappearance of companies like Enron, Parmalat, or WorldCom also involved inappropriate behavior and cover-ups. And, of course, regulators have recently been levying significant fines – in excess of $1 billion – on companies such as Pfizer, Johnson & Johnson, and Siemens, to name but a few.

Conversely, some organizations have managed to create significant goodwill for themselves with their key constituencies, and this goodwill can help reduce the negative consequences of an isolated breakdown.

In the short term, however, there can be clear conflicts between the requirements of performance and integrity: a price increase can destroy customer goodwill but lead to a short-term improvement in profit; employee goodwill can be depleted by withholding compensation or investments in safety or training, with negative longer-term consequences but a positive impact on short-term financial performance.

Arguably, though, these trade-offs are at the heart of managers' jobs. Most managerial decisions involve some form of trade-off: more focus on one activity leaves less time for another activity; more attention to the urgent can lead to less attention to the important; an investment here will preclude an investment there. Managers operate under conditions of scarce resources – time, attention, money – and hence need to allocate these resources in one direction at the expense of another.

Similarly, a great many managerial decisions involve trade-offs between short-term and long-term financial performance. In the absence of short-term financial objectives, most organizations could make much larger (and not necessarily long-term NPV enhancing) investments in increasing the satisfaction of their customers, suppliers, employees, or to improve their social and environmental performance.

These trade-offs are hard to assess because of the same two reasons that make possible the divergence between short- and long-term performance: (a) imperfect – i.e., incomplete – measures of performance that capture the impact of managerial decisions with some lag and (b) managers' always imperfect understanding of cause–effect relationships underlying long-term performance. The divergence can then be magnified by offering managers incentives that are tied to short-term performance measures.

Management accounting and control scholars have proposed ways of addressing this challenge. Many authors have emphasized different aspects, but overall the prescription would include:

- The development and use of multiple and more sophisticated indicators of performance to
  - help managers better understand over time the cause–effect relationships involved (and thus make better short-term and long-term trade-offs);
  - propose incentives tied to more complete and more long-term friendly objectives.
- The development of a "control system" that following the work of Simons (2000) and others, might include
  - a belief system featuring corporate values, leadership competencies, and/or a Code of Conduct, to inspire employees to "do the right thing";
  - boundary controls featuring significant numbers of regulations and policies, accompanied by employee training and certification, with audits to check compliance by an independent group of specialized employees.

I will argue below that these prescriptions are insufficient and, in some cases, can be counterproductive.

In summary, the challenge of creating high-performance–high-integrity organizations and corporate cultures requires top management to strive to create a common way of thinking that drives a common way of acting, such that (tens of) thousands of managers and employees – coming to the matter with different personal norms and often different cultural norms – will become willing and able to respect the law and, more generally, to "do the right thing" for a complex set of stakeholders sometimes presenting divergent needs.

INTEGRITY BREAKDOWNS

Conscious Breakdowns

In order to understand how to create such a culture, let us first examine why the employees of an organization would fail to "do the right thing."

One obvious reason is that employees face some motivation to misbehave. There are two major drivers: fear and greed.

The greed factor is clear: Employees sometimes try to take advantage of the imperfections of performance measures and/or of the control system to take actions that enhance their own welfare (e.g., expected compensation) at the expense of other stakeholders and potentially of the organization's long-term reputation and/or performance. While one can never be sure of individual motivations, the behavior of David Sokol (a Berkshire Hathaway senior executive who bought a company's stock very close to the time he recommended the stock to his employer and then realized a multimillion dollar profit when Berkshire Hathaway bought the company a few weeks later), as described in Barrett (2011), could reasonably be attributed to a desire for personal enrichment.

The other major driver of "unethical behavior" is fear. This fear – of losing relationships, money, promotion opportunities, or personal reputation if one does not engage in the unethical behavior – is often self-imposed by the employee. It is rooted in the individual's understanding of what the boss expects or desires, and what the boss, the organization, or relevant others are likely to value.

Of course, in some cases the boss or relevant others can reinforce the employee's perception of their preferences through their behavior. Employees tend to be sensitive to what their bosses say and do (e.g., explicit references to a clear link between achievement of some performance level and some reward or punishment). But employees also notice what their bosses do *not* say or do (e.g., the failure to mention that these performance objectives ought to be met through behavior that meets certain ethical criteria).

Managers and organizations must understand just how little encouragement is needed for most reasonable human beings to perform acts that they know to be problematic. This human propensity was vividly demonstrated by Milgram's famous (1963, 1974) experiments, in which the vast majority of subjects – with minimal pressure being exerted on them – agreed to administer electrical shocks of very high intensity to other individuals, regardless of the apparent pain experienced by recipients. These experiments have since been replicated in numerous countries over the years, including over the last few years on TV programs in the United Kingdom, France, and the United States, with surprisingly consistent results across countries and decades.

The lesson is: Human beings the world over have a strong propensity to conform to the wishes of credible and legitimate authority figures. In most national and corporate cultures, bosses are seen as such figures and must hence be very mindful of the signals they send – and fail to send.

Having succumbed to fear and/or greed and committed actions that they know to be problematic, employees then need to rationalize these actions, i.e., to explain to themselves that while the action would normally be inappropriate, it is actually permissible under the circumstances. This rationalization process can take a number of forms, including:

- Minimizing the magnitude of the breach: "Yes it was inappropriate, but it's a very small breach."
- Comparing to other offenders: "Others are doing it too – and started doing so before me."
- Attributing the breach to an unavoidable/external cause: "I did not really have a choice; 'they'/this made me do it."
- Attributing the breach to a higher purpose: "If I had not done this, worse outcomes would have occurred – e.g., the organization would have gone bankrupt and many people would have lost their jobs."

Research shows – and our own human experience every day confirms – that the human tendency to rationalize is considerable. This propensity is based on the human need to reduce cognitive dissonance, i.e., to reduce or eliminate the tension that people experience when they hold two cognitions that are psychologically inconsistent, such as "I'm a highly ethical person" and "I just committed an action that I would normally regards as inappropriate."

Human beings' propensity to reduce cognitive dissonance was first studied and analyzed by Festinger (1957). Since then, a considerable number of studies have documented the pervasiveness and robustness of this phenomenon. In particular, many studies show that human beings can be remarkably resistant to the existence of accurate and relevant disconfirming data. This empirical regularity is explained by recent neuroscience studies. MRIs of the brain show that when people are exposed to dissonant information, reasoning areas of the brain shut down. By contrast, when exposed to consonant information, emotion circuits of the brain light up as if the subject was experiencing a pleasurable sensation (Westen, Blagov, Harenski, Kilts, & Hamann, 2006).

Studies of individual and corporate wrongdoing also suggest that misbehavior tends to increase over time. The individual engages on a "slippery slope" by committing a small ethical breach, which leads to a larger one and another. As Cynthia Cooper, the whistleblower in the WorldCom accounting misreporting, explained it in her (2008) account of the debacle: "People don't wake up one day and say, 'I think I'll become a criminal today.' Instead, it's often a slippery slope and we lose our footing one step at a time."

Cooper's (2008) account offers a fascinating insight into the incremental and social nature of such corporate misbehavior. At the time, WorldCom was a fast-growing and exciting place to work. The darling of Wall Street, it had built on very humble beginnings and was in the process of conquering the United States' telecom world. Headquartered in a small town in the south of the United States, it was the most glamorous and desirable employer around as well as a source of local pride.

The end of a quarter revealed an unexpected shortfall in quarterly profits. Communication of that shortfall would likely have a significant negative impact on a stock price that rested on very lofty growth expectations by the market. A reduction in the stock price would in turn hamper the continued growth of WorldCom, which was financing most of its acquisitions by issuing more stock. Three accountants were asked to investigate the spike in expenses and could not explain it. They were then asked by their charismatic boss – a smart, energetic, hard-working, engaged, and engaging professional – to "make it go away."

The individuals knew that what they were being asked to do was inappropriate. They considered refusing to comply, but ultimately relented. They "sharpened the pencil" and reduced various reserve accounts to make up for the shortfall. They also prepared their letter of resignation, which they were prepared to hand in if their boss did not agree to issue a clear indication to Wall Street that future growth would be slightly reduced. The CFO made the announcement and the letters of resignation stayed in the drawers. The individuals figured that this was a small breach – it was almost like a "bridge loan," which would be repaid next quarter – committed for a good cause – the protection of a great company's growth prospects ... a small price to pay not to let down one's friends and "corporate family." As the next quarter progressed, memories of the emotional tension experienced at the time started to fade.

Unfortunately, the end of the next quarter proved problematic as well. The quarterly profit was again below expectations. The three accountants again agreed to find the required amounts in various reserve accounts. In time, they exhausted all the reasonable reserve accounts and had to resort to more aggressive techniques to cover up the misreporting: they started capitalizing items that should have been considered current expenses.

Of course, the individuals' discomfort increased with the magnitude and severity of the cheating; they grew increasingly uncomfortable with their actions. But by then, their initial motivation to collude – their desire not to let down their "corporate family" and in the process lose a job they loved in a company they loved – was now supported by another motivation. If they stopped now, their past ethical breaches would become public and would have devastating consequences for them and their families. They were – and clearly felt – trapped.

Cooper (2008) cites two classic thinkers on this notion of a slippery slope:

- From Scottish author Samuel Smiles: "Sow a thought, reap an action; sow an action, reap a habit; sow a habit, reap a character; sow a character, reap a destiny."
- From C. S. Lewis: "Every time you make a choice, you are turning the central part of you, the part that chooses, into something a little different than it was before .... Good and evil both increase at compound interest. That is why the little decisions you and I make every day are of such importance."

Cooper concludes that "the foundation of our character is laid brick by brick, decision by decision, in how we choose to live our lives" (2008, p. 364).

One could argue that these individuals lacked courage. Gebler (2006) indeed reports quotes from them suggesting that they would agree with that statement.

- "I'm sorry for the hurt that has been caused by my cowardly behavior."
  - *Scott Sullivan, CFO*
- "At the time I consider the single most critical character defining moment of my life, I failed. It's something I'll take with me the rest of my life."
  - *David Myers, controller*
- "Faced with a decision that required strong moral courage, I took the easy way out .... There are no words to describe my shame."
  - *Buford Yates, director of general accounting*

One response would hence be that organizations need to hire individuals who because of their character and personality are more likely to display ethical and courageous behavior. Many studies of high performing organizations indeed tend to suggest that great companies do invest more time, effort, and energy in selecting the "right people for them."

But in fairness, the individuals engaged in the WorldCom debacle were not known to be especially weak. As described by Cooper, who knew them well, they were basically "good people" who made bad decisions. They took a first action that breached their principles and rationalized it away. This first ethical breach made it easier to commit a second one, at which point they became almost condemned to continue lying in order to conceal their initial transgressions.

Unconscious Breakdowns

Fear of hypothesized consequences and/or a desire to improve one's welfare can hence lead employees and managers to set aside their principles and knowingly fail to do "the right thing." In many cases, however, employees and managers fail to do the right thing because they simply do not know what is "the right thing to do."

In some cases, people can identify that they face a dilemma but find it difficult to weigh the various parameters of the decision. But recent research shows that in many cases, people do not even realize they are facing an ethical issue; as a result, they often fail to recognize that they (and/or others) are behaving unethically, which also leads them to overestimate their own (and their favored others') integrity.

There are at least four major causes to this blindness, which are described in the following sections.

"Ethical Fading"

Ethical fading refers to the process by which the ethical aspects and/or implications of the decision fade away from our mind; we fail to notice them. Research (see Bazerman & Tenbrunsel, 2011a) shows that ethical fading is more likely to happen when:

- The "business" implications of the decisions are heavily emphasized (as they would typically be in any resource-constrained environment).
- Language euphemisms are used (e.g., referring to "aggressive" accounting practices, "let go" for "fire," or "right sizing" for "mass sackings").

- Potential victims are numerous and anonymous (as opposed to an individual who can be identified).

More worryingly, research shows that some of the countermeasures proposed in "mainstream ethical programs" may also contribute to ethical fading. Studies suggest that the following conditions can trigger an *increase*, rather than decrease in ethical fading and unethical behavior:

- Small financial penalties associated with transgression
  - Several laboratory experiments (e.g., students facing an environmental decision, in Bazerman & Tenbrunsel, 2011a) and field experiments (e.g., parents showing up late to pick up their child at day care, in Frey & Jegen, 2001) show that the presence of a penalty leads people to feel that paying the penalty discharges them from any other responsibility. The decision is no longer an ethical one, it becomes a business decision.
- The existence of a large number of rules and regulations
  - Many "ethics and compliance" programs feature large numbers of rules and regulations. This number also tends to increase over time, as new challenges get identified in some part of the organization and the organization feels the need to regulate this point throughout its ranks.
  - Unfortunately, research in psychology (Self-Determination Theory, Deci & Ryan, 1985) and economics (Motivation Crowding Theory, Frey & Jegen, 2001) shows that this tactic is likely to backfire. Several experiments indeed document an association between a large number of rules and regulations and an increase in transgressions. It seems that large numbers of rules and regulations lead individuals to feel that they no longer have to self-monitor; the monitoring is externalized to the (seemingly exhaustive) rules and regulations. The question then ceases to be, "Is this the right thing to do?" and rather becomes, "Is this allowed?"
- Disclosure of the "conflict of interest"
  - The idea of requiring individuals to disclose conflicts of interest seems quite reasonable. Awareness of this conflict of interests should make other interested parties more cautious in their dealings and more wary of the conflicted party's estimates. In an interesting experiment, however, Cain, Loewenstein, and Moore (2005) showed that third parties are not always able to discount the effects of the conflict of interest appropriately. More interesting still, they found that individuals asked to disclose their conflict of interest provided estimates that were more biased than those who faced the same incentives but did not

disclose this conflict of interest to the estimates' recipients. Again, this result is consistent with a decrease in self-monitoring by individuals who feel that disclosing their conflict of interest discharges them of their responsibility to be accurate.

Motivated Blindness

Motivated blindness refers to human beings' tendency to selectively disregard information that would be costly for them to notice, and to interpret available information in ways that are supportive of their best interest. This includes observation and interpretation of others' behavior, i.e., human beings' propensity not to notice others' unethical behavior when noticing it would potentially harm the observer.

Moore, Tanlu, and Bazerman (2010) ran an interesting experiment in this area, when they asked four groups of individuals to estimate the value of a fictional company. Each group was given a role in the potential transaction: buyer, seller, buyer's auditor, and seller's auditor. The two groups of auditors were asked to act as independent experts and were rewarded based on the accuracy of their estimate. As expected, "sellers" provided higher estimates of company value than "buyers." More surprisingly, sellers' auditors also provided higher estimates – 30% higher on average! – than buyers' auditors. This study was replicated with actual auditors from one of the Big Four. The outcome was similar, confirming that even in the presence of incentives encouraging the impartial behavior they were being asked to adopt, the "auditors" were unconsciously influenced by the point of view of the party by whom they were supposedly being hired.

Confirmatory Biases

A considerable body of work in the psychology literature documents how human beings' perception of reality is shaped by their beliefs and expectations. In particular, we are prone to selecting, interpreting, and even remembering events in ways that are consistent with our beliefs. At a conscious level, this process is a manifestation of Festinger's (1957) observation that human beings need to reduce cognitive dissonance.

But very often this filtering process will be totally unconscious, i.e., will occur without the individual noticing it. The dissonant information (dissonant because contradicting our view or simply because unexpected) will simply not print on our radar screen! If it does appear on the radar screen, it will often – and in complete good faith – be interpreted in ways that are consistent with our belief/expectation.

And while for decades we thought that memory, at least, was a reliable companion, recent research shows that memory is far less reliable than previously thought. As Tavris and Aronson (2007, p. 6) put it: "Memories are often pruned and shaped by an ego-enhancing bias that blurs the edges of past events, softens culpability and distorts what really happened." Evidence on the imperfections of human memory has grown to the point that the US police and justice system is actively reexamining the role and influence of eyewitnesses (see Liptak, 2011).

So going back to the David Sokol example mentioned above, Bazerman and Tenbrunsel (2011b) examine Warren Buffett's failure to understand the inappropriateness of David Sokol's action and attribute it to motivated blindness. I think confirmatory biases are at least as good an explanation. David Sokol did tell Warren Buffett that he owned some stock in the company they were discussing. And I am absolutely willing to buy that Warren Buffet never thought of asking Sokol, "And when did you buy that stock?" I think it did not even cross Buffett's mind that Sokol bought the stock a few days before! Sokol was one of his trusted lieutenants; ethical behavior is an important pillar at Berkshire Hathaway; Buffett had no reason to suspect foul play and likely made an immediate and unconscious assumption on the timing of the purchase.

The unconscious nature of this process is important, because unconscious processes are harder for human beings to observe and hence correct. Unconscious biases are also a lot more pervasive than we tend to think. For example, when asked if they are prejudiced based on others' race, age, or weight, most individuals will respond negatively. When tested in ways that do not rely on introspection and instead tap the individuals' unconscious processes, however, many individuals turn out to be more prejudiced than they thought (see 1998 study by Greenwald, McGhee, and Schwartz on the Implicit Association Test, and the body of work that it has spawned over the last 13 years).

Self-Serving Bias/Identity Protection

Much evidence supports the idea that most human beings want to have a positive image of themselves. Studies have identified a "better-than-average effect," also called "illusory superiority," which leads a vast majority of people to rank themselves as above average in a variety of dimensions including driving skill, social sensitivity, leadership ability, and many other attributes (see, e.g., Kruger, 1999; Roese & Olson, 2007; Suls, Lemos, & Stewart, 2002). Human beings are also vulnerable to the widespread self-serving bias, which leads them to attribute their success to their merits and internal dispositions but their failures to unfavorable external factors (see Miller & Ross, 1975).

This desire to maintain a positive image can also lead us to regard as acceptable, when performed by us, actions that we otherwise know to be problematic. DeSteno and Valdesolo (2011) thus report on a study that asked people to allocate between themselves and another party two tasks: a short and rather pleasant one, and a longer, less pleasant one. The other party would not know who assigned the task to them.

In a first test, a group of individuals was asked what would be the fair and ethical thing to do. A vast majority of people indicated that the decision makers should allocate the long and difficult talk to themselves. Another (similar) group of individuals was then faced with that choice, and a vast majority of people chose to allocate the short, easy task to themselves. They then rated their actions as perfectly fair and ethical, thus displaying a clear double-standard.

The study then went one step further. Individuals were given wristbands of one of two colors, and asked to watch an individual from their group or from the other group, go through the exercise and assign him or herself the easy task. Even though observers had no personal stake in the outcome, the color of their wristband was enough to bias their judgment. Their assessment of the action was colored – literally! – by the wristband, with observers tending to rate the action as appropriate when the wristband was the same color as theirs, but decidedly less ethical if the individual wore a different color wristband.

So just as human beings have a tendency to extend the self-serving bias to their "in-group," we also seem to have a tendency to give the benefit of the doubt to people who are – or at least seem to be – "in our camp."

Going one step further, DeSteno and Valdesolo focused another study on the rationalization process (I am/you are a good person, so if I/you do this, it must be OK). They hypothesized that this rationalization process, while unconscious, probably consumed mental energy. So they asked another group of subjects to memorize a series of numbers and retain it in their heads as they were answering questions on the experiment and their actions. When thus kept cognitively busy, participants no longer rated their action as ethical. The rationalization process hence does seem to require processing and mental energy.

One last comment on this front: Sachdeva, Iliev, and Medin (2009) primed individuals to think of themselves as virtuous (generous, caring) or not (greedy, disloyal). In a subsequent, apparently disconnected step of the experiment, they asked participants whether they wanted to make a donation to charity. The people who wrote about themselves as virtuous ended up giving less – five times less! – than the people who were prompted to write about themselves in negative terms. Belonging to a group or an organization that has historically been morally upstanding might hence lead people to overestimate their ethical "nature" and underestimate how much effort they need to make to behave ethically. I wonder to what extent this phenomenon might have contributed to the recent series of mishaps, quality problems, and associated cover-ups observed at Johnson & Johnson, a company historically recognized as a paragon of virtue (see Voreacos, Nussbaum, & Farrell, 2011).

IMPLICATIONS FOR RESEARCH AND PRACTICE

This discussion suggests a number of possible implications for practitioners and management educators.

Implications for Practice

- Be careful with intuitively appealing responses such as increasing the number of rules and regulations, imposing penalties for transgressions and relying on the requirement to acknowledge conflicts of interests. While reasonable on paper, these practices can backfire and lead to more rather than less inappropriate behavior, as the presence of external control can lead individuals to exercise less self-control.
- Invest time and energy in a massive "managerial judgment education program." Each employee and manager brings to their work their individual programming, in part shaped by their national culture and their personal prior experiences. Yet they will make decisions that engage the name and reputation of the organization. In an ideal world, they would hence be trained to internalize the set of principles and trade-offs that the organization would like them to make.
- Combat ethical fading, by making sure that managers – starting with top management – regularly emphasize the importance of all stakeholders and the fact that financial performance is the result of a job well done, rather than an end in itself.
- Create a climate where employees can speak up and express their uneasiness about the decision that is being contemplated, thereby

increasing the likelihood that possible transgressions will be identified and intercepted.

Implications and Questions for Management Educators

Working on this chapter has led me to question whether I do enough in my teaching to increase managers' capabilities on the ethics/integrity front. For example, instructors might ask themselves the following questions.

- Do I highlight clearly and frequently enough for my students and executive education participants the organizational and personal costs of ethical transgressions, and the slippery slope that the initial transgression triggers?
- Does my teaching sufficiently help increase managers' and auditors' awareness of various important concepts and mechanisms discussed above, such as:
  - Ethical fading
  - Motivational blindness and confirmation biases
  - Self-serving bias
  - The dangers of intuitively appealing responses featuring more rules, regulations, and incentives.

REFERENCES

Barrett, P. M. (2011). Sokol's choice. *Bloomberg Businessweek*, May 2, pp. 50–55.

Barsoux, J.-L., & Manzoni, J.-F. (2008). *Tesco: Delivering the goods (A) and (B)*. Cases # IMD-3-1955 and IMD-3-1956. Lausanne: IMD.

Bazerman, M. H., & Tenbrunsel, A. E. (2011a). *Blind spots: Why we fail to do what's right, and what to do about it*. Princeton, NJ: Princeton University Press.

Bazerman, M. H., & Tenbrunsel, A. E. (2011b). *A lesson from Warren Buffett about ethical blind spots. HBR Blog*, May 31. Retrieved from http://blogs.hbr.org/hbsfaculty/2011/05/a-lesson-from-warren-buffet-ab.html. Accessed on February 9, 2012.

Cain, D. M., Loewenstein, G., & Moore, D. A. (2005). The dirt on coming clean: Perverse effects of disclosing conflicts of interest. *Journal of Legal Studies*, *34*, 1–25.

Collins, J. C., & Porras, J. I. (1994). *Built to last: Successful habits of visionary companies*. New York, NY: Harper Business.

Cooper, C. (2008). *Extraordinary circumstances: The journey of a corporate whistleblower*. Hoboken, NJ: Wiley.

Deci, E. L., & Ryan, R. M. (1985). *Intrinsic motivation and self-determination in human behavior*. New York, NY: Plenum.

DeSteno, D., & Valdesolo, P. (2011). *Out of character: Surprising truths about the liar, cheat, sinner (and saint) lurking in all of us*. New York, NY: Crown Publishing.

Festinger, L. (1957). *A theory of cognitive dissonance*. Stanford, CA: Stanford University Press.

Frey, B. S., & Jegen, R. (2001). Motivation crowding theory. *Journal of Economic Surveys*, *15*(5), 589–611.

Galbraith, J. R. (1995). *Designing organizations*. San Francisco, CA: Jossey-Bass.

Gebler, D. (2006). Creating an ethical culture. *Management Accounting*, *87*(11), 29–34.

General Electric. (2012). *Our culture*. Retrieved from GE website http://www.ge.com/company/culture/index.html. Accessed on February 9, 2012.

Goffee, R., & Jones, G. R. (1998). *The character of a corporation: How your company's culture can make or break your business*. New York, NY: Harper Business.

Greenwald, A. G., McGhee, D. E., & Schwartz, J. K. L. (1998). Measuring individual differences in implicit cognition: The implicit association test. *Journal of Personality and Social Psychology*, *74*, 1464–1480.

Hill, C. W. L., & Jones, G. R. (2001). *Strategic management theory: An integrated approach* (5th ed.). Boston, MA: Houghton Mifflin.

Kotter, J. P., & Heskett, J. L. (1992). *Corporate culture and performance*. New York, NY: Free Press.

Kruger, J. (1999). Lake Wobegon be gone! The "below-average effect" and the egocentric nature of comparative ability judgments. *Journal of Personality and Social Psychology*, *77*(2), 221–232.

Liptak, A. (2011). 34 years later, Supreme Court will revisit eyewitness IDs. *New York Times.com*, August 22. Retrieved from http://www.nytimes.com/2011/08/23/us/23bar.html?_r=1&src=me&ref=us. Accessed on February 9, 2012.

Manzoni, J.-F. (2002). Management control: Toward a new paradigm? In M. J. Epstein & J.-F. Manzoni (Eds.), *Performance measurement and management control: A compendium of research* (pp. 15–46). Oxford, UK: Elsevier Science.

Manzoni, J.-F. (2004). From high performance organizations to an organizational excellence framework. In M. J. Epstein & J.-F. Manzoni (Eds.), *Performance measurement and management control: Superior organizational performance* (pp. 19–45). Oxford, UK: Elsevier Science/JAI Press.

Manzoni, J.-F. (2006). Reflections on the human implications of the search for performance. In M. J. Epstein & J.-F. Manzoni (Eds.), *Performance measurement and management control: Improving organizations and society* (pp. 19–47). Oxford, UK: Elsevier Science/JAI Press.

Manzoni, J.-F. (2008). On the folly of hoping for A, simply because you're (trying to) pay for A. In M. J. Epstein & J.-F. Manzoni (Eds.), *Performance measurement and management control: Measuring and rewarding performance* (pp. 19–41). Bingley, UK: JAI Press/Emerald.

Milgram, S. (1963). Behavioral study of obedience. *Journal of Abnormal and Social Psychology*, *67*(4), 371–378. Retrieved as PDF from http://content.apa.org/journals/abn/67/4/371. Accessed on February 9, 2012.

Milgram, S. (1974). *Obedience to authority: An experimental view*. New York, NY: Harper & Row.

Miller, D. T., & Ross, M. (1975). Self-serving biases in the attribution of causality: Fact or fiction? *Psychological Bulletin*, *82*(2), 213–225.

Moore, D. A., Tanlu, L., & Bazerman, M. H. (2010). Conflict of interest and the intrusion of bias. *Judgment and Decision Making*, *5*(1), 37–53.

O'Reilly, C. A., & Pfeffer, J. (2000). *Hidden value: How great companies achieve extraordinary results with ordinary people*. Boston, MA: Harvard Business School Press.

Pascale, R. T., & Athos, A. G. (1981). *The art of Japanese management*. New York, NY: Simon & Schuster.

Ravasi, D., & Schultz, M. (2006). Responding to organizational identity threats: Exploring the role of organizational culture. *Academy of Management Journal, 49*(3), 433–458.

Roese, N. J., & Olson, J. M. (2007). Better, stronger, faster: Self-serving judgment, affect regulation, and the optimal vigilance hypothesis. *Perspectives on Psychological Science, 2*, 124–141.

Sachdeva, S., Iliev, R., & Medin, D. L. (2009). Sinning saints and saintly sinners: The paradox of moral self-regulation. *Psychological Science, 20*(4), 523–528.

Schein, E. (1992). *Organizational culture and leadership: A dynamic view*. San Francisco, CA: Jossey-Bass.

Schein, E. H. (2008). Creating and managing a learning culture: The essence of leadership. In J. V. Gallos (Ed.), *Business leadership: A Jossey-Bass reader* (2nd ed., pp. 362–369). San Francisco, CA: Jossey-Bass.

Simons, R. (2000). *Performance measurement and control systems for implementing strategy*. Upper Saddle River, NJ: Prentice-Hall.

Suls, J., Lemos, K., & Stewart, H. L. (2002). Self-esteem, construal, and comparisons with the self, friends and peers. *Journal of Personality and Social Psychology, 82*(2), 252–261.

Tavris, C., & Aronson, E. (2007). *Mistakes were made (but not by me): Why we justify foolish beliefs, bad decisions, and hurtful acts*. Orlando, FL: Harcourt.

Voreacos, D., Nussbaum, A., & Farrell, G. (2011). Johnson & Johnson's quality catastrophe. *Bloomberg Businessweek*, March 31, pp. 64–71.

Westen, D., Blagov, P. S., Harenski, K., Kilts, C., & Hamann, S. (2006). Neural bases of motivated reasoning: An fMRI study of emotional constraints on partisan political judgment in the 2004 U.S. Presidential Election. *Journal of Cognitive Neuroscience, 18*(11), 1947–1958.

APPENDIX: TESCO'S MODEL (ADAPTED FROM VARIOUS DOCUMENTS, PARTICULARLY TESCO'S ANNUAL REPORT 2009 AND TESCO'S CORPORATE RESPONSIBILITY REPORT 2011)

Tesco's long-term strategy:

- To be a successful international retailer
- To grow the core UK business
- To be as strong in nonfood as in food
- To develop retailing services
- To put community at the heart of what we do

The basic model:

- Our **core purpose** is to create value for customers to earn their lifetime loyalty.
- **Our success depends on people**: the people who shop with us and the people who work with us.
  - If our customers like what we offer, they are more likely to come back and shop with us again.
  - If the Tesco team find what we do rewarding, they are more likely to go that extra mile to help our customers.
- Hence our **two values**:
  - *No one tries harder for customers.*
  - *Treat people as we like to be treated.*

This model rests on four major thrusts:

A great place to work

Our staffs have told us what is important to them:

- To be treated with respect
- Having a manager who helps them
- Having an interesting job
- Having an opportunity to get on

Shopping trip

Customers have told us they want:

- Clear aisles
- To be able to get what they want
- Good prices
- No queues
- Great staff

Treat people how we like to be treated.

The way we work

- We deliver consistently every day
- We try to get it right first time
- We make jobs easy to do
- We know how vital our jobs are
- We always save time and money

Better for customers, simpler for staff, and cheaper for Tesco.

No one tries harder for customers.

Our community

- Actively supporting local communities
- Buying and selling our products responsibly
- Caring for the environment
- Giving customers healthy choices
- Good jobs for local people

Together we can make a difference.

Tesco's "Steering Wheel" (the measurement tool the company uses to report, analyze, and reward performance) reflects these four thrusts as well as Tesco's long-term strategy. Retrieved from http://www.tescoplc.com/corporate-responsibility/our-approach/ (accessed on February 9, 2012).

NEW TRENDS IN PERFORMANCE MEASUREMENT AND MANAGEMENT CONTROL

Antonio Davila

ABSTRACT

Purpose – *The purpose of this chapter is to present a state of the art of performance measurement and management control systems based on the papers presented in the 6th Conference on Performance Measurement and Management Control in Nice, France, 2011. The chapter is structured around two main topics: performance measurement and control systems.*

Methodology/approach – *The chapter is based on a thorough review of all the papers presented at the conference. Each paper was classified according to the topic it dealt with and the methodology used in the paper.*

Findings – *We structure the papers around the following topics: cost measurement and management, mapping the landscape, financial measures, nonfinancial performance frameworks, performance measurement design, sector-specific measures, and marketing measures within performance measurement. The topics identified in management control systems are compensation, management control systems' design, motivational aspects, strategy, risk management and enterprise software, governance, nonprofit organizations, and innovation.*

Performance Measurement and Management Control: Global Issues
Studies in Managerial and Financial Accounting, Volume 25, 65–87

ISSN: 1479-3512/doi:10.1108/S1479-3512(2012)0000025006

Research limitations/implications – *The chapter provides a comprehensive review of these topics. The review is limited to the selection bias implicit in the papers submitted to the conference. However, with almost 200 papers, the conference includes a large set of papers and topics. The analysis indicates that the field is dynamic with close connections to practice and promising research lines.*

Practical implications (if applicable) – *This state-of-the-art review reveals relevant research findings for companies and organizations. It also indicates that the research community is tackling important issues to society and the evolution of research promises to offer interesting alternatives to tackle these problems.*

Social implications – *Performance measurement and management control systems are important topics for management. Recent work is extending the implications of these fields beyond companies to include nonprofit organizations and public sector departments. Advances will definitely shape the way societies are designed.*

Originality/value of chapter – *The chapter provides a review of the field based on the contributions at the conference. As such it portrays the state of the field and gives researchers as well as practitioners a quick way to update their knowledge of the topics that are dominating the field.*

Performance measurement and management control is a dynamic field addressing the new challenges that emerge in management practice. The Performance Measurement and Management Control Conference provides the opportunity to gauge this evolution. It showcases new ideas and identifies new trends that shape the future of this field of knowledge. This chapter presents trends from the conference at the forefront of performance measurement and management control.

Performance measurement is progressing in various fronts. New ideas are emerging in traditional fields. Financial performance measures and cost measurement, topics that have characterized the field for decades, are still being fruitfully explored to reflect new management needs. Another aspect where ideas are emerging is around extending performance measures beyond financial to include nonfinancial measures; a trend that dates back to the 1960s with the concept of Key Performance Indicators from GE and then given a big push in the 1990s with concepts such as the Balanced Scorecard. New concepts and frameworks are extending performance measurement to provide information on innovation opportunities and specific industries. Performance measurement models are also extending their scope beyond traditional functions such as finance or manufacturing to go deep into functions such as marketing and sales and R&D where intangibles play more of a role.

Management control is also extending its reach into new and attractive areas. Budgeting, the original starting point of this field back in the 1960s (Anthony, 1965), is still a central topic. But the field is extending to embrace the complexity around management control in organizations. Research effort is progressing on issues such as incentives, compensation, and control systems' design. New topics are being researched. For instance, management control presents challenges in public administration and nonprofit organizations that are very different from those of traditional for-profit companies. These organizations do not have a dominant logic, the logic of economic value, but work with objectives that are both more numerous, harder to conceptualize and summarize in measures, and more dynamic than economic value. Management control in these organizations balances a wider range of interests and works with a more diverse set of motivators. Management control is also exploring governance. Traditionally, management control was centered on the implementation of strategies with little or no attention to the way in which agency relationships at the top shaped the performance of companies. This attention on governance examines how stakeholders relate to the management of companies through their governance mechanisms. Finally, research is also extending into the role of ethics and risk management. Internal controls often appear in various models of management control systems (Simons, 2000), but it is only now that they are being studied within the broader context of risk and ethics.

The objective of this chapter is to present the trends based on the latest research presented in the Performance Measurement and Management Control Conference, 2011. The chapter is structured into two parts. The first part presents the evolution of topics around performance measurement. The second part focuses on topics in management control.

NEW TRENDS IN PERFORMANCE MEASUREMENT

Ideas on performance measurement are advancing quickly. For the purpose of this chapter, the discussion of the evolution of these ideas is structured around the following topics: (1) cost measurement and management,

(2) mapping the landscape, (3) financial measures, (4) nonfinancial performance frameworks, (5) performance measurement systems' design, (6) sector-specific measures, and (7) marketing measures.

Cost Measurement and Management

Cost measurement is often studied from a technical perspective, and cost allocation is addressed as an engineering challenge. The challenge of cost measurement is how to best reflect resource consumption through the design of a cost system. Activity-based costing is based on this logic and claims that an activity perspective does a better job of reflecting this resource consumption (Kaplan & Porter, 2011). Further research has extended this perspective to address cost allocation challenges including an agency perspective (Baldenius, Dutta, & Reichelstein, 2007).

But the issue of cost measurement is also being examined from a different perspective. Researchers are looking at how cost system design impacts the behavior of users. Depending on whether resources are shared and how performance for the various managers is measured, distorting actual costs can be a fruitful approach to entice a particular behavior. For instance, managers of different product development projects can chose to either share a common part or design a specific one for their product. The benefits of the common part are associated with lower costs, yet these benefits do not happen unless enough projects use these common parts. But projects face lower performance compared to unique parts that are tailor-made for the specific product. One alternative is to force managers to use common parts and remove the decision out of these people who often have better specific information. Another alternative is to lower the perceived cost of the common part to get enough adopters to realize the cost savings from part commonality.

The role of cost measurement also varies across countries and this line of research has provided interesting insights; for instance, a comparative study of German and Japanese firms found that the former have more complex and timely systems than the latter due to cultural and institutional reasons (Kajuter, Moeschler, & Wada, 2011). Cost measurement does not appear as just an engineering problem to be solved but also as a social construction where power and tradition lead to different solutions and different amount of resources devoted to designing a cost system. Another cost measurement aspect that is gaining relevance is the increasing weight of industries with cost structures that are very different from the traditional manufacturing ones. Healthcare (Kaplan & Porter, 2011) and services more broadly have economic structures where fixed costs play a much larger role. However, new research is examining settings that will be central to future competitiveness including creative and knowledge workers. These settings present several challenges not only in terms of measuring resource consumption but also in measuring value creation (Harri, Jaaskelainen, Lonnqvist, & Ruostela, 2011).

In this volume, Cescon (2012) examines the relationship between Advanced Manufacturing Techniques (AMT) and organizational characteristics including environmental uncertainty, organizational size, and level of integration. The chapter then moves to study the impact of the presence of AMT on these companies' use of more advance cost and control techniques such as activity-based costing, strategic costing, target costing, and life cycle costing. The study also looks into the quality of nonfinancial performance measures and the adoption of innovative managerial practices, just-in-time manufacturing, total quality management, and activity-based management.

Mapping the Landscape

Traditional performance measurement systems have focused on tracking the progress of the organization. Most performance measures take an internal perspective on how processes within the company are progressing. They inform about quality, productivity, uptime, or budget performance. Other measures take an outside-in perspective. They look at the organization from the perspective of external stakeholders. For instance, customers' perspective is a traditional way of outside-in view of the organization. Another stakeholders' view that is common is partners' view. The underlying assumption is that the organization is at the center of the performance measurement system.

However, the volatility of the environment and competitive threats coming from a global market have enhanced the importance of taking an inside-out perspective. The idea is to look at an organization's environment and its landscape from the perspective of the company. The company is not anymore the center of the measurement system but rather a perspective to look into the world. These landscape scorecards map the landscape of a company to identify, measure, and track those actors that are potential sources of opportunities or threats. For instance, universities can become important actors if the company believes that new technologies may come from them. Venture capitalists can be important to track new promising start-ups. Governments can change the rules of the game creating both opportunities and threats. These measurement systems look at the world from the perspective of the organization, rather than looking at the organization from the perspective of different external stakeholders.

Financial Measures

Financial measures have been at the core of performance measurement. Still, technology and conceptual advances are making this topic interesting. For instance, technology is facilitating the communication of financial information through the Internet. New information communication protocols are easing the transfer of financial information. Lowering the cost of information transfer facilitates the integration of new players to the financial markets but also simplifies the use of intelligent software to make performance evaluation decisions for investment purposes.

Another emerging issue is new financial measures such as comprehensive income but also a focus on accounting for sustainability. Concepts such as the triple bottom line or one report (Eccles & Krzus, 2010) are efforts to integrate into financial measures of performance costs that so far have been ignored. The impact of organizational decisions on the environment and society at large are coming into the picture with force. These effects had traditionally been ignored because there is no cash flow associated with them coming out of the company. Environmental and social costs were absorbed by society and the public sector with organizations failing to recognize them as part of their production function. The transition to recognizing these costs as part of the cost of doing business cannot happen until they are properly measured. Financial measurement is working on generating adequate measures to be able to integrate these costs.

The recent financial crisis has also provided the opportunity to use financial information to identify financial characteristics of those companies that have excelled versus those that did not do as well (Needles, Shigaev, Powers, & Frigo, 2011).

Financial measures are also a central aspect of budgeting and behavioral implications or budgets are still central to management research. As discussed later on when talking about management control systems, the recent crisis has put a lot of stress into management tools such as the budget.

Nonfinancial Performance Frameworks

The complexity of reality makes it impossible to summarize all the information in a single performance measure. This complexity is translating into higher disclosure requirements in capital markets. It is also translating into more sophisticated analysis of how to structure measurement systems within companies. Frameworks themselves are progressing. For instance, the idea of structuring measures around the concept of value-based management is gaining some interest. However, the concept itself still requires some additional theoretical refinement to be widely adopted. Companies, especially in Northern Europe, are starting to experiment with the idea of organizing measurement around the value creation process.

There is also progress in looking at existing measurement frameworks, notably the Balanced Scorecard, from different perspectives. The original approach to these measurement frameworks was functionalist, giving almost an engineering view to the challenge of having adequate information for management. The success of the Balanced Scorecard relied heavily on its logic and simplicity, appealing to all organizations and managers. On the research front, knowledge is advancing through looking at the phenomena from different perspectives. The history of the Balanced Scorecard in early adopters offers the possibility of having long-term case studies that look at implications over long horizons of this tool (Bescos & Foulquier, 2011). Also there are contributions on the role of this framework coming from the fields of psychology and sociology.

Performance measurement frameworks' research is also looking at the deployment of these frameworks at all levels of the organization. For instance, the sophistication of the performance measurement system at the top of the organization permeates and influences the quality of systems lower in the organization. The quality of these lower-level systems is also found to affect the translation of value drivers into performance (Burkert, Davila, & Oyon, 2011).

Performance Measurement Design

Another fruitful research line in performance measurement is explaining the variability of performance measurement design through contingency factors of the environment, the organization, and the characteristics of top management. For instance, managers have different attitudes toward the use of performance measurement systems. Roughly, they align between analytical managers who rely heavily on them and intuitive ones who use their informal networks and their own intuition to a larger extent. These two types of managers need different systems. The former feels more comfortable with detailed information, while this same level of information can be overwhelming for the latter type (Burkert, Davila, & Lueg, 2011). Experiments have also found that the level of knowledge also affects the design of performance measurement systems. Knowledge of more financial measures is associated with a preference for numbers, while managers with less knowledge prefer graphical information.

Performance measurement design also depends on the environment. This line of research is not only limited to the traditional contingency paradigm looking at the covariance of certain design characteristics and environmental changes, but it is expanding to in-depth case studies using sociological paradigms. For instance, network theory and network analysis techniques are being used frequently to understand the particular design of a performance measurement system. This network perspective is especially effective in looking at interorganizational performance measurement systems, a phenomenon that is increasing in importance as the complexity of business requires the cooperation and integration of network of companies to deliver services.

Sector-Specific Measures

A new trend in performance measurement research is studying sector-specific measures. So far, most research had a general perspective providing the logic of the particular framework and leaving the manager with the task of finding measures relevant to her or his particular industry. This new trend is filling this void. Healthcare industry is the most noticeable example of this trend. Researchers are developing measures specific to healthcare delivery that will help better understand and manage this industry.

Energy is another sector where specific measures are being developed around concepts such as efficiency and externalities. The trigger factor appears to be renewable energies. The discussion around these alternative sources of energy is how they compare to traditional sources in terms of efficiency but also in terms of a lower environmental impact. In this book, Virtanen, Tuomaala, and Pentti (2012) address the issue of measuring sustainability in the context of energy consumption. In particular, they examine the issue of measuring energy efficiency. This issue that may be simple from a conceptual perspective poses significant challenges when brought into practice as the various sources and uses of energy need to be identified and valued to have a reliable measure of efficiency. The case study reveals how existing measures do not meet the criteria for a good performance measurement system and how its use as a management tool and the target-setting process also present some unique challenges.

Performance measurement is also taking a relevant role in nonprofit organizations and the public sector. Organizations, for which economic performance is not the main objective, require alternative approaches to measuring performance. Hirtz and Guernaccini (2012) in this book offer an interesting application. The chapter is not only relevant for its conceptual contribution bridging theory of constraints with output measures in the public sector, it is also relevant as an example of applying theory to improve practice. The authors document how the application of performance measurement concepts allowed a public sector organization to have the right information for decision making and improve its performance.

Marketing Measures

Traditional performance measurement has not looked at the marketing function. Marketing was often associated with intuition, creativity, and numbers-free. However, the marketing function has evolved dramatically over the last decade. Information technology has probably affected marketing more than any other management discipline. Now, marketing is deeply embedded with large databases of consumer behavior, customer relationships, and transaction information. Marketing is now a highly quantitative discipline. As such, performance measurement has become a relevant aspect of the marketing function. The performance measurement community has been slow at adapting to these new circumstances and a lot of the advances have come from marketing itself. Concepts such as customer equity are financial measures that using discounting techniques estimate the value of a customer. The cash flow predictions are based on customer demographics and transaction history to profile the customer within a particular segment with a cash flow profile. Customer equity is a life cycle value measure and as such the question of relevant cash flows becomes significant. Incremental cash flow estimates assume that each customer is marginal and can lead to a situation where all customers have positive equity but the company itself does not. On the other hand, full cost estimates can lead to rejecting customers attractive at the margin. These are traditional issues in financial performance measurement that marketing is now starting to face.

Performance measurement is integrating marketing in various ways. For instance, looking at customer satisfaction with management control systems is a new perspective to assess the quality of these systems. Social media and social marketing also require carefully designed performance measurement systems. Finally, financial measures can be important sources of information to value brands.

NEW TRENDS IN MANAGEMENT CONTROL

Management control is a much broader field than performance measurement. This diversity is reflected in a larger number of papers at the conference around this topic addressing a large variety of topics using different perspectives and research methods. For the purposes of discussion and identifying trends, I have identified the following topics: (1) compensation, (2) management control systems' design, (3) motivational aspects of management control systems, (4) strategy, (5) risk management and enterprise software, (6) governance, (7) nonprofit organizations, and (8) innovation.

Compensation

Economic incentives have been a traditional way of addressing the goal divergence issue at the heart of the control problem. Agency theory has made important contributions to theory moving from single- to multiperiod models and from explicit to implicit contracts. This theory has helped in sharpening our concepts and arguments. It has also given a strong foundation for empirical research and researchers from accounting, but finance and strategy have also contributed significantly to testing whether existing contracts for top managers align with agency predictions.

Management control research is still devoting attention to top management compensation and the problems of balancing short- and long-term incentives. Stock prices that traditionally were argued to take care of this problem have proven to be less powerful than expected. The level of noise in this performance measure makes it expensive to use beyond top management, and even there it is expensive. Furthermore, it creates strong pressure to influence stock price and therefore encourages behavior that is often at the limit or beyond legal or ethical boundaries. Finally, different investors have different time horizons and the timing of stock price maximization is a controversial aspect.

Budgeting research, a research line with a long tradition in management control, is also integrating new evidence. The short-term perspective inherent in budgeting needs to be balanced with longer-term perspective to balance incentives for myopic decisions. Budget setting is also being studied beyond the issue of negotiation and integrating information from comparable divisions. Relative performance evaluation has often been presented as a fruitful approach to reduce noise and the cost of using noisy performance measures in compensation contracts. However, contracts are sticky over time and they seldom change from one year to the next. Relative target setting refers to using performance from comparable companies or organizational units to set targets. In contrast to the stickiness of compensation contracts, budget targets are more flexible and the weight that supervisors put on different measures can radically change from one year to the next or from one person to another.

Recent research is extending the reach of compensation beyond top management and beyond the economic perspective. One issue is how to replicate market mechanisms within the organization (Tamas, Friis, & Hansen, 2011) to enhance the resource allocation process. Another issue is the role of subjective evaluation. Bringing in qualitative assessments of performance gives an additional degree of freedom to compensation mechanisms but also the option of misusing this freedom. Centrality and leniency biases are both challenges in subjective performance evaluation that only now are receiving attention (Nicola, Hartmann, & Collini, 2011). Compensation is also studied from a psychological perspective in terms of understanding how managers perceive different types of incentives and relative performance evaluation issues (Alvesi, Emmanuel, & Kominis, 2011). Ruggeri (2012) uses an experiment to contrast agency predictions and the controllability principle in settings with high or low interdependence. She finds that agency theory performs better in low interdependence, while managers rely on the controllability principle to a larger extent in high interdependence settings.

Management Control Systems' Design

Contingency theory has been the most common approach to study management control systems' design. The accumulated knowledge from all this work is well established with a few consolidated findings but also with a complex picture where design depends on the interaction of multiple contingency factors. Simons' work dating back to his finding that management control systems were used intensively in dynamic environments (when theory predicted the opposite) opened up the field to qualitative, ethnographic studies that provide the richness to understand the underlying forces. This research approach complements contingency theory using questionnaire-based methods and has provided interesting insights.

An aspect of management control design that is receiving attention recently is design in interorganizational settings. This topic is important because organizations as value chains migrate to networks. Mineev (2012) takes an institutional perspective to analyze a case of a network in North-West Russia and how the network and its systems are formed. The chapter is interesting in documenting the emergence of systems and how local networks are created. The chapter shows how path dependency is as important as contingency factors.

Florez, Ramon, Velez, Araujo, and Sanchez (2012) address another important aspect of interorganizational control mechanisms. They analyze the relationship between exporters and their partners in different parts of the world. Their findings indicate that control systems enhance the relationship and social control is the type of control that has a more significant impact on performance.

Another important topic in interorganizational relationships is open book accounting. Sharing of information can lead to important savings; however, power struggles can bias the appropriation of these savings to one party. Understanding the implications of open book accounting and its impact on power relationships is important to better structure value networks (Piontkowski, Hoffjan, Schuchardt, & Lachmann, 2012).

Management control systems' design is a fruitful line of research. New research approaches are complementing traditional questionnaire-based contingency studies. This transition is welcome. Understanding this phenomenon requires in-depth knowledge that cannot be accessed through questionnaires. This new trend is best appreciated in interorganizational research. Yet, other aspects of management control systems' design are also benefiting from these complementary approaches ranging from better understanding interactive systems, the systems' perspective on performance measurement systems, and the role of systems in innovation and creativity.

Motivational Aspects of Management Control Systems

Management control systems affect behavior through their impact on the motivational structure of people. Incentives speak to extrinsic motivation and people's economic needs. But control systems also affect other types of motivation. For instance, Kunz and Linder (2012) use an experiment to study the impact of low controllability on extrinsic and intrinsic motivation. Consistent with psychological predictions, low controllability is associated with lower extrinsic motivation. Interestingly, it also reduces intrinsic motivation, suggesting that the selection of performance measures is important not only when they are linked to incentives but also in settings where intrinsic motivation is more important.

Control systems also influence the perception that employees have about procedural fairness and the effect on employee commitment. He and Lau (2011) find that the use of nonfinancial information has a positive impact on job relevant information, role clarity, and procedural fairness. Therefore, the use of more advanced performance measurement systems not only help in having more relevant information for decision making but also enhance people's perception of fairness within the organization.

Motivational aspects beyond economic incentives include issues associated with social identity. The concepts of belief systems highlight the importance of vision and missions not as much as isolated statements but as values that permeate the organization. Values that appeal to higher needs and self-actualization will reinforce intrinsic motivation and lower the need to rely on extrinsic motivation. This motivational structure is associated with creativity, commitment, and effort; all of them characteristics that are needed in a market where differentiation is an important source of competitive advantage. The concept of interactive systems has also come to the forefront of research (see next subsection). Interactive systems have an important role to exchange information about emerging opportunities and threats but they may also affect motivation. Interactive systems provide a communication mechanism to involve lower levels of the organization in strategic discussions around opportunities. The integration in the strategic thought of the company can reinforce intrinsic motivation and self-actualization to the benefit of coming up with sources of differentiation.

Strategy and Control Systems

Management control systems were originally designed as tools to implement the strategy that came out of the formulation stage. Our understanding of the interplay between strategy and control systems has evolved significantly since then. The separation between formulation and implementation led to the input–process–output model or personnel–action–results model. As the strategy field progressed, our conceptualization of management control systems also advanced. The concepts of intended, emergent, and realized strategies saw the development of Simons' levers of control model where traditional controls came as diagnostic systems and beliefs, boundaries, and interactive systems captured the idea of search and discovery of new strategies. This model is currently dominating the literature with significant work going on understanding the workings of interactive systems.

Meanwhile, the strategy field and the strategic process subfield have evolved from the idea of emerging strategies to autonomous and induced strategic actions. The latter refers to the traditional idea of emerging strategies and interactive systems where people on the ground adapt the intended strategy of top management to the reality of their day-to-day work. The former captures the development of radically new strategies by people outside the top management team. Autonomous strategic actions are not about adapting the existing strategy but about creating a totally new one and rather than coming from top management, coming from operating people. In most companies, existing management control systems are designed to ignore ideas that do not fit with the current strategy. However, leading ones are creating structures and systems to capture these radical ideas spread out inside and outside the organization. The concept of corporate entrepreneurship and corporate venture capital are exploring these new strategic issues (Cardenas, Oyon, & Davila, 2011).

Management control systems' research is also looking into issues associated with strategy implementation. For instance, research is examining the role and evolution of these systems in mergers and acquisitions (Cording & Christmann, 2011), multinational companies (Kornacker, 2011), green field investments, and strategy formulation.

The strategic field has diversified into several lines of research ranging from top management teams to networks, industry dynamics, technology transfer, or government relationships (lobbies). Almost each strategic issue has a management control challenge behind it. The progress of management control research and strategy needs to collaborate closely with the development of strategic field. So far, the lag between progress in strategy and progress in management control has been significant. For research in this intersection to be more relevant, it needs to work closer to the development in strategy.

Risk Management and Enterprise Software

The developments of the last few years with the 2008 crisis as their starting point have brought the issue of risk management to the forefront. The failures leading to the financial crisis have been often associated with risk management mistakes. Risk is intimately associated with control as traditionally interpreted: mechanisms to limit self-interested behaviors from agents that undermine the position of the principal. Failures in the financial sector were associated with faulty control systems that did not trigger action when the unintended consequences of incentive systems created problems.

Risk management is being approached from different angles. One aspect is to modify existing incentive mechanisms to include risk measures. Making risk measures explicit helps improve people awareness about risks and influences organizational culture attitude toward risk (Aureli & Salvatori, 2012). Another angle is looking at Enterprise Risk Management (ERM) and how it is deployed in organizations. ERM are triggered by regulation, stakeholders' demands, and business competitiveness. These systems combine quantitative approaches to evaluate and manage risk exposure traditional of the financial world with qualitative approaches that emphasize sensitivity toward risk and ethical criteria in decision making (Silva de Souza, da Silva Gomes, Garcia de Oliveria, Santos Sampaio, & Almeida de Faria, 2012).

A topic closely related to risk management is enterprise software. Risk management requires extensive information, and advances in enterprise software are making this information available. Of course, this latter topic is much broader than just risk management and crucial to management control. However, the attention it has received so far is limited. Several reasons can explain this lack of research. First, it is a complex issue at the intersection of technology progress and control theory. As such it falls in between technology and control research, which is dealt by few researchers. Good research on this field requires up-to-date knowledge on technology advances and then in-depth analysis of its consequences. The possibility to the Foucauldian concept of "seeing through" becomes central to the advances in technology. On the one hand, communication technology allows people to coordinate and communicate skipping the traditional hierarchies. This possibility has been shown in the recent events in Arab countries where grassroots organizations have been able to bring down governments. This ability to self-organize is also present in larger organizations exploring new ways of doing businesses. The flip side is the possibility of controlling through the tracking of information exchanges and low-level performance measures. This flip side not only presents control issues but also presents ethical issues involving privacy, access to personal communication, or disclosure of confidential information (and insider trading).

Governance

Management control systems range all the way from the implementation of strategy at the bottom of the organization to the monitoring at the board of directors. While the former has been the focus of research since the inception of management research in the early part of the 20th century, the focus on boards is more recent. The high-profile failures in the early part of the 21st century turned research attention to control at the very top of the organization, the governance as implemented through the board of directors (Larcker & Tayan, 2011).

Governance presents many challenges. The board of directors' main role is to represent the shareholders who choose them. As such, its members have a fiduciary responsibility. It is the ultimate body in charge of supervising the direction that the company takes. Board members' main duties are to monitor the work of the CEO and her or his team and to contribute and approve strategic decisions such as large investments both in terms of money and human resources. While the mandate is clear, its implementation is challenging. There are issues of independence associated with whether board members have shareholders' interest and the long-term growth of the company as their objective or rather they are loyal to the CEO who appointed them. Another issue is whether a part-time dedication to a company is enough to grasp the complexity of today's businesses. These issues come to the forefront when high-profile breakdowns of governance catch the headlines.

Epstein (2012) in a chapter in this book highlights another relevant issue which is how governance has very different realizations across the world. Governance in the United States is a very different phenomenon from governance in Germany, Japan, or Spain. Merino, Manzaneque, and Banegas (2012) examine governance in this latter country and, in particular, board members' compensation. Their conclusions are interesting and critical suggesting various changes that need to be implemented for governance to be more effective.

Most research on governance has so far relied on quantitative analysis of board characteristics using publicly available data. Yet, the complexity of this phenomenon requires a more in-depth analysis of what is actually going on in the process of governance relying on qualitative research methods.

Nonprofit Organizations

The recent pressure on the public sector in Western economies is putting additional demands on nonprofit organizations. They are asked to use management control systems with the objective of being more effective and efficient in achieving its goals (Jaaskelainen, 2011). But these organizations are qualitatively different from traditional for-profit companies, and simply translating the tools from profit-seeking companies does not work. New Public Sector Management was developed mostly in the United Kingdom over the last decade with precisely this objective. The translation of management control systems proved to be much harder than anticipated with mixed results.

Nonprofit and public sector organizations have a much more complex set of objectives. Profit-seeking companies have a clear purpose: maximize profits over time within a set of constraints that are different across companies depending on how they structure their relationship with society and the environment. Even these companies face important challenges in balancing the various time horizons. Nonprofit organizations often have a range of objectives that are not always well defined or easy to measure. In addition, the motivational structure of their people is more complex with noneconomic motives playing a much larger role.

These organizations face multiple challenges of a higher complexity than those companies face. For instance, the definition of objectives can be a controversial issue. Certain government functions see different logics competing to have their rationales dominate government action. Competition for ideas affects the objectives and the design of the institution. This competition often happens in the public domain and external control systems have a role that is seldom seen in for-profit organizations. Research on this area is emerging. Some of the work is looking into the negotiation process among stakeholders of nonprofit organizations and how this negotiation gets reflected in the information and evaluation systems (Francisco & Alves, 2012).

Nonprofit organizations also face challenges in implementing their strategies. Challenges range from the people employed who can be at-risk groups, limited economic resources, and, in some cases, the use of volunteers, and using the economic logic embedded in existing management control systems in an environment where this logic is marginal. Research in public sector and nonprofit and management control systems is looking into how the control challenge is being addressed. These studies are using frameworks and concepts developed in the profit-seeking sector and the question remains whether they will be powerful enough to translate into this new reality, or theory will need to advance to develop new concepts (Sedysheva, 2011; Wallstedt, 2011; Verbeeeten & Spekle, 2011).

Research on this field is examining a variety of issues. Case studies are looking at the institutional changes going on in nonprofit organizations in their transition to incorporate some of the economic logic of efficiency. Longitudinal case studies are important in documenting these transitions. Leotta and Ruggeri (2012) use as a framework to structure these episodes around the object (performance measurement and evaluation systems), the subjects, the place and time, and the how and why change happens. Their setting is a hospital. The healthcare sector is an important part of public sector services in most countries where a sizeable part of this sector is public. Management control research in hospitals started a decade ago and is now a well-established line of research. Research questions are structured around the interaction between formal and informal control mechanisms, the adoption of management reforms on performance, and the impact of institutional changes (Frances & Widener, 2011).

Innovation

Management control systems' research is also moving from traditional settings characterized by repetitive processes where unexpected deviations are seen as problems rather than opportunities into creative settings. Innovation has taken central stage in today's competitive world. Value creation is intimately related with the ability to differentiate. Creativity and the ability to transform it into value is one of the main sources of differentiation (others include privileged access to resources either through regulation or ownership). Creativity literature constantly reinforces that creative activities require stability in terms of objectives that are stable over a certain period of time and through clear boundaries that limit the search space. Management control systems play an important role in these two aspects. Economic incentives are not the way to motivate creative activities. Creative people have to perceive economic rewards as fair to forget them and get their intrinsic motivation to act. They also need inspiration, an attractive reason to deploy their creative instincts. Management systems play a role in inspiring systems. These are inspirational systems (Davila & Ditillo, 2011). Management control systems are also important elements in setting the boundaries of creative activities. They set the limits on the budget available, time to market, type of products to develop, or interaction with other parts of the companies. These systems are aspirational systems that manage the aspirations of the creative teams.

Management control systems also play important roles in the process of moving ideas into value. This process is configured with stages that require these systems. Selection of ideas to invest in is a resource allocation process intimately associated with strategic control systems and budgeting. Monitoring of innovation projects is also associated with control. Incremental innovation relies on stage gate systems that have detail project plans describing the characteristics of the project going forward and its development. Gates are control events where deviations from the plan are analyzed and used for decision making. Radical innovation uses a less-structured mechanism inspired in the control mechanisms used in start-up companies through staged financing. Integration of different projects requires portfolio management tools to have an overall view of the strategy of the company. The use of partners during development also relies heavily on interorganizational control practices, while product release requires close coordination with production and marketing.

CONCLUSIONS

The previous sections have detailed trends identified through the articles at the conference and this book. Yet, this research field also includes researchers interested in the evolution of knowledge (Libby & Lindsay, 2011), the process of research (Kruis, 2011), and education (MacDonald & Kam, 2011). The outcomes of this work are not directly associated with management practice but they are very important to help researchers reflect on how to approach their projects and to help educators improve the learning process. These papers provide tools to improve the validity of research processes or identify those articles that have had more influence and from which we can learn to do better research. They propose new concepts that make sense of findings difficult to explain with existing theories.

This review of trends based on the sixth conference on performance measurement and control systems identifies a rich and diverse field. An important aspect is the intimate relationship between researchers and the challenges facing managers. Most of the research is empirical and often field based with an in-depth relationship between academia and practice. This relationship is important to the relevance of the field and to translate the progress in basic sciences such as economics, sociology, and psychology into practice.

The papers at the conference cluster around traditional research questions such as financial performance measures, incentives, motivation, or budgeting and emerging fields where practice requires the insights from careful analysis. Emerging topics include performance measures specific to certain industries and functions including sustainability, healthcare, or marketing. Emerging topics in management control include risk management, governance, nonprofit organizations, or innovation. All these topics are central to the challenges facing society today, speaking of the relevance of this field.

REFERENCES

Alvesi, F. C., Emmanuel, R., & Kominis, G. (2011). *Exploring the reward perceptions across organizational levels: A study on middle managers*. Working paper. University of Venice, Italy.

Anthony, R. N. (1965). *The management control function*. Boston, MA: Harvard Business School Press.

Aureli, S., & Salvatori, F. (2012). Are performance-dependent rewards a viable tool to assure managers' commitment toward firm's goals about risk management? In A. Davila, M. Epstein & J. F. Manzoni (Eds.), *Performance measurement and management control: New global perspectives to advance theory and practice*. Bingley, UK: Emerald.

Baldenius, T., Dutta, S., & Reichelstein, S. (2007). Cost allocation for capital budgeting decisions. *Accounting Review*, *82*(4), 837–867.

Bescos, P. L., & Foulquier, P. (2011). *Improvement in the implementation of balanced scorecards: The case of an insurance company*. Working paper. EDHEC Business School, France.

Burkert, M., Davila, A., & Lueg, R. (2011). *Complementarity effects of performance measurement systems' design and top managers' use on managerial performance*. Working paper. HEC Lausanne, Switzerland.

Burkert, M., Davila, A., & Oyon, D. (2011). *How top management measurement systems' design affects company-wide performance measurement and the relationship between value drivers and performance*. Working paper. HEC Lausanne, Switzerland.

Cardenas, F., Oyon, D., & Davila, A. (2011). *Innovation approaches to value creation*. Working paper. HEC Lausanne, Switzerland.

Cescon, F. (2012). The impact of AMT firm characteristics on innovative costing techniques and non-financial measures of performance. In A. Davila, M. Epstein & J. F. Manzoni (Eds.), *Performance measurement and management control: New global perspectives to advance theory and practice*. Bingley, UK: Emerald.

Cording, M., & Christmann, P. (2011). *Form over substance: The differential effects of embedded versus non-embedded integration practices on acquisition performance*. Working paper. Rice University, Italy.

Davila, A., & Ditillo, A. (2011). *Controlling creativity: Design and use of management accounting and control systems in creative environments*. Working paper. Bocconi University, Italy.

Eccles, R. G., & Krzus, M. P. (2010). *One report: Integrated reporting for a sustainable strategy*. Hoboken, NJ: John Wiley.

Epstein, M. (2012). Governing globally: Convergence, differentiation, or bridging. In A. Davila, M. Epstein & J. F. Manzoni (Eds.), *Performance measurement and management control: New global perspectives to advance theory and practice*. Bingley, UK: Emerald.

Florez, R., Ramon, J. M., Velez, M., Araujo, P., & Sanchez, J. M. (2012). The role of management control systems on interorganizational efficiency: An analysis of export performance. In A. Davila, M. Epstein & J. F. Manzoni (Eds.), *Performance measurement and management control: New global perspectives to advance theory and practice*. Bingley, UK: Emerald.

Frances, K., & Widener, S. (2011). *The power of organic control: A case study of a lean implementation in a health clinic*. Working paper. Clemson University, USA.

Francisco, L., & Alves, M. C. (2012). Accounting information and performance measurement in a non-profit organization. In A. Davila, M. Epstein & J. F. Manzoni (Eds.), *Performance measurement and management control: New global perspectives to advance theory and practice*. Bingley, UK: Emerald.

Harri, L., Jaaskelainen, A., Lonnqvist, A., & Ruostela, J. (2011). *New ways of working: Measuring the impacts of knowledge work productivity*. Working paper. Tampere University of Technology, Finland.

He, J. Q., & Lau, C. M. (2011). Does the reliance on nonfinancial measures for performance measurement enhance managers' perceptions of procedural fairness? In A. Davila, M. Epstein & J. F. Manzoni (Eds.), *Performance measurement and management control: New global perspectives to advance theory and practice*. Bingley, UK: Emerald.

Hirtz, T., & Guernaccini, P. (2012). Toward a model of performance measurement of output based on the theory of constraints. In A. Davila, M. Epstein & J. F. Manzoni (Eds.), *Performance measurement and management control: New global perspectives to advance theory and practice*. Bingley, UK: Emerald.

Jaaskelainen, A. (2011). *A novel approach to measuring and managing productivity in large public service organizations*. Working paper. Tampere University, Finland.

Kajuter, P., Moeschler, M., & Wada, S. (2011). *Conceptual design of cost accounting systems: A cross-national study of German and Japanese firms*. Working paper. University of Munster, Germany.

Kaplan, R. S., & Porter, M. E. (2011). How to solve the cost crisis in health care. *Harvard Business Review* (September), 47–64.

Kornacker, J. (2011). *Budgeting and complementary management control in high growth countries: Evidence from German multinational enterprises in China*. Working paper. Dortmund University, Germany.

Kruis, A. M. (2011). *Validity-feedback interviews. Assessing the construct validity of MAC survey instruments*. Working paper. Nyenrode Business Universiteit, The Netherlands.

Kunz, J., & Linder, S. (2012). Buy one, get one free: Benefits of following controllability principle for intrinsic motivation. In A. Davila, M. Epstein & J. F. Manzoni (Eds.), *Performance measurement and management control: New global perspectives to advance theory and practice*. Bingley, UK: Emerald.

Larcker, D., & Tayan, B. (2011). *Corporate governance matters: A closer look at organizational choices and their consequences*. Prentice Hall.

Leotta, A., & Ruggeri, D. (2012). Changes in performance measurement and evaluation systems as institutional processes: The case of an Italian teaching hospital. In A. Davila, M. Epstein & J. F. Manzoni (Eds.), *Performance measurement and management control: New global perspectives to advance theory and practice*. Bingley, UK: Emerald.

Libby, T., & Lindsay, R. M. (2011). *Life after RAPM: First steps in the development of a new construct*. Working paper. University of Waterloo, Waterloo, Ontario, Canada.

MacDonald, S., & Kam, J. (2011). *The skewed few: People and papers of quality in management studies*. Working paper. University of Sheffield, Sheffield, South Yorkshire, UK.

Merino, E., Manzaneque, M., & Banegas, R. (2012). Control of directors' compensation in Spanish companies. Corporate governance and firm performance. In A. Davila, M. Epstein & J. F. Manzoni (Eds.), *Performance measurement and management control: New global perspectives to advance theory and practice*. Bingley, UK: Emerald.

Mineev, A. (2012). Building of a network and the construction of controls. In A. Davila, M. Epstein & J. F. Manzoni (Eds.), *Performance measurement and management control: New global perspectives to advance theory and practice*. Bingley, UK: Emerald.

Needles, B. E., Shigaev, A., Powers, M., & Frigo, M. L. (2011). *The operating performance of high performance companies during a period of financial crisis: Risks and opportunities*. Working paper. DePaul University, USA.

Nicola, D. V., Hartmann, F., & Collini, P. (2011). *Long-term incentives, managerial effort and supervisor evaluation bias*. Working paper. University of Venice, Italy.

Piontkowski, J. O., Hoffjan, A., Schuchardt, L. D., & Lachmann, M. (2012). Determinants of the implementation and long-term use of inter-organizational cost management tools – an experimental investigation. In A. Davila, M. Epstein & J. F. Manzoni (Eds.), *Performance measurement and management control: New global perspectives to advance theory and practice*. Bingley, UK: Emerald.

Ruggeri, D. (2012). Subjective weighting under organizational interdependence conditions: An experimental investigation. In A. Davila, M. Epstein & J. F. Manzoni (Eds.), *Performance measurement and management control: New global perspectives to advance theory and practice*. Bingley, UK: Emerald.

Sedysheva, M. (2011). *Strategic management systems and methods of controlling as key elements of military expenditure policy making process*. Working paper. Estonian Business School, Estonia.

Silva de Souza, R., da Silva Gomes, S. M., Garcia de Oliveria, G., Santos Sampaio, M., & Almeida de Faria, J. (2012). Enterprise risk management and performance improvement: A study with Brazilian non-financial firms. In A. Davila, M. Epstein & J. F. Manzoni (Eds.), *Performance measurement and management control: New global perspectives to advance theory and practice*. Bingley, UK: Emerald.

Simons, R. (2000). *Performance measurement and control systems for implementing strategy*. Upper Saddle River, NJ: Prentice Hall.

Tamas, V., Friis, I., & Hansen, A. (2011). *Bringing the market inside the firm: Information and incentive problems*. Working paper. Copenhagen Business School, Copenhagen, Denmark.

Verbeeeten, F., & Spekle, R. F. (2011). *Management control systems, results oriented culture and performance: Evidence from Dutch municipalities*. Working paper. Erasmus University, The Netherlands.

Virtanen, T., Tuomaala, M., & Pentti, E. (2012). Challenges in energy efficiency performance measurement in the process industry. In A. Davila, M. Epstein & J. F. Manzoni (Eds.), *Performance measurement and management control: New global perspectives to advance theory and practice*. Bingley, UK: Emerald.

Wallstedt, N. (2011). *Management control in the public sector: The professionalization of public sector workers*. Working paper. Stockholm University, Stockholm, Sweden.

PART II
DEVELOPMENTS IN PERFORMANCE MEASUREMENT

THE IMPACT OF AMT FIRM CHARACTERISTICS ON INNOVATIVE COSTING TECHNIQUES AND NON-FINANCIAL MEASURES OF PERFORMANCE

Franco Cescon

ABSTRACT

This chapter proposes a contingency model that examines the most relevant contingent variables for understanding the factors that explain innovative costing techniques in a sample of firms investing in advanced manufacturing technologies (AMT firms). Furthermore, the aim of this study was also to ascertain whether AMT firms use significantly innovative managerial practices (IMPs).

The chapter uses material from a survey investigation and interviews with financial directors and the survey was conducted using a sample of AMT firms selected from an Italian industry. The method differs from previous studies in that it considers the relationship between a relevant contingency variable of AMT firms (i.e. the levels of integration, but

Performance Measurement and Management Control: Global Issues
Studies in Managerial and Financial Accounting, Volume 25, 91–115

ISSN: 1479-3512/doi:10.1108/S1479-3512(2012)0000025007

also environmental uncertainty and size) and various innovative costing, a relationship that has not been previously explored. The research was developed from the relevant literature.

The results indicate that there is no association between innovative costing and AMT firms in general, however the findings show that activity-based costing (ABC) is positively associated with fully integrated AMT firms. Large AMT firms have the highest percentage of innovative costing usage, such as ABC and strategic costing (SC). The relationship between AMT firms that perceive a high degree of environmental uncertainty and innovative costing was supported by the data in case of strategic dimension (target costing (TC) and life cycle costing (LCC)). Expectations of a relationship between AMT firms and IMPs, such as just-in-time (JIT), total quality management (TQM) and activity-based management (ABM), were not supported by the data.

We recognise that specific research limitations might reduce their generalisation, especially the number of statistical observations.

INTRODUCTION

Global markets required a change in the system by which firms are directed and controlled. Commentators such as Abdel-Maksound, Dugdale, and Luther (2005, p. 261) argue that 'many companies have responded to these challenges by implementing innovative managerial practices such as JIT, investing in advanced manufacturing technologies such as CAD/CAM, and emphasising quality, delivery, innovation and flexibility in meeting customer needs'.

In response to these changes and together with other firm change circumstances, several management accounting (MA) innovations have been advanced. Abdel-Kader and Luther (2008, p. 3) report that 'these new accounting techniques have been designed to support modern technologies and new management processes, such as total quality management (TQM) and just-in-time (JIT) production systems, in the search for a competitive advantage to meet the challenge of global competition'.

Following the contingency theory literature (e.g. Covaleski, Dirsmith, & Samuel, 1996; Otley, 1995; Reid & Smith, 2000), this chapter presents the results of a survey investigation into innovative management accounting practices (MAPs) in firms investing in advanced manufacturing technologies (AMT firms). Our main aim is to determine which specific contingency characteristics of AMT firms explain the adoption of different innovative cost techniques. Moreover, the chapter contributes to contingency theory research by considering a number of relevant innovative costing techniques and concentrating on key contingency factors, rather than considering a plurality of AMT firm characteristics. Three key contingency factors of firms investing in AMT are examined: (1) an external factor, i.e. the difference in the dynamism of product life cycles as a key external characteristic of environmental uncertainty; (2) an internal organisational factor, i.e. the difference in size as a characteristic of complexity and diversity and (3) an internal technology factor, i.e. the difference in the levels of integration as a key internal characteristic of new manufacturing systems.

This chapter further contributes to contingency theory research by considering a wide range of non-financial information as a part of innovative MA techniques for testing the association between the levels of integration and the non-financial measures of performance. An additional aim of the research was to ascertain whether AMT firms use innovative managerial practices (IMPs) such as JIT and ABM to reduce defect rates while improving on-time delivery. Fig. 1 portrays the research model proposed for analysing the MAPs.

In the top half of Fig. 1, the framework depicts the two variables (external and internal) of AMT firm characteristics and their potential relationship (impact) with the firms' MA innovations and IMPs. The bottom half of Fig. 1 illustrates that MA innovations will affect all of the various internal reports, including innovative costing approaches (techniques), non-financial

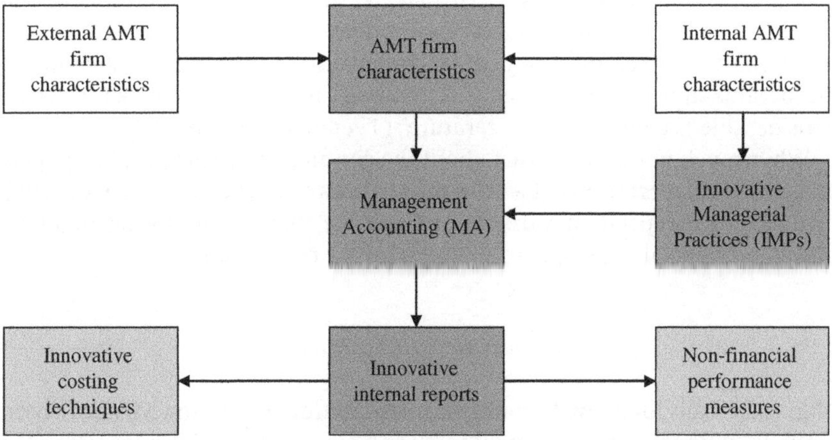

Fig. 1. AMT Firm Characteristics and Their Relationships with Management Accounting (MA) Innovations (and IMPs).

measures of performance for performance evaluations and, in some circumstances, decision-making.

The remainder of the chapter is organised as follows. Section 'The literature: A theoretical framework' provides an overview of the relevant literature. Section 'Development and hypotheses' develops a hypotheses derived from the literature. Section 'Research design and data collection' outlines the research design employed for this study. The survey results are then presented and discussed in section 'Survey results and hypothesis testing'. The final section contains a summary and conclusion.

THE LITERATURE: A THEORETICAL FRAMEWORK

Contingency-Based Research

It has been argued that 'contingency-based research has a long tradition in the study of management control systems (MCS). Researchers have attempted to explain the effectiveness of MCS by examining designs that best suit the nature of the environment, technology, size, structure, strategy, and national culture' (Chenhall, 2003, p. 127). However, 'contingency literature in the accounting control area has for a long time been criticized for being fragmentary and contradictory as results of methodological limitation' (Gerdin & Greve, 2004, p. 303). Although the identification of theoretical contributions remains difficult for academics (see Zimmerman, 2001), some commentators suggest that the significant amount of empirical research signals the importance of contingency theory. In particular, 'the question of which contingency factors influence the level of sophistication of an organization's management accounting system (MAS) has received considerable attention in the literature' (Tilemma, 2005, p. 103).

While much has been written about the contingency approach, few papers consider the expectations of a relationship between the contingency of AMT firms as an independent variable and specific innovative costing practices and non-financial measures of manufacturing performance.

Innovative Costing Models

Numerous calls for a revolutionary transformation of cost and MA have been made to support advanced manufacturing environments by integrating the financial and strategic dimensions (e.g. Kato, 1993; Raffish & Turney, 1991).

In particular, some innovative cost models have been developed into a theoretical framework. In the current study, within the financial dimension of innovative costing techniques, we include activity-based costing (ABC) and time-driven activity-based costing (TDABC), while the strategic dimension includes strategic costing (SC), target costing (TC) and life cycle costing (LCC).

Many authors (e.g. Johnson & Kaplan, 1987; Kaplan, 1983) have highlighted the strong interest in implementing ABC, especially when challenges posed by a changing business environment influence the adoption of AMT and marketing strategies and, consequently, the cost structure of the firms. In particular, the normative literature suggests that ABC provides a more accurate product-costing framework for decision-making, performance evaluation and external (financial) reports.

In the accounting literature, SC has been proposed as a means of addressing the limitations of internal cost and developing a strategic dimension with external information, especially for strategy formulation. Bromwich and Bhimani (1994, p. 126) assert that the aims of strategic MA are (a) 'to focus managerial efforts more on their markets, where customers have to be won and retained and competitors repulsed, and on the costs of these market activities' and (b) 'to ascertain the enterprise's cost positioning relative to its rivals'. Using primarily the case study method, the leading proponents of SC (Shank & Govindarajan, 1989) sought to explain the importance of analysing the value chains of competitors and the value chain of the firm as a set of cost drivers to understand variations in costs as a source of competitive advantage.

Like SC, TC has also been proposed as a means of addressing the limitation of internal cost and developing a strategic dimension with external information. TC, which is one of the most important accounting practices in large Japanese firms, is a market-orientated tool with a clear strategic dimension of the cost information regarding competitors' strategies. As Bromwich and Bhimani (1994, p. 175) note, 'It indicates for each product (and each cost element of that product) what cost reductions (over time) must be achieved by the firm and its suppliers in order to ensure that the company's strategy is realised throughout the product's life cycle'.

LCC is emerging as an approach that is related to the firm's budgeting process. It is described as 'the period that starts with the initial product specification and ends with the withdrawal of the product from the market place' (Raffish & Turney, 1991, p. 61). In particular, LCC not only has clear potential from a marketing perspective in terms of different phases, but also has potential for costing purposes.

Non-financial Measures of Performance

While some authors (e.g. Datta, Sambasivarao, Kodali, & Deshmukh, 1992; O'Brien & Smith, 1993) suggest that intangible or strategic benefits are associated with investments in AMT and others (e.g. Zimmerman, 1995) include both (management) accounting systems and non-financial information systems in the performance evaluation system (PMS), the majority (e.g. Chenhall & Langfield-Smith, 1998) suggest that non-financial information systems are an extension of the established role that innovative MA plays in evaluating managerial and corporate performances.

The literature has expanded non-financial measures of performance, which are particularly effective in controlling production that makes use of advanced technologies and management innovations. As Abdel-Maksoud et al. (2005, p. 270) explain, 'non-financial performance measures have attained greater theoretical prominence since the promotion of the "balanced scorecard" (Kaplan & Norton, 1992) and other related ideas such as the "performance prism" (Neely, Adams, & Kennerly, 2002) and the "value chain scorecard" (Lev, 2001)'.

In this study, we focus on a wide range of non-financial information for testing the associations between contingency factors of AMT firms, such as level of integration, and non-financial performance measures. Although the use of non-financial measures of performance such as productivity, inventory costs, quality and customer satisfaction measures has been a major concern in almost all previous surveys, we have focused on the following 10 non-financial measures of manufacturing (quality) performance in AMT firms: (1) number of set-ups, (2) lead time (cycle time), (3) number of product defects, (4) labour productivity, (5) capacity utilisation, (6) schedule adherence, (7) on-time delivery to customers, (8) numbers of customer returns, (9) number of complaints from customers and (10) proportion of overtime worked.

DEVELOPMENT OF HYPOTHESES

Environmental Uncertainty and Innovative Costing Models

Since the 1980s, there has been an important change in the external environment of firms in many sectors of the economy.

Under conditions of significant change, environmental uncertainty is an external factor that reportedly affects strategy, organisation design and technology (Baines & Langfield-Smith, 2003). It has been argued that a business strategy is a response to an environment. For example, Chong and Chong (1997) found a positive relationship between perceived environmental uncertainty and strategy. They also discovered that more sophisticated MAS reports can reduce uncertainty and improve managerial decision-making. These research findings lead us to include that environmental uncertainty as a variable will potentially have an impact on innovative costing practices.

These considerations lead to the following hypothesis:

Hypothesis 1. AMT firms perceiving a higher degree of environmental uncertainty, which is measured by the competition intensity variable given by product life cycle dynamism, employ more innovative costing practices than AMT firms that perceive lower environmental uncertainty.

Organisation Size and Innovative Costing Models

Since the 1980s, industry in Western countries has been compelled to respond to some international competitive challenges through changes in strategy. As a result, there has been an emphasis on more customer-oriented variables (Sim & Killough, 1998) and changes in organisation design. Changes need an appropriate matching of strategy, organisational design and size, especially when size is related to mergers and acquisitions. Otley (1995, p. 54) suggests that 'organizational size is an important factor affecting both structure and other control arrangements'. Previous surveys found a significant relationship between firm size and the use of sophisticated cost accounting practices (e.g. Al-Omiri & Drury, 2007). Thus, it might be expected that large AMT firms will tend to increase investments in fully integrated systems.

As this survey includes a sample of large firms, the second hypothesis is as follows:

Hypothesis 2. Large AMT firms use more innovative costing techniques than small AMT firms by investing in fully integrated systems

Level of Integration and Innovative Costing Models

Considerable attention has been paid to incorporating technology as a contingency factor of MA and control system design. Otley (1995, p. 53) notes that 'one of the longest-established relationships between a contingent variable and control systems design has been that associated with production technology'.

Thus, it can be argued that the use of new technologies will affect the cost structure, costing system design and the development of non-financial measures in manufacturing firms. For example, Haldma and Lääts (2002, p. 393) note that 'the tightening global competition and increasing fixed costs associated with the use of advanced manufacturing technologies have prompted the need to analyse, allocate and manage fixed cost better'.

It is useful to note that, with reference to AMT firms, some conceptual frameworks (see e.g. Meredith & Hill, 1987) suggest that each level of integration has a different purpose, objective, benefits, scope of effects, organisational impact and risk. It is also important to note that many authors examine the appropriateness of different justification techniques, such as financial, non-financial and risk analysis techniques, for each level of integration (e.g. Alkaraan & Northcott, 2006).

Similarly, we maintain that there are implications for cost accounting and believe that each level of integration potentially justifies the use of different costing techniques. More recent work has indicated a positive relationship between the degree of automation in the production process and the formality of innovative costing techniques. However, as indicated previously, there are different levels of integration in AMT firms; thus, we classify the companies in three groups (stand-alone, less- and fully integrated AMT firms).

Considering the characteristics of their high level of integration, it might be expected that fully integrated AMT firms tend to result in a higher level of costing technique sophistication. The following hypothesis is set to test the association between innovative costing techniques and fully integrated AMT firms:

Hypothesis 3. AMT firms investing in fully integrated systems use more innovative costing techniques than stand-alone AMT firms and less-integrated AMT firms.

Levels of Integration and Non-financial Measures of Performance

As discussed above, the firm's accounting systems are used for decision-making, performance evaluation and external reporting.

While some authors (e.g. Kaplan, 1986) suggest that traditional managerial accounting may distort cost and that traditional product costing does not reflect the pattern of cost causation in firms using AMT and management innovations, other authors believe that a number of intangible (or strategic) benefits are associated with AMT investments.

Recent surveys have shown a strong interest in the topic of certain strategic benefits with non-financial investment criteria in AMT investments. There is empirical evidence that some intangible (or strategic) benefits, such as quality of output, manufacturing flexibility, reduced lead-times and reduced inventory level, are important in the justification and decision of AMT investments.

At the same time, many authoritative commentators agree with the report's view that internal reporting, which is used for performance evaluation, should emphasise MA innovations, such as innovative costing approaches and non-financial systems. While some authors suggest that performance measurement systems based on financial information are insufficient, Bromwich and Bhimani (1994, p. 12) note 'that monitoring performance and indeed the panning of performance may be improved if non-financial information is utilised together with financial information. This is especially helpful where some costs may be difficult to define and measure though a wide array of real time non-financial information is available'.

These considerations regarding innovative production systems lead to the following hypothesis:

Hypothesis 4. A range of non-financial information is important in the control of organisational problems, such as performance evaluations, in AMT firms regardless of the level of integration.

RESEARCH DESIGN AND DATA COLLECTION

Empirical data were collected in the following two ways: (a) a survey investigation into the specific MA innovations in relation to AMT firms and (b) interviews. The survey was conducted using a mailed questionnaire so as to facilitate access to a large number of finance directors and controllers. In addition, the survey respondents were asked to indicate whether they would be willing to participate in a follow-up interview. The aim of the follow-up interviews was to analyse questionnaire responses in greater depth.

The interviews were semi-structured and conducted with a degree of flexibility. A list of the main questions was sent in advance to facilitate the interviews.

The sample included 277 firms selected from the list supplied by the Italian Confederation of Industry (ICI). This ICI list was a representative database of the companies that adopt advanced manufacturing systems. The selected companies were classified in eight different manufacturing sub-sectors and satisfied, in particular, a minimum turnover and a minimum number of employees (30).

The questionnaires were sent to the financial directors and controllers by e-mail. A letter was attached that explained the purpose of the survey and assured the confidentiality of the information supplied by each firm.

After three e-mails and telephone reminders to those who had been sent one of the 277 questionnaires, 39 were returned blank due to errors in the address and 31 were returned for various other reasons. Of the remaining 207 questionnaires, we received 54 usable responses, giving a response rate of 26.0%. (54 completed questionnaires from 207 potential respondents)

The possibility of a non-response bias was examined by comparing the 54 responding companies to the sample according to their manufacturing groups.

The results of Fischer's Exact Test for Count Data (P-value = 0.267) indicated that, with regards to sub-sector composition, there was no statistically significant difference between the two groups of companies (54 complete questionnaires and 207 potential respondents). In addition, when considering the eight manufacturing sub-sectors, cross-tabulation indicated that the composition of the 54 responding firms differed from the sample of 207 firms in so much as there were proportionally fewer firms in the following groups: (1) motor vehicles, tyres and related components; (2) textiles, clothing and related components; (3) computers, information technology and related components. Furthermore, the results indicated that the responding firms, especially the engineering companies, were smaller than other firms with respect to turnover.[1]

To determine the turnover range of the large firms, the responding companies were measured in three groups by turnover (and number of personnel): (a) small units, (b) medium-sized units and (c) large units.[2]

Fifty-two of the responding companies were classified as AMT firms. The composition, according to size, is summarised in Table 1.

Table 1. Size of AMT Firms.

	No (52)	%
Small AMT firms	8	15.4
Medium AMT firms	38	73.0
Large AMT firms	6	11.6

The measurement instrument used in the questionnaire to estimate the importance of certain suggested issues was a five-item Likert scale.

Moreover, this analysis facilitated the use of a Pearson chi-square test to determine whether, in AMT firms, there was a statistical correlation between innovative costing techniques and contingency variables. In the test, the *P*-value was displayed. A 10% significance level was used in this study.

SURVEY RESULTS AND HYPOTHESIS TESTING

Classification of AMT Firms by Levels of Integration

To test the hypotheses, responding firms were categorised according to their types of AMT systems. Consistent with the literature (e.g. Abdel-Kader & Dugdale, 1998), these AMT firms were categorised into three levels of integration: (1) stand-alone AMT firms, (2) less-integrated AMT firms and (3) fully integrated AMT firms (see Table 2).

Table 2 indicates that the second level and the first level (in particular, robotics and/or computer numerical control (CNC)) were most prevalent, while the lowest percentages applied to the third level (computer integrated manufacturing (CIM) and/or flexible manufacturing systems (FMS)).

Considering Meredith and Hill's (1987) justification of new manufacturing systems, this categorisation allowed us to understand that the majority of AMT firms tend to invest in the first level of automation (stand-alone) for the purpose of saving costs and in the second level of automation (less integrated) to create significant tangible and intangible benefits, such as less work-in-process, reduced lead time, greater flexibility, better quality and a set of information packages (master scheduling, material-requirements planning, inventory, plant monitoring). In contrast, AMT firms tend to invest in the third level of automation (fully integrated) with the purpose of implementing a new production strategy and becoming a true competitive lever in the market. Consequently, strategic benefits are the primary focus,

Table 2. Levels of Integration in the AMT Firms.

	No (52)	%
Stand-alone AMT firms	18	34.6
Less-integrated AMT firms	23	44.2
Fully integrated AMT firms	11	21.2

and everything is changed at once, which includes purchasing, manufacturing, marketing, finance and top management functions.

Levels of Integration and Costing Innovations

The survey respondents were asked to indicate their use of the four innovative costing techniques to provide the cost information required for decision-making and control.

Table 3 summarises the number, percentage and Pearson test calculated for the innovative costing techniques used according to the levels of integration in AMT firms.

In general, few AMT firms use only innovative cost techniques as an internal source of data for several purposes, such as decision-making, controlling the behaviour of lower-level managers and reporting top managers' performance.

Furthermore, the results reveal a certain variability in the use of these innovative costing techniques at different levels of integration.

In particular, these techniques are less likely to be implemented in stand-alone AMT firms probably because the main focus at the first level of automation is an increase in efficiency to produce cost savings, i.e. decreased direct labour costs, lower levels of inventories, fewer inspections, increased safety, reduced reworking and scrap. Consequently, traditional costing may be appropriate in controlling elements of cost as costs are identified and converted into measures with predetermined criteria.

In addition, innovative cost techniques are not likely to be implemented in less-integrated AMT firms. It should be noted that the emphasis is not

Table 3. Innovative Costing Techniques According to Levels of Integration in AMT Firms.

Techniques	Stand-alone AMT Firms (N=18)		Less-Integrated AMT Firms (N=23)		Fully Integrated AMT Firms (N=11)		Pearson Test
	No	%	No	%	No	%	*P*-value
Activity-based costing	1	5.5	2	8.6	4	36.3	**0.04**
Strategic costing	1	5.5	2	8.6	3	27.2	0.17
Target costing	–	–	3	13.0	2	18.1	0.20
Life cycle costing	–	–	1	4.3	1	9.0	0.45

exclusively on efficiency, as in the case of stand-alone AMT firms, but also on tangible and intangible benefits. Therefore, it can be said that this new manufacturing system requires some changes in the firm's production, such as product mix, capacity and lead time, in its organisational design and cost structure. In this case, the activity accounting focuses on the management of activities that consume resources rather than on controlling elements of cost, which is more appropriate, especially for service department costs, such as set-up (set-up times), scheduling (machine utilisation) and engineering (engineering changes).

In general, although this second level of automation does not ignore strategic benefits, it is not the management's primary focus in the new manufacturing strategy. Therefore, as the results of less-integrated AMT firms reveal, the thinking in this case is that the strategic dimension of the innovative costing techniques, such as SC, TC and LCC, could not be justified.

When strategic benefits are the management's primary focus and everything is changing, as is the case with fully integrated AMT firms, moving beyond its traditional costing method and introducing a given innovative costing technique could be justified. The research findings indicate that fully integrated AMT firms tend to adopt higher percentages of innovative costing techniques compared to stand-alone and less-integrated firms in the case of the financial dimension (ABC) and the strategic dimension (SC).

Two finance directors working in fully integrated AMT firms suggest that SC and TC might be considered important as they were more directly linked to strategic benefits in firms with AMT. 'Strategic benefits are the primary focus of managerial decision-making. We adopt strategic costing (SC) using external information to develop competitive advantage comparisons and target costing (TC) to capture cost information about competitors'.

LCC was considered less important in fully integrated AMT firms by the respondents in this study, and the follow-up interviews suggest that financial directors perceive this technique as difficult to apply.

A financial director of a large electrical household appliances firm illustrated this point by stating, 'I think that LCC is theoretically a correct approach, but in practical terms it is not feasible in our business. Moreover, the management accounting practices at the corporate level do not consider LCC as a tool for the firm's budgeting process'.

The Pearson test (Table 3), which was adopted to examine the difference in use of innovative costing techniques for stand-alone, less-integrated and fully integrated AMT firms, indicates that no statistically significant difference was found in SC, TC and LCC use. The results of the test are P-value = 0.17, P-value = 0.20 and P-value = 0.45, respectively.

However, there is a statistically significant association between fully integrated AMT firms and ABC (P-value = 0.04). Hypothesis 3 is, therefore, supported for ABC. Consequently, our data suggest that the use of fully integrated AMT systems is a contextual factor influencing the implementation of innovative costing techniques, especially ABC.

Our study is consistent with that of Abdel-Kader and Luther (2008), who examined in the United Kingdom the impact of firm characteristics on MAPs, only in the case of fully integrated AMT firms.

Size and Costing Innovations

According to cost accounting research, organisation size is an important contextual factor that influences innovative costing techniques as larger companies have relatively greater access to resources than do smaller organisations. A particular feature of this study is the division of responding AMT firms into three groups. These groups include those employing AMT in small organisations, those using AMT in medium-sized organisations and those using AMT in large organisations.

In Table 4, the numbers of innovative costing practices are shown separately for each category of firm size. While the findings indicate that large AMT firms tend to adopt a high percentage of innovative costing techniques for both the financial dimension (ABC, 83.3%) and the strategic dimension (SC, 50.0%), the use of TC was limited (16.6%) and LCC was not used at all.

Table 4. Innovative Costing Techniques According to Size in AMT Firms.

Techniques	Small AMT Firms (N = 8)		Medium AMT Firms (N = 38)		Large AMT Firms (N = 6)		Pearson Test
	No	%	No	%	No	%	*P*-value
Activity-based costing	–	–	2	5.2	5	83.3	**<0.01**
Strategic costing	1	12.5	2	5.2	3	50.0	**<0.01**
Target costing	–	–	4	10.5	1	16.6	0.54
Life cycle costing	–	–	2	5.2	–	–	0.68

In contrast, the analysis reveals that medium-sized AMT firms do not emphasise innovative costing techniques. Overall, innovative costing techniques were absent in small AMT firms.

A Pearson chi-square test was used to identify the relationship between size and innovative costing techniques. The Pearson test (Table 4) supports Hypothesis 3 with regard to the financial dimension with a *P*-value for ABC being <0.01 and one strategic dimension and the *P*-value of SC being <0.01. In both cases, large firms have the highest percentage of innovative costing usage.

In general, the results support the findings of Haldma and Lääts (2002) quite clearly. Furthermore, this research is consistent with that of Al-Omiri and Drury (2007) across all three groups of firm size (small, medium and large). Their results found that higher levels of cost systems sophistication are positively associated with size as a factor influencing the choice of product costing systems in UK organisations.

In addition, there is a basis for comparison when addressing the question of ABC adoption in large firms (e.g. Innes, Mitchell, & Sinclair, 2000). Despite the results, international comparisons tend to indicate that ABC plays a smaller role in Italy than in the United Kingdom and the United States (see e.g. Bright, Davies, Downes, & Sweeting, 1992; Foster & Swenson, 1997).

Levels of Perceived Environmental Uncertainty and Costing Innovations

Environmental uncertainty is an important external factor that influences innovative costing techniques. Following Gul and Chia (1994), when the perceived environmental uncertainty (PEU) is high, the decision-maker tends to require additional ex-post and ex-ante internal and external information, more sophisticated reporting from management and a cost accounting system that is able to cope with environment complexity. Also, Chong and Chong (1997) suggest that more sophisticated reporting from the MAS can help to reduce uncertainty and enhance managerial decision-making.

Once again, a particular feature of this study is the division of responding AMT firms into three groups: those that perceive a low degree of environmental uncertainty, those that perceive a medium degree of environmental uncertainty and those that perceive a high degree of environmental uncertainty. To measure environmental uncertainty, we use the dynamism of product life cycles because it is considered a key variable in relation to the intensity of competition.

Table 5. Innovative Costing Techniques for Each Level of PEU in AMT Firms.

Techniques	PEU: Low Dynamism of Product Life Cycle (N = 15)		PEU: Medium Dynamism of Product Life Cycle (N = 33)		PEU: High Dynamism of Product Life Cycle (N = 4)		Pearson Test
	No	%	No	%	No	%	*P*-value
Activity-based costing	2	13.3	5	15.1	1	25.0	0.70
Strategic costing	1	6.6	4	12.1	1	25.0	0.58
Target costing	–	–	3	9.1	2	50.0	0.01
Life cycle costing	–	–	–	–	2	50.0	**< 0.01**

The numbers for innovative costing practices in AMT firms are shown separately for each level of PEU in Table 5.

The results, as presented in Table 5, indicate that 72% of the respondents perceived a medium to high degree of environmental uncertainty. However, AMT firms perceiving high and medium degrees of environmental uncertainty tend to adopt a limited percentage of innovative costing techniques with respect to the financial dimension ABC and a relatively high percentage in the case of strategic dimension.

We also sought to identify a relationship between the levels of PEU and innovative costing techniques in AMT firms. For this external factor, we found a significant relationship between the degree of PEU, as measured by the dynamism of product life cycles, and innovative costing practices in AMT firms.

Namely, Pearson test results indicate that AMT firms perceiving a higher degree of environmental uncertainty use more innovative costing practices than AMT firms that perceive lower environmental uncertainty. The *P*-value = 0.01 for TC, and the *P*-value < 0.01 for LCC. Hypothesis 1 is, therefore, supported for both TC and LCC.

Levels of Integration and Non-financial Measures of Performance

It was Kaplan and Norton (1992) who first promoted the balanced scorecard as a framework to incorporate non-financial measures and provide some balance to the more traditional financial measures used to evaluate corporate performance.

Table 6. Non-financial Measures of Performance.

Non-financial Measures of Performance	Stand-alone		Less Integrated		Fully Integrated		*P*-value
	Mean	Std. Dev.	Mean	Std. Dev.	Mean	Std. Dev.	
1. Number of set-ups (% batches)	4.00	0.84	3.18	1.05	3.10	1.59	**0.06**
2. Manufacturing cycle efficiency	3.53	1.35	3.32	1.71	3.00	1.34	0.57
3. Defects (% of total production)	4.25	1.12	3.52	1.27	3.09	1.64	**0.07**
4. Efficiency (standard hours produced/hours worked)	4.12	1.31	3.81	1.36	3.55	1.44	**0.00**
5. Capacity utilization (hours worked/budgeted hours)	3.75	1.06	3.63	0.84	3.00	1.70	0.23
6. Schedule adherence %	4.00	1.03	4.13	0.77	3.55	1.50	0.32
7. On-time delivery to customers	4.37	0.80	3.95	1.17	3.72	1.48	0.32
8. Proportion of overtime worked	3.13	1.58	3.04	1.32	3.50	1.65	0.70
9. Number of complaints from customers	2.50	1.36	2.70	1.08	3.20	1.31	0.37
10. Number of customer returns	4.31	0.94	3.47	1.08	3.90	1.37	**0.07**

Our findings reveal that non-financial measures, which are used in controlling organisational problems (performance evaluation), are extensively employed in Italian AMT firms (see Table 6).

Importance was measured on a five-point scale ranging from 1 (no importance) to 5 (critical importance), and the 10 measures are shown in Table 6 according to levels of integration. A 10% significance level (*P*-value) was used in this analysis.

An analysis of non-financial measures of performance suggests that such factors are of critical importance in Italian AMT firms and significantly so with respect to the (1) number of set-ups, (2) defects as a percent of total production, (3) efficiency and (4) number of customer returns.

Although Hypothesis 4 is supported, the number of significant non-financial measures of performance is restricted. While four non-financial measures of manufacturing performance were identified in this survey, the other six factors in the statistical test failed to reveal any significant differences in the behaviour of the three levels of integration of AMT firms.

Table 7. IMPs in AMT Firms (% of Respondents).

Innovative Managerial Practices (IMPs)	% of AMT Firms Using IMPs	*P*-Value
JIT (just-in-time)	40.3	0.24
TQM (total quality management)	36.5	0.77
ABM (activity-based management)	9.6	0.96

In general, considering both Hypotheses 3 and 4, our results confirm those of Abdel-Kader and Luther (2008, p. 20) who found that 'organization that adopt AMT need a multidimensional performance (financial and non-financial) measurement system to provide managers with continuous signals as to what is most important in their day-to-day activities and where efforts must be directed'. Overall, the practices of Italian manufacturing AMT firms support other UK surveys (e.g. Bhimani, 1993) and place emphasis on non-financial performance measures.

Further Analysis of the Innovative Managerial Practices of AMT Firms

The practices of Italian AMT firms were further investigated through a question that aimed to identify which IMPs they implement. A list of three relevant management innovations was provided, and respondents were asked to indicate if each management innovation was implemented together with AMT investments to improve the contemporary manufacturing environment.

While AMT firms tend to use JIT (40.3%) and TQM (36.5%), we found that only a minority of AMT firms adopted ABM (9.6%).

Furthermore, Table 7 shows that there are no significant differences, at the 10% level of significance, among the three groups (stand-alone, less integrated, fully integrated) of AMT firms as regards IMPs. The results are as follows: JIT with a *P*-value = 0.24, TQM with a *P*-value = 0.77 and ABM with a *P*-value = 0.96.

DISCUSSION

The survey indicates that MA innovations are affected by some (internal and external) AMT firms' characteristics, such as levels of environmental uncertainty, size and levels of integration. Considering a contingency approach for each AMT firm characteristic, this conclusion is based on the importance of the two relevant groups of MA innovations: (i) innovative costing techniques and (ii) non-financial measures of performance. The following discussion is structured around these two groups of MA innovations.

Innovative Costing Techniques

1. The findings, derived from the questionnaire, indicate that differences in innovative costing techniques are significantly explained for each level of PEU, i.e. lower PEU, medium PEU and higher PEU. Thus, the findings support Hypothesis 1, which states that AMT firms perceiving a higher degree of environmental uncertainty use more innovative costing practices than AMT firms that perceive lower environmental uncertainty. However, the hypothesis is only supported with respect to two innovative costing techniques: TC and LCC. Such techniques are important to strategic cost analysis.

   The data confirm that in specific manufacturing companies (AMT firms) environmental uncertainty should be considered as an external variable in the contingency theory paradigm.
2. The survey results support Hypothesis 2, which states that large AMT firms use more innovative costing techniques than small AMT firms investing in fully integrated systems. In fact, a significant relationship was found between large AMT firms and two innovative costing practices: (1) ABC, which is a more accurate accounting cost system that enhances managers' focus on the internal activities of the organisation and (2) SC, which is viewed as sophisticated costing that provides information of strategic value.

   This survey generally supports previous findings that the innovation of costing (and budgeting) increases in relation to firm size. The data confirm that size should be considered as an internal variable in the contingency theory paradigm. In particular, this study provides evidence of the impact of AMT firm size on innovative cost techniques, especially with regard to ABC and SC.
3. Traditional cost systems, such as absorption costing and variable costing, are still important in firms investing in AMT. Conversely, in AMT firms, all of the innovative costing techniques, such as ABC, SC, TC and LCC, appear to have fewer applications than traditional cost systems.

Although the results reveal considerable variability in the use of these innovative costing techniques in the three levels of integration of AMT firms, the statistical test is consistent with Hypothesis 3. That is, AMT firms investing in fully integrated systems use more innovative costing techniques than stand-alone and less-integrated AMT firms. It would be logical to conclude that each level of integration in AMT firms affects MAS and, consequently, costing techniques. This conclusion is open to speculation as formal and technically sophisticated approaches tend to be viewed as weaker with respect to fully integrated systems. However, this survey reveals a statistically significant association between fully integrated AMT firms and ABC. The remaining three new costing techniques were not considered important as there were non-significant differences between the three groups of AMT firms.

The firms investing in fully integrated systems viewed ABC, despite being expensive and time-consuming, as a technique that emphasised the categorisation of activities and enhanced manager focus on the management of activities that consume resources rather than control elements of cost.

The data confirm that the level of integration should be considered as an internal variable in the contingency theory paradigm. In particular, this study shows that differences in innovative cost techniques are significantly explained by the levels of integration, which are manufacturing characteristics of AMT.

In general, it would be logical to conclude that there is evidence that AMT firms with fully integrated systems adopt ABC and that this cost system is seen as marginally more important than the other innovative costing techniques.

Non-financial Measures of Performance

4. The results appear to support Hypothesis 4: A range of non-financial information is important in the control of organisational problems (performance evaluation) in AMT firms regardless of the level of integration. However, caution should be exercised in drawing conclusions as three non-financial measures (manufacturing cycle efficiency, proportion of overtime worked, number of complaints from customers) were not considered to be important by the respondents. Similarly, caution should be exercised in assessing the importance of three non-financial measures (on-time delivery to customers, schedule adherence percentage, capacity utilisation) as no significant differences were

observed at the 10% level among the three levels of integration of AMT firms.

In this study, the number of set-ups, defects, efficiency and the number of customer returns were the four non-financial performance measures that showed a significant association with AMT levels of integration.

5. There is evidence to suggest that AMT firms tend to use IMPs. However, expectations of a statistical significant relationship between AMT firms and IMPs, such as JIT, TQM and ABM, were not supported by the data.

   In addition, there are no significant differences among firms using IMPs and MA innovations, such as innovative costing techniques and non-financial performance measures.

CONCLUSION

This chapter applied contingent approaches to respond empirically to the considerations of Otley (1995) and Chenhall (2007) when theorising contingencies in management control system research. In particular, using the contingent theory of MA, we examined the impact of a few potentially relevant contingency variables on a set of innovative costing practices in a sample of selected Italian manufacturing companies that have invested in AMT.

The main aim was to increase the understanding of the external and internal factors that explain the (potential) usefulness of innovative costing techniques as part of MA sophistication in firms that have invested in AMT. This study thus sought to contribute to the contingency theory of MA by considering a number of innovative costing techniques, rather than a single popular emerging cost approach, and by concentrating on a single relevant contingent variable, rather than considering a plurality of firm characteristics. The method differs from previous studies in Europe in that it is the first to examine the relationship between different levels of integration in AMT firms, innovative costing practices and non-financial performance measures. Furthermore, this study includes the relationship between the level of environmental uncertainty and size, as firm characteristics, and innovative costing techniques.

Questionnaire survey results indicate that there is no association between innovative costing and AMT firms in general. Consequently, the present study in Italy confirms some reservations (e.g. Bromwich & Hong, 1999; Gosselin, 1997; Malmi, 1999) that have been expressed concerning the use of

ABC. However, when explaining why firms adopt different innovative costing techniques, the results show that ABC is positively associated with fully integrated AMT firms.

In accordance with numerous international empirical studies some expectations of a relationship between large AMT firms and innovative costing techniques were supported by the data.

The relationship between environmental uncertainty and innovative costing techniques was supported by the data in the case of strategic dimensions, including TC and LCC.

A range of non-financial performance measures are important in the control of organisational problems in AMT firms. However, there are four significant criteria focusing specifically on product quality and marketing activities; therefore, caution should be exercised in assessing the importance of all non-financial performance measures among the three levels of integration.

Finally, expectations of a relationship between AMT firms and IMPs, such as JIT, TQM and ABM, were not supported by the data.

In reflecting upon the results, we recognise that specific research limitations might reduce their generalisation, especially the number of statistical observations.

NOTES

1. The potential respondents operate in the following eight manufacturing sub-sectors: (1) motor vehicles, tyres and components; (2) engineering, machinery, electrical household appliances and related products; (3) food products and beverages; (4) textiles and clothing products; (5) chemicals and chemical products; (6) computers and related products; (7) wood, furniture and related products and (8) others (e.g. paper, construction and related products).
2. If the turnover was less than 25.8 million euro, the unit was classified as a small firm; a turnover of 25.9–2582.2 million euro was regarded as a medium-sized firm, whereas a turnover of more than 2582.2 million euro was classified as large.

ACKNOWLEDGMENTS

My particular thanks to Luca Grassetti for his advice on statistical analysis. Also, I gratefully acknowledge the valuable comments received from Professor Al Bhimani, the participants at the IMAP workshop (Bocconi University, Milan, 2009) and the participants at the 6th Conference on Performance Measurement and Management Control (EIASM), Nice,

France, September 2011. Permission has been received from the Department of Management, Institute of Strategy (Bocconi University, Milan, Italy), to use certain questions and answers on innovative costing techniques from past Institute of Strategy examinations.

REFERENCES

Abdel-Kader, M., & Dugdale, D. (1998). Investment in advanced manufacturing technology: A study of practice in large U.K. companies. *Management Accounting Research*, *9*, 261–284.

Abdel-Kader, M., & Luther, R. (2008). The impact of firm characteristics on management accounting practices: A UK – based empirical analysis. *The British Accounting Review*, *40*, 2–27.

Abdel-Maksound, A., Dugdale, D., & Luther, R. (2005). Non-financial performance measurement in manufacturing companies. *The British Accounting Review*, *37*, 261–297.

Alkaraan, F., & Northcott, D. (2006). Strategic capital investment decision-making: A role for emergent analysis tools? A study of practice in large UK manufacturing companies. *The British Accounting Review*, *38*, 149–173.

Al-Omiri, M., & Drury, C. (2007). A survey of factors influencing the choice of product costing system in UK organizations. *Management Accounting Research*, *18*, 399–424.

Baines, A., & Langfield-Smith, K. (2003). Antecedents to management accounting change: A structural equation approach. *Accounting, Organization and Society*, *28*, 675–698.

Bhimani, A. (1993). Performance measures in UK manufacturing companies: The state of play. *Management Accounting*, *71*(December), 20–23.

Bright, J., Davies, R. E., Downes, C. A., & Sweeting, R. C. (1992). The deployment of costing techniques and practices: A UK study. *Management Accounting Research*, *3*(3), 201–211.

Bromwich, M., & Bhimani, A. (1994). *Management accounting: Pathways to progress*. London: Chartered Institute of Management Accountants.

Bromwich, M., & Hong, C. (1999). Activity-based costing systems and incremental costs. *Management Accounting Research*, *10*, 39–60.

Chenhall, R. (2003). Management control systems design within its organizational context: Findings from contingency-based research and directions for the future. *Accounting, Organization and Society*, *28*, 127–168.

Chenhall, R. (2007). Theorizing contingencies in management control systems research. In C. Chapman, A. Hopwood & M. Shields (Eds.), *Handbook of management accounting research* (pp. 163–205). Amsterdam: Elsevier.

Chenhall, R., & Langfield-Smith, K. (1998). Factors influencing the role of management accounting in the development of performance measures within organizational change programs. *Management Accounting Research*, *9*(4), 361–386.

Chong, V., & Chong, K. (1997). Strategic choices, environmental uncertainty and SBU performance: A note on the intervening role of management accounting systems. *Accounting and Business Research*, *27*(4), 268–276.

Covaleski, M., Dirsmith, M., & Samuel, S. (1996). Managerial accounting research: The contribution of organizational and sociological theories. *Journal of Management Accounting Research*, *8*, 1–35.

Datta, V., Sambasivarao, K. V., Kodali, R., & Deshmukh, S. G. (1992). Multi-attribute decision model using the analytic hierarchy process for the justification of manufacturing systems. *International Journal of Production Economics*, *28*, 227–234.

Foster, G., & Swenson, D. G. (1997). Measuring the success of activity-based cost management and its determinants. *Journal of Management Accounting Research*, *9*, 109–142.

Gerdin, J., & Greve, J. (2004). Forms of contingency fit in management accounting research – a critical review. *Accounting, Organization and Society*, *29*, 303–326.

Gosselin, M. (1997). Bandwagon theories: Some explanation for the activity based costing paradox. Paper presented at the EIASM Workshop on Manufacturing Accounting, Edinburgh, UK.

Gul, F., & Chia, Y. (1994). The effect of management accounting systems, perceived environmental uncertainty and decentralization on managerial performance: A test of three- way interaction. *Accounting, Organization and Society*, *19*(4/5), 413–426.

Haldma, Z., & Lääts, K. (2002). Contingencies influencing the management practices of Estonian manufacturing companies. *Management Accounting Research*, *13*, 379–400.

Innes, J., Mitchell, F., & Sinclair, D. (2000). Activity-based costing in the U.K.'s largest companies: A comparison of 1994 and 1999 survey results. *Management Accounting Research*, *11*, 349–362.

Johnson, H. T., & Kaplan, R. S. (1987). *Relevance lost: The rise and fall of management accounting*. Boston, MA: Harvard Business School Press.

Kaplan, R. S. (1983). Measuring manufacturing performance: A new challenge for managerial accounting research. *The Accounting Review*, *LVIII*(4), 686–705.

Kaplan, R. S. (1986). Must CIM be justified by faith alone? *Harvard Business Review*, *March–April*, 87–95.

Kaplan, R. S., & Norton, D. P. (1992). The balanced scorecard: Measures that drive performance. *Harvard Business Review* (January-February), 71–79.

Kato, Y. (1993). Target costing support system: Lessons from leading Japanese companies. *Management Accounting Research*, *4*(March), 33–47.

Lev, B. (2001). *Intangibles: Management, measurement and reporting*. Washington, DC: Brooking Institution Press.

Malmi, T. (1999). Activity-based costing diffusion across organizations: An explanatory empirical analysis of Finnish. *Accounting Organization and Society*, *24*(November), 649–672.

Meredith, J. R., & Hill, M. M. (1987). Justifying new manufacturing systems: A managerial approach. *Sloan Management Review*, *Summer*, 49–61.

Neely, A., Adams, C., & Kennerly, M. (2002). *The performance prism: The scorecard for measuring and managing business success*. London: Prentice Hall.

O'Brien, C., & Smith, S. J. E. (1993). Design of the decision process for strategic investment in advanced manufacturing systems. *International Journal of Production Economics*, *30–31*, 309–322.

Otley, D. (1995). Management control, organizational design and accounting informations systems. In D. Aston, T. Hopper & R. W. Scapens (Eds.), *Issue in management accounting* (pp. 45–63). London: Prentice Hall.

Raffish, N., & Turney, P. B. B. (1991). Glossary of activity-based management. *Journal of Cost Management for the Manufacturing Industry* (Fall), 53–63.

Reid, G., & Smith, J. (2000). The impact of contingencies on management accounting system development. *Management Accounting Research, 11*, 427–450.

Shank, K., & Govindarajan, V. (1989). *Strategic cost analysis: The evolution from managerial to strategic accounting*. Homewood, IL: Irwin.

Sim, K., & Killough, L. (1998). The performance effects of complementarities between manufacturing practices and management accounting systems. *Journal of Management Accounting Research, 10*, 325–346.

Tilemma, S. (2005). Toward an integrated contingency framework for MAS sophistication: Case studies on the scope of accounting instruments in Dutch power and gas companies. *Management Accounting Research, 16*, 101–129.

Zimmerman, J. L. (1995). *Accounting for decision-making and control*. Chicago, IL: Richard D. Irwin.

Zimmerman, J. L. (2001). Conjectures regarding empirical managerial accounting research. *Journal of Accounting and Economics, 32*, 411–427.

TOWARD A MODEL OF PERFORMANCE MEASUREMENT OF OUTPUT BASED ON THE THEORY OF CONSTRAINTS

Thierry Hirtz and Pierre Guernaccini

ABSTRACT

Purpose – *The notion of output (Bouckaert, 2005; OECD, 2007) is more and more widespread in the performance measurement systems in the public sector. More generally, several reforms are promoting the implementation of key indicators such as management by objectives and quality management.*

Our study introduces a first reflection about a performance measurement model of output based on the Theory of Constraints (TOC) in one public administration. The particularity of our model is the combination of theoretical design and practical implementation. Thus, an adjustment stage has been achieved in order to reach a realistic and sturdy model. It is also an opportunity to test the TOC indicators system, which is becoming prevalent in research activities (Davies & Mabin, 2009).

Methodology/approach – *First of all, we present a literature review, which associates the concept of output, the TOC, and the notion of*

Performance Measurement and Management Control: Global Issues
Studies in Managerial and Financial Accounting, Volume 25, 117–137

ISSN: 1479-3512/doi:10.1108/S1479-3512(2012)0000025008

performance in a public administration. After that, we explain the design stage to define the key performance indicators.

During a two-month test phase, this measurement system was implemented in one public administration in Luxembourg; data were collected to draw up a dashboard. Subsequently, the model was adjusted in order to fit with the practice and observed cases.

Findings – *We consider the limits of this measurement system, especially on certain aspects like the compliance with law. Furthermore, we show the major contribution of our model based on the TOC, which is to move from one local optimum to a global optimum.*

Originality/value of paper – *Thus, this publication highlights a model of performance measurement in a public administration, and the results of its implementation. It emphasizes the practical value of these indicators, the decision-making efficiency, and the appropriate level of use. The measurement model based on the TOC had a high interest for operational as well as for the managerial staff.*

INTRODUCTION

Generalities

The concept of output (Bouckaert, 2005; OECD, 2007) is frequently applied in the performance measurement systems in the public sector. More generally, several reforms are promoting the implementation of key indicators such as management by objectives and quality management.

In the context of reengineering the operational activities in one public administration, a model of performance measurement was designed and implemented. This study introduces a first reflection about a model, which combines the concept of output and the Theory of Constraints (TOC). The second particularity of the model is the combination of theoretical design and practical implementation. Thus, an adjustment stage was achieved in order to reach a realistic and sturdy model. It is also an opportunity to test the TOC indicator system, which is becoming more prevalent in research activities (Davies & Mabin, 2009).

In a literature review, we introduced the notion of performance in the public sector, the concept of output, and the TOC and its indicators. Then, we explained the design stage to define the key performance indicators in a collaborative approach.

During a two-month test phase, this measurement system was implemented in one public administration in Luxembourg; data were collected to draw up a dashboard. Subsequently, the model was adjusted in order to fit with the practice. Thus, this publication is a first reflection about a model of performance measurement in a public administration, and the results of its implementation.

Context

In the framework of reengineering activities in a public administration in Luxembourg, a model performance measurement was designed and implemented. The reengineering project deals with increasing the efficiency and the quality of the operational processing, which concerns the requests of family allowances. The top management wished to decrease the processing delay and to have a high level of satisfaction of the beneficiary.

In this way, the new project of reengineering was launched in 2010 with the support of the Public Research Centre Henri Tudor. The first step includes a design of the new operational processes with a test stage. This stage, which concerned a team of three persons from the administration, was important in order to test the new concept and to readjust it if necessary. This test technique has also been used to facilitate the change and its acceptance.

The value analysis method has been applied to define the new organization with a high participation of both employees and management during four workshops of one day. In back office, the Centre elaborated the mapping of the new processes and the implementation planning.

The second step is the planning definition of the new structure deployment followed by its implementation.

In parallel to the new organization design, the performance of the requests in processing should be measured in order to assess the efficiency of the new organization. The question was to define a tool allowing the measurement, creating performance indicators as well as dashboards.

After a review of the theoretical concept of the performance measurement in the public sector, we focused the research through the concept output and the TOC and its indicators.

Literature Review

The TOC is a holistic management philosophy put forward by Eliyahu Goldratt in 1984. It is based on five steps that focus on identifying the system constraints, exploiting them, subordinating the rest of the system to the needs of the bottleneck, improving the constraints, and repeating the process continually (Goldratt, 1990). In essence, the TOC (Wu, Blos, & Wee, 2010) is about change and how best to effect change; it is a set of management principles that help to identify obstacles to one's goal(s) and effects the changes essential to eliminate them. The changes at these leverage points would deliver a positive (or negative) effect on overall system performance considerably out of proportion to the magnitude of the change effort. Goldratt (1990) referred to these leverage points as constraints, because inaction at these locations prevented the system from realizing better.

Davies and Mabin (2009) resume the several approaches and tools of the TOC, and they conclude that the TOC has reached a sufficient level of maturity and provides benefits in terms of results and thinking processes, such that it warrants an additional look.

Goldratt (1990) has introduced three performance measures through which the impact of any local decision can be judged against the global goal. These measurement indicators are Throughput, Inventory, and Operating Expense.

Throughput (Dettmer, 1997): The rate at which the entire system generates money through sales (products or services). In a nonprofit organization or a government agency, the concept of sales may not apply: In cases where an organization's Throughput may not be easily expressed in money, it might be defined in terms of the delivery of product or service to a customer. A variety of alternatives have been suggested to modify expressions. *Inventory*: All the money the system invests in things it intends to sell. *Operating Expense*: All the money the system spends turning Inventory into Throughput.

Although it has been interchangeably used with the term output (Üstün, 2010), Goldratt's definition of Throughput is definitely not what is generally referred to as output in a traditional manufacturing environment. Üstün (2010) presents a study, which adapts the tools and principles of the TOC to the service of University and defines the performance measures of the system, which are Throughput, Inventory, and Operating Expense.

While the Throughput, Inventory, and Operating Expense are operational indicators, the TOC introduces a tactical or a decision level with the Throughput Dollars Days (TDD) and Inventory Dollars Days (IDD) (Goldratt, 1990; Cox & Spenser, 1998). A study (Ho & Li, 2004) shows that use of these performance indicators improves decision making. Lockamy and Spencer (1998) underline that the performance measurement system associated with the TOC has been advocated by some researchers as a mechanism for linking operational objectives to the global goals of an organization, with a simplification of the problem and reduction in the number of measures and areas actively and continuously monitored by the management. However, there is a lack in empirical research in determining and testing the relevance of the proposed performance measurement system presented in the TOC (Steven, Douglas, & Morgan, 2004).

Nevertheless, the TOC is more and more prevalent in the private sector both in the research world and among practitioners; the research studies of the TOC system measurement in the public sector are very limited.

In the public sector, the notion of performance measurement is widely discussed. Van Dooren and Van de Walle (2008a) introduce the main streams of the performance measurement in the public sector, the differences between the objective and subjective measures, and the Popperienne versus the Kuhnienne tradition. Today, they underline that the performance measurement is clearly subjective due to its design, which is based on a social construction. Schachter (2010) analyzes the dichotomization of objective and subjective measures in the public sector performance measurement literature and the relationship of that dichotomy to the field's perennial science wars. The analysis also examines the potential impact of the dichotomization on the role accorded to citizens in helping develop public sector performance measures. Finally, for both types of measures, the key question is the same. How well does the indicator help the agency move toward attaining the underlying conceptual goal?

The same authors, Van Dooren and Van de Walle (2008b), wonder about the use of performance information by the managers in the public sector. The information on the performance should not be used by the manager to blame or punish, but to find solutions triggering improvements. The priority should be the managerial action rather than the measuring or reporting finality. The question of the performance information use in the public sector by politicians and citizens is raised. Although the political use of this information varies according to the countries studied, it is primarily used to observe and almost never explain, and it does not identify the success factors.

According to Brewer and Selden (2000), the performance is regarded as a multidimensional concept that includes at least six dimensions (internal efficiency, internal efficiency, internal equity, external efficiency, external efficiency, external equity). The new public management (Osborne & Gaebler, 1992) has played an important role in the dissemination of the performance measurement. According to this concept, the best way to improve performance is to interest the person. One of the main ways to seek self-interest is compensation to performance. The empirical research by Hondeghem and Perry (2009) shows that the relationship between motivation and performance is positive, but the links between public service motivation and performance are more nuanced and complex than previous work.

OECD (2007) underlines trends toward output and outcome performance management; nevertheless, the governments continue to make use of evaluations. The performance is directly in link with the government results, defined as outputs and outcomes of their activities. They aim at improving the efficiency, effectiveness, and value for money of public activities. The evidence about performance that is collected and used systematically – called performance information – may be quantitative (numerical) or qualitative (descriptive). Over the past two decades, there has been a renewed emphasis on performance measures (principally of outputs and outcomes) in budgeting and management. Countries appear to have recognized the dangers of focusing only on outputs. It can lead to goal displacement as agencies risk losing sight of the intended impact of their programs on wider society, and concentrate on quantifiable activities at the expense of those that are less measurable. It can also result in less attention being paid to crosscutting issues. While outcomes incorporate a wider focus on the impact of programs on society and have greater appeal to politicians and the public, some are difficult to measure. Of the countries that developed performance measures, the majority produce a combination of outputs and outcomes. The outputs consist in the combined results of the activities and the upstream inputs. They may be products or services which can be objects of direct consumption (garbage collection, delivery of a passport) (Bouckaert, 2005). The quantitative aspect may be expressed in financial or in physical terms (Hatry, 1999). They could be also assessed on the quality aspect, can it be quantified or not. Outputs are never an end in themselves for the public sector. The effects or intermediate and final subsequent outcomes are the focus (Fig. 1).

According to Festré and Garrouste (2008), there isn't always a perfectly identified output in the public sector. A same agency can produce different outputs depending on its missions. Finally, the outputs can be complementary or substitute. For example, the alcohol fight is dependent on the police, the social services, the markets, and the pubs.

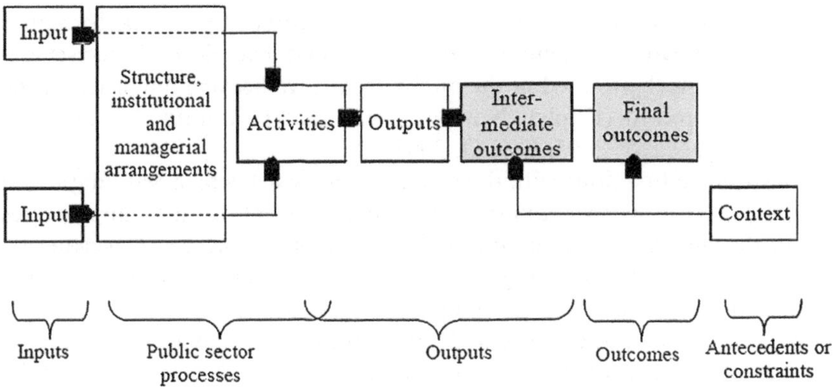

Fig. 1. The Concept of Output (OECD, 2007).

METHOD

Our research is positioned in the field of Research Engineering (Chanal, Lesca, & Martinet, 1997), which in some ways is akin to intervention research. It differs, however, by imagining a new status termed "researcher-engineer" who designs the tool to support his research, builds it, and acts both as a facilitator and as an evaluator in its implementation in the organization, thereby contributing to the emergence of representations and new scientific knowledge.

The objective of the project is to measure the performance of the operational process in one public administration in the framework of a reengineering process. We start with a theoretical approach in order to select and to design the performance measurement system. One of the first reflections was to identify the most appropriate tools among the diversity of methods (the approach is detailed below).

Then, this system was implemented for two months during a test stage in the public administration mentioned above. During and after the experimental step, the measurement system was monitored and the results analyzed. The theoretical system was improved on the basis of the experimental feedback.

Theoretical Design

Selection of the Method

The first question was to identify the appropriate method in order to measure the performance in the framework of the reengineering.

The research team makes a state of art on the topic of public performance measurement with a scientific watch. In a first overview of the several approaches which are used in the public sector, we became interested in the balanced scorecard (Kaplan Norton, 1992) and the output and outcome concept (Bouckaert, 2005; OECD, 2007).

We started a first draft with the balanced scorecard, which was adapted in the public sector in other countries. The dimensions of the internal processes and of the customer were suitable with the aim of our project. Nevertheless, the financial and the learning dimensions weren't directly adapted to the performance measurement of the operational process.

In another way, the concept of the output/outcome is the emergent concept of the measurement in the public sector which is in line with the operational activities and with a global approach of the performance in the public organizations. The framework of the operational reengineering fitted with the concept, like the numerous examples of applications in the public sector attested. Moreover, the efficiency was easy to understand with the ratio output/input. In our case, the inputs (resources) were steady during the complete project. In consequence, a positive or a negative variation of the output/outcome will be a reflection of the performance.

The next issue was to build the performance indicators of the output/outcome on a scientific approach, which calls for the operationalization of the concept that is linked to the TOC.

The method of the TOC was known by our team and previously used in an industrial production with interesting results. A case of application of indicators was made in public service by our team, without making the link with the concept of output or being the object of research. Finally, we chose this TOC approach to design the output and the outcome indicators because of the common approach of the "global performance," the previous positive experience, and its use in an operational environment. Moreover, this was a first opportunity to test the combination of the output/outcome and the TOC in the public sector from a research perspective.

Definition of the Output and the Outcome

The next step was to build an indicator of the output and the outcome. During a first workshop with the top management and the middle management of this public organization, we defined the goal in accordance with the TOC. Every member of the group expressed and wrote a formulation of its goal on a paper and after a round table, a common goal was defined after the consensus meeting. The classical confusion of the "goal formulation" is to give a necessary condition instead of the true finality. For example, the processing of an allowance request is a necessary condition, and its payment is the real finality. It is very usual to have as many definitions of a goal as persons in the group.

The goal formulation must include a "goal unit," which could be measured or evaluated – it represents the Throughput. The goal should be in line with a sustainable development in order to avoid the short-term performance.

After the consensus meeting, the definition of the allowance service goal, which was explained, was "to pay the family allowances in the shortest possible time to ensure the compensation of family charges in a sustainable way." In this result, we distinguished the concepts of both the output and the intermediate outcome.

In the case of the family allowance service, we quickly noticed that the intermediate outcome and the final outcome will be difficult to measure. There are different reasons to explain it; the outcomes of the allowance family are multiple and these outcomes come themselves from the effects of several organizations. By a literature review (OECD, 2003; Sundström, 2001), we identified, on the one hand, the outcomes of the family allowance payment and, on the other hand, we showed the role of the other organizations on the outcomes (Fig. 2). At first, the family allowance is a political instrument in order to increase the birthrate, and later in order to give access to education. It has also been used for the economic goals, that is, to promote the consumption of direct beneficiaries and to avoid savings among wealthy families. The birthrate is a function of several factors like the family allowance payment; the structure of child care, the possibility of parental leave, the tax benefits, and the amount of the allowances are other key factors. In a global context, the economic, social and political context has a strong impact on the birthrate. Another problem is to know the relative proportion of the impact of the administration on the outcome in question.

Based on these elements, the final outcome measurement seems to be far removed from the performance of the studied administration. The level of the outcome could be important to study in a political level in order to know

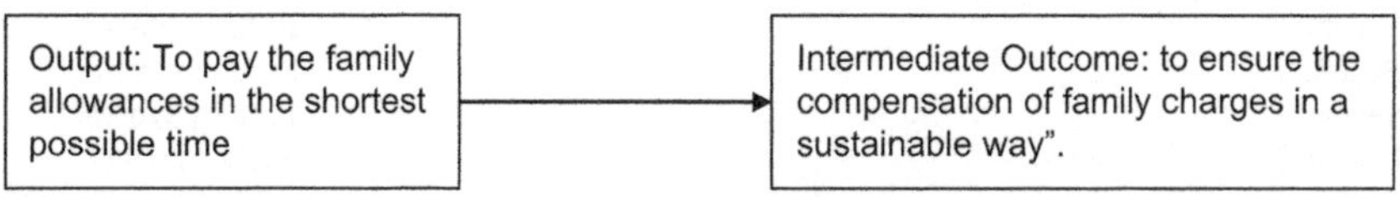

Fig. 2. Output and Outcome Definitions.

the most effective lever, but in our case we limit the perimeter of the performance on the output measurement which is "to pay the family allowances in the shortest possible time."

Definition of the Throughput and of the Inventory

The throughput is the production of the goal's unit, which is focused on the output, "To pay the family allowances in the shortest possible time." In consequence, the throughput equals the number of payment of the family allowance requests. To be precise, the throughput includes the refusal responses of the requests.

Then, the throughput is the number of payment of the family allowance requests and the number of refusal responses to the requests, which are more globally called the number of processed (NP) requests. The inventory is the number of pending requests in processing (NPR).

Definition of the "Throughput Dollars Days" and the "Inventory Dollars Days"

The notion of "dollars" corresponds to the Throughput value. In the private sector, this notion refers to money, but in the case of the public administration the value is based on the nature of the results. In the case of the public administration, we consider that the value corresponds to the kinds of requests. In consequence, we identify the different kinds of requests, which are processed in this public administration, but we do not give the financial value of these requests. The administration is in charge of six different kinds of family benefits, which are themselves composed by different requests. In Table 1, the different requests for the first kind of family benefits are the requests for the initial payment, the periodic payment, the rights change, or the stop payment.

Table 1. Example of Requests.

Benefit 1	Periodic payment
Benefit 1	Stop payment
Benefit 1	Initial payment
Benefit 1	Rights change
Benefit 2	Stop payment
Benefit 2	Initial payment
Benefit 2	Rights change
Benefit 3	Allowance payment

Table 2. Description of the Indicators Following the Theory of Constraints.

Indicators of the Theory of Constraints	Application of the Indicators
Throughput	Number of processed (NP) – Number of payment of the family allowance requests. – Number of refusal responses to the requests.
Inventory	– Number of pending requests in processing (NPR)
Throughput Dollars Days	– Number of Processed by kinds of requests by month (NPM)
Inventory Dollars Days	– Number of pending requests by kinds of requests by month (NPRM).

The notion of "days" refers to the periodicity of the activities. In our case, the majority of the payments are made monthly on a fixed date (liquidation).

The TDD was called in the project the number of processed by month (NPM), and the IDD was the number of pending requests in processing by month (NPRM).

The selection of the indicators following the TOC is described in Table 2. These indicators have been declined for each type of family benefits requests.

Definition of the Calculation Method and of the Rules

What information do we dispose of to calculate these indicators?

The first level of information is to identify the type of the requests based on the types of family allowances. The management team of this public administration has provided us with the different appellations of the requests.

The second level of information is to know if the request is closed or not. If the request is closed, the NPR increases by one unit, and if the request is not closed, the NPRM increases by one unit.

A huge difficulty was to define clearly when a request is closed.

For example, in the case of an initial payment request, the steps are:

- the request reception,
- the processing,
- the decision (acceptance or refusal),

– the sending of decision letter,
– and the monthly liquidation.

We considered that a request is closed when the payment is made. In the other example, in the case of an application for change of bank account, is the request closed when the change is made or when the payment on the new account is made? The management chose the second option.

For each request, we needed to define the specific conditions of closing to allow the calculation of the indicator. The definition of these conditions was developed and validated in collaboration with the management and staff concerned.

At the end of each month, the number of processed by month and number of pending requests by month are added in order to know the NPM and the NPRM.

Data Collection Method

During two months, we needed to collect information concerning the processing of requests in a team of the public administration.

The team was composed of three people, who are the operational files managers. Their job is to process applications for family benefits, which vary between 15 and 60 requests per day for three people. The process starts with the reception of the request, its analysis, and the final decision. Some kind of files (about 30% of the total files) should be controlled by a fourth person, who check the compliance of the decision.

In the case of incomplete request (lack of documentary evidence), a letter is sent to the applicant in order to receive the missing documents. It's very usual to have several exchanges between the applicant and the file manager.

The three team members were trained to fill in the date table (excel table), which was designed, in order to collect the data given in Table 3.

Depending on the application type, some files could be controlled by the function controller.

Table 3. Excel Table for the Operational Team.

Affiliates Number	Type of request	Date of reception	Date of processing	Date control	Action performed (status)	Closed (date)

In this case, the date control is added to the file by the controller itself.

During the processing of a request, there are several exchanges of letters between the affiliated and the public administration because of incomplete requests (e.g., lack of civil status documents). The principle of the TOC is to assess the overall performance of the system. In this way, the letters for a same request are identified and grouped in order to have the global processes.

Monitoring during the Two Months Test Stage

During the test stage, the Centre's team was regularly present beside the operational team. We want to guarantee the quality of the database filling in order to calculate the performance indicators. A custom coaching was given to the three persons of the operational team to avoid some mistakes of filling in and to bring an answer of any questions.

Another objective of this monitoring is to collect the qualitative information (usability, problem, remark, suggestion). The research team plans the observations with a daily presence the first week and a weekly presence at the end of the project.

Analysis of the Database, Adjustment of the Model of Measurement and Calculation

After the two months test stage, the table was analyzed by the Centre's team in order to calculate the performance indicators.

In a first time, we rejected the unusable lines (17 lines) of the datebase: information that weren't clear or partial. After two months, the database contained 1,280 lines, which correspond to the number of received letters by the pool of three persons.

The researchers corrected the identification of the type of requests according to the different designations for the same type of requests.

Moreover, we discovered new activities, which weren't mentioned by the top management during the designation of the activities. We identified several new requests by type of family benefits, then we had to redefine the expectations for each request in order to calculate the indicators. We added the information of the type of requests in the table of family benefits. This was an indirect and an important result to formalize the type of requests.

Thereafter the research team grouped the different letters for a same request to finally obtain 1,003 requests (for 1,280 letters). During the theoretical design, we noticed that several letters could address the same request, but in the test stage, we identified, in addition, that a letter could contain several requests.

Lastly, we could calculate the indicators for both months, if the requirements were reached; the NPM is increased by one unit; if not, the NPRM is increased.

RESULTS

After two months of collecting and processing data, we pointed out the following indicators. To maintain confidentiality, we present the results, but we have used the generic appellation of "benefit" to protect the administration.

The first results give a picture of the processing time for the mails received (Fig. 3). This figure shows that 80% (month 1) or 67% (month 2) of the mails are processed in two days. The reason for the decrease in the processing time is the increase in the activity (the number of mails).

This representation shows the processing time of mails, but it does not show if the requests were processed. For example, if there are missing pieces, the mail is processed when the manager asks complementary pieces, but the request remains pending.

This representation is a classical view of the performance indicators, which are used in the public administration in general, but it only gives a local view of the processing, not the entire view of the processing.

The next part represents the results in terms of request, which is focused on the entire processing (global view) and not only in terms of mails (local view).

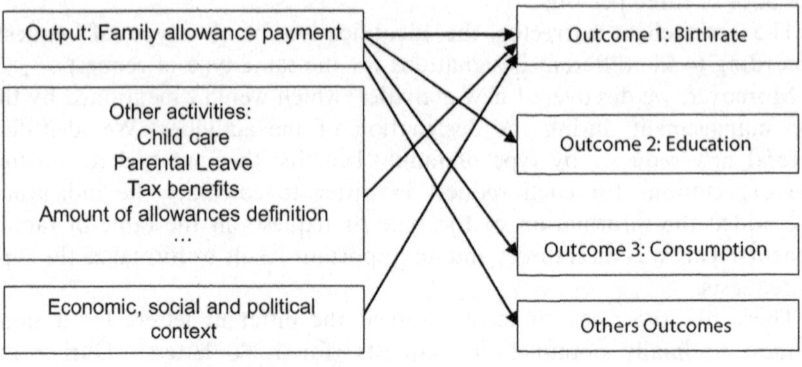

Fig. 3. Final Outcomes of the Family Allowances Payment.

Table 4. Number of Received Requests During the Two Months.

Family Benefits	Type of Request	Number	Percent
Benefit 1	Periodic payment	155	15.5
Benefit 1	Stop payment	148	14.8
Benefit 1	Initial payment	138	13.8
Benefit 1	Rights change	111	11.1
Benefit 2	Stop payment	42	4.2
Benefit 2	Initial payment	50	5.0
Benefit 2	Rights change	38	3.8
Benefit 3	Allowance payment	45	4.5
Benefit 3	Rights change	35	3.5
Benefit 4	Stop payment	22	2.2
Benefit 4	Initial payment	19	1.9
Benefit 4	Rights change	9	0.9
Benefit 5	Stop payment	40	4.0
Benefit 5	Initial payment	55	5.5
Benefit 5	Rights change	25	2.5
Benefit 6	Stop payment	33	3.3
Benefit 6	Initial payment	37	3.7
Benefit 6	Rights change	1	0.1

For each family benefit, a first result of our study was to identify and formalize the different types of requests for these types of family benefits, which represent the outputs of the public administration (Table 4).

Table 5 shows the performance indicators, NPM and NPRM, following the types of request and family benefits. In the global view, the dashboard informs us how many requests are closed and how many requests are pending.

DISCUSSION

This study introduces a performance measurement model of output based on the TOC, which is one of the first experimentation in the public sector.

To sum up the model (Fig. 4), we decline the output "To pay the family allowances in the shortest possible time" on the notion of goal "to pay the family allowances in the shortest possible time to ensure the compensation of family charges in a sustainable way," which gives more accuracy and avoid the confusion with the outcome. At the level of the case studied, we

Table 5. Number of Processed/Months and Number of Pending Requests/Months for Each Type of Request and Family Benefits.

Family Benefits	Type of Request	Number	Number of Processed/Months	Number of Pending Requests/Months
Benefit 1	Periodic payment	155	145	10
Benefit 1	Stop payment	148	130	18
Benefit 1	Initial payment	138	120	18
Benefit 1	Rights change	111	99	12
Benefit 2	Stop payment	42	40	2
Benefit 2	Initial payment	50	45	5
Benefit 2	Rights change	38	38	0
Benefit 3	Allowance payment	45	41	4
Benefit 3	Rights change	35	32	3
Benefit 4	Stop payment	22	20	2
Benefit 4	Initial payment	19	15	4
Benefit 4	Rights change	9	1	8
Benefit 5	Stop payment	40	20	20
Benefit 5	Initial payment	55	45	10
Benefit 5	Rights change	25	12	13
Benefit 6	Stop payment	33	22	11
Benefit 6	Initial payment	37	22	15
Benefit 6	Rights change	1	1	
	Total	1,003	848	155

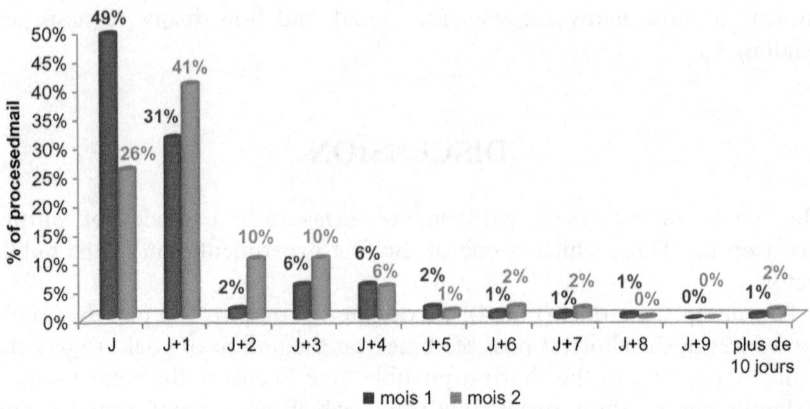

Fig. 4. Processing Time for Received Mail.

identified several final outcomes, which themselves depend on other organizations (public administration, political decision). We could ask ourselves about the link between output and outcome as well as the nature of the link. Concretely, what is the real impact of performance in the payment of family allowances on birthrate? On consumption? This could be an issue for a next research by another study.

The output is very similar to the throughput "NP," which explains the goal unit and allows the measurement. In parallel, we define the Inventory "NPR."

Then we used the notion of "TDD" and "IDD" in order to identify all the different kinds of requests (notion of value) and to define the appropriate periodicity. We propose to replace these appellations by "Throughput Value Periodic" and "Inventory Value Periodic," which is more suitable for the public sector. Several kinds of requests were identified during the theoretical design stage, which were completed during the test stage. For each one, we define the precise conditions to calculate the results (Fig. 5).

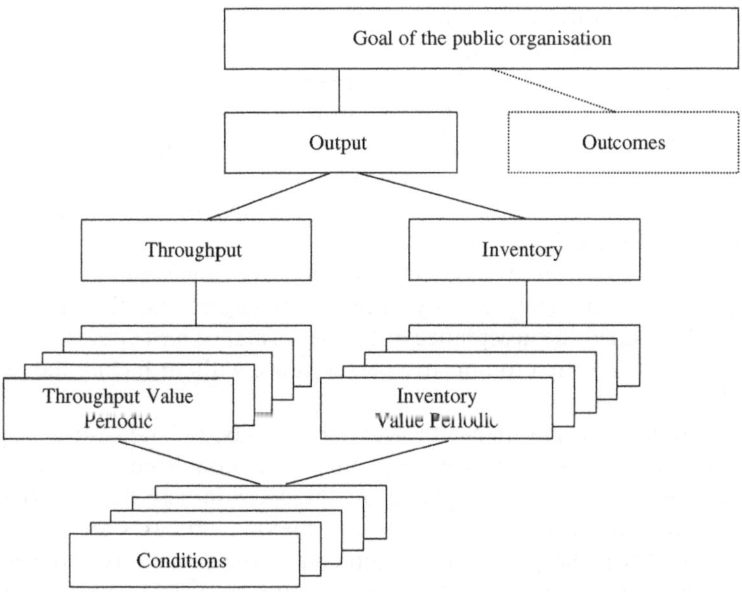

Fig. 5. Output Measurement with the Theory of Constraints Indicators.

Concerning the information use and contrary to the local view of the mail in processing, which informs if the request is really closed or pending, these indicators give a clear information on the processing of the requests. More than an internal performance, this clearly indicates if the "customer's" request was satisfied.

These indicators give a general picture of the process to the management team. On the basis of these data, the managers have better visibility of their business, which leads to decisions simplifying the processing of certain requests, and increasing the number of processed requests. The decisions of new IT developments were taken after the analysis of these indicators.

The performance by type of requests was not known before, even if the managers had an idea of their business. The reality was more understandable with these figures, since it helped the managers to take right decisions. Some new priority was given by the management to increase some kinds of requests.

Thus, the decisions seek to reach the global optimum and not the local optimum, which could have been the case with the indicator of mail processed.

On the operational level, these indicators quickly give the information concerning the status of the request. Previously, the operational staff had to read all the previous letters in order to understand the status, which takes a lot of time and is exhausting. In a way, the indicator simplifies the reality to provide useful information to the operator.

The method also forces to identify all the different kinds of outputs (with 18 types of requests), which were more numerous than we first thought. The indirect result of this study was to clarify the activity and to specify the appellation of the different requests.

In terms of approach, the collective definition of the goal, of the output, and of the measurement rules by both the managers and the operational staff has facilitated the acceptance of the performance measurement system. The teams quickly tried to improve performance without having the feeling of being controlled.

Concerning the performance measurement limits, the indicators measure the processing capacity; nevertheless, some dimensions aren't directly taken into account, such as the compliance of the decision with the law. If some cases are controlled by another person, a majority of cases aren't directly controlled. When the processing is controlled, the request is sent back to the operational staff in case of nonconformity in order to reconsider the decision, which takes into account the performance measurement. But in the other cases, these indicators don't give the assurance of the right decision in conformity with the law. In consequence, the research team suggested adding a random control of some requests.

The quality of the information (letters), which is sent in parallel to the payment isn't directly measured. Sometimes, some applicants send a new mail because of the clear information lack, which is integrated in the performance measurement. But if there are no questions, this criterion isn't evaluated. The complementary option could be to make a customer satisfaction inquiry in order to measure the quality of the service.

CONCLUSION

The study's objective is to introduce a performance measurement model of output based on the TOC in a public administration, which is one of the first experimentation in the public sector. It shows the mutual contribution of the TOC and of the concept of output. On the one hand, the operationalization of the output concept is easier; on the other hand, it facilitates the adaptation of the TOC in the public sector. The identification of the output and outcome clarifies the "generic goal" of the TOC. Nevertheless, in the case we studied, the measurement of outcome is complex due to the multiple outcomes and their dependences with other organizations. We can also wonder about the nature of the link between the output of this administration and the final outcomes. If the link exists, the output and the outcome seem to be on different levels (political and operational); this issue could be the object of future research.

The indicators of the throughput and the inventory clarify the notion of the output, which allows the identification of a goal's unit. Nevertheless, the tactical indicator of the TDD and IDD needs to be suitable to the public sector; we suggest the appellation of the Throughput Value Periodic and Inventory Value Periodic. These indicators, which were implemented during the test stage, provide clear information on the overall system performance. This system facilitates the decision-making process for both the management and the operational staff.

The participation of the administration's members in defining indicators is a key factor of the acceptance of the measurement, and could avoid creating the feeling of being controlled. Moreover, the establishment of these indicators has led to some improvement in the process. The first objective was to evaluate the performance of the new organization by the establishment of indicators. The implementation of these indicators has facilitated the understanding of the organization and also has determined the actions for improvement.

After this first integration of the output and the TOC, we will attempt at developing this model in other public administrations. A future application of the outcome and the TOC should be explored.

REFERENCES

Bouckaert, G. (2005). Un nouvel examen de la mesure de la performance dans le secteur public. *Téléscope*, *12*, 12–25.

Brewer, A., & Selden, C. (2000). Why elephants gallop: Assessing and predicting organizational performance in federal agencies. *Journal of Public Administration Research and Theory*, *10*(4), 685–711.

Chanal, V., Lesca, H., & Martinet, A.-C. (1997). Vers une ingénierie de la recherche en sciences de gestion. *Revue Française de Gestion*, *116*, 41–51.

Cox, J. F., & Spenser, M. S. (1998). *The constraints management handbook*. New York, NY: St. Lucie Press.

Davies, J., & Mabin, V. (2009). Theory of constraints. *Decision Line*, *40*, 9–15.

Dettmer, H. W. (1997). *Goldratt's theory of constraints: A systems approach to continuous improvement*. Milwaukee, WI: ASQ Quality Press.

Festré, A., & Garrouste, P. (2008). *Motivation, incentives and performance in the public sector*. Working Paper.

Goldratt, M. E. (1990). *What is this thing called the theory of constraints and how should it be implemented?* Great Barrington, MA: North River Press.

Hatry, H. (1999). *Performance measurement: Getting results*. Washington, DC: The Urban Institute Press.

Ho, T. F., & Li, R. K. (2004). Heuristic dispatching rule to maximize TDD and IDD performance. *International Journal of Production Research*. Retrieved from http://www.informaworld.com/smpp/title~db=all~content=t713696255~tab=issueslist~branches=42-v4242,5133-5147

Hondeghem, A., & Perry, J. L. (2009). Introduction. *Revue Internationale des Sciences Administratives*, *75*, 5–10.

Kaplan, R. S., & Norton, D. P. (1992). The balanced scorecard-measures that drive performance. *Harvard Business Review*, *83*(7/8), 172–180.

Lockamy, A., III., & Spencer, M. S. (1998). Performance measurement in a theory of constraints environment. *International Journal of Production Research*, *36*(8), 2045–2060.

OECD. (2003). *Panorama de la société: Les indicateurs sociaux de l'OCDE*. Éditions OCDE, ISBN 9264297979.

OECD. (2007). *Mieux mesurer l'administration publique, Documents de travail sur la gouvernance publique*, 2007/1, Éditions OCDE. doi:10.1787/247538122856

Osborne, D., & Gaebler, T. (1992). *Reinventing government: How the entrepreneurial spirit is transforming the public sector*. A William Patrick Books, Reading, MA.

Schachter, H. L. (2010). Objective and subjective performance measures: A note on terminology. *Administration & Society*, *42*(5), 550–567.

Steven, A. M., Douglas, M. S., & Morgan, S. (2004). Metrics and performance measurement in operations management: Dealing with the metrics maze original research article. *Journal of Operations Management*, *22*(3), 209–218.

Sundström, K. (2001). *Fécondité et changement social.* L'Observatoire de l'OCDE, 229.

Üstün, P. (2010). *Application of the theory of constraints to an elective course registration system.* A thesis submitted to the Graduate School of Social Sciences of Middle East Technical University. Orta Dogu Teknik Üniversitesi.

Van Dooren, W., & Van de Walle, S. (2008a). La réalité est simplement une illusion, quoique très persistante. Introduction au numéro spécial sur la mesure de la performance. *Revue Internationale des Sciences Administratives*, *4*(74), 559–562.

Van Dooren, W., & Van de Walle, S. (2008b). *Performance information in the public sector – How it is used* (p. 296). Basingstoke: Palgrave Macmillan.

Wu, S., Blos, M., & Wee, H. (2010). Can the Toyota way overcome the recent Toyota set back? A study based on the theory of constraints. *Journal of Advanced Manufacturing System*, *9*(2), 145.

CHALLENGES IN ENERGY EFFICIENCY PERFORMANCE MEASUREMENT IN THE PROCESS INDUSTRY

Tuija Virtanen, Mari Tuomaala and Emilia Pentti

ABSTRACT

Purpose – *The accounting literature has demonstrated increased concern over issues of sustainability. One of the most critical issues that corporations face at the moment is climate change. This especially concerns companies that are substantial consumers of materials and energy. Corporations have increasingly acknowledged that to address this issue the existing ways of managing business must be changed.*

This study examines the challenges posed by measurement of energy efficiency performance in an energy-intensive industry. It covers the challenges encountered with respect to the energy efficiency indicator and its use as a management tool.

Methodology – *The case study method was used to conduct the research, which took place in a single firm in an energy-intensive process industry.*

Findings – *The results indicate that the current energy efficiency indicator, specific energy consumption (SEC), does not meet the criteria for good performance measurement. In particular, the controllability of*

Performance Measurement and Management Control: Global Issues
Studies in Managerial and Financial Accounting, Volume 25, 139–166

ISSN: 1479-3512/doi:10.1108/S1479-3512(2012)0000025009

energy efficiency seems problematic. Another major challenge is target setting. Measurable targets are needed to identify and prioritise areas where consumption and emission can be reduced.

Practical implications – *The study provides practical knowledge on what is happening in organisations that pursue sustainable development and in particular, environmental efficiency.*

Originality/value – *Although the conceptual challenges in energy efficiency measurement are well known in the technical literature, discussions dealing with its management have been few in number. This study is a cross-disciplinary work and combines technical energy efficiency literature and management accounting research on performance management.*

INTRODUCTION

The world's energy system is at a crossroads. The International Energy Agency (IEA) states that the "current global trends in energy supply and consumption are patently unsustainable – environmentally, economically, socially". The projected rise in greenhouse gas emissions suggests that the concentration of those gases in the atmosphere by the end of this century will result in a rise in global average temperature of up to 6°C. Reaching the targets set for emission reduction requires innovative policies, an appropriate regulatory framework, rapid development of a global carbon market and increased investment in energy research, development and demonstration (IEA, 2008a). The European Commission implemented a strategy for Climate Action in 2008. One of the targets is to reduce primary energy consumption by 20% compared with the energy consumption projected for 2020. This stresses the need to increase energy efficiency. Process industry companies can promote sustainability effectively by increasing energy efficiency.

To reach the target of improved energy efficiency, companies are carrying out energy efficiency programs that can be structured on the basis of voluntary energy efficiency agreements (see Motiva, 2007). Environmental management systems (EMS) like ISO 14001 or energy management systems like ISO 50001 provide insights into the general implementation process. Nevertheless, these procedures merely develop general guidelines for energy efficiency monitoring (EK, 2011). Academic literature has presented frameworks for describing the drivers of corporate social performance, the actions that managers can take to affect performance and the consequences of those actions for both corporate social and financial performance (Epstein & Roy, 2001). Accordingly, the key issue would be to develop performance indicators to monitor and assess the value of the sustainability actions undertaken.

It has also been reported that due to the number and the complexity of industrial processes and product end-uses, designing consistent and comparable energy efficiency indicators can be difficult (IEA, 2008b). There is evidence that energy use in firms today is not managed as effectively as it should be (Sivill, 2011, p. 15). It has been noticed that the reasons for promoting energy efficiency may lie in organisational and managerial factors and that energy management is more about people than technology (Caffall, 1995). Several organisational and behavioural factors have been identified as barriers to or drivers of corporate energy efficiency (Sardianou, 2008; Sola & Xavier, 2007; Thollander & Ottosson, 2008). However, discussions dealing with the challenges involved in using energy efficiency indicators as management tools have been few in number.

In 2005, industry accounted for one-third of global primary energy use and around 25% of global energy and process CO_2 emissions (IEA, 2008c). In Finland, manufacturing has accounted for as much as 50% of final energy use (Statistics Finland, 2008). Consequently, improvements are needed in the tools required for energy management.

The purpose of the paper is to identify the challenges posed by measurement of energy efficiency performance. This covers two aspects: (1) challenges due to the energy efficiency indicator and (2) challenges encountered in its use as a management tool. This purpose was achieved by studying the measurement of technical energy efficiency against theories of performance measurement and management. Hence, the paper provides an understanding of the complexity of energy efficiency management in the process industry. The case study method was used in conducting the research and the case company represented the petrochemical industry.

The rest of the paper is structured as follows. The following section presents energy efficiency concepts and indicators and reviews earlier research on challenges concerning the indicator and its use. Next, the research method is described. After that, the empirical results from the case study are presented. The next section analyses and discusses the challenges posed by current measurement of energy performance in the case company. Finally, the conclusions end the paper.

EARLIER RESEARCH

Energy Efficiency Concepts and Indicators

Energy-using products Directive 2006/32/EC (Article 3b) defines energy efficiency as "a ratio between an output of performance, service, goods or energy, and an input of energy" (EC, 2006). In general, energy efficiency refers to using less energy to produce the same amount of useful output.

The most typical energy efficiency indicator in industry is *specific energy consumption* (SEC). It is the ratio between the total energy used and the useful output of the process measured in physical units, most commonly tons of products (for example GJ/t):

$$\mathrm{SEC} = \frac{\text{energy used}}{\text{products produced}} \tag{1}$$

The total energy used is the difference between energy inputs and energy outputs. In industrial processes, the total energy used normally covers fuel, electricity and steam (Ahtila, Holmberg, Tuomaala, & Turunen, 2010, p. 340). Each of these has a different *energy value* (*exergy*). All the energy used should therefore be converted into primary energy, that is energy before conversion. This is done by taking into account the efficiencies of the conversion processes. SEC is then calculated using Eq. (1) (EC, 2009 modified):

$$\mathrm{SEC} = \frac{E_{\text{Fuels}} + (E_{\text{Steam}}/\eta_{\text{Steam}}) + (E_{\text{Electricity}}/\eta_{\text{Electricity}})}{\text{products produced}} \tag{2}$$

where η_{steam} is the efficiency of steam generation and $\eta_{\text{electricity}}$ is the efficiency of electricity generation. As the energy use is summarised in a single equation, it can no longer be decomposed into original vectors. The single value now indicates a summary of the changes in the various energy vectors. An example of energy efficiency measurement is illustrated in Fig. 1.

Current energy efficiency can be compared with a reference situation in a dimensionless *energy efficiency index* (EC, 2009):

$$\mathrm{EEI} = \frac{\mathrm{SEC}_{\text{ref}}}{\mathrm{SEC}} \tag{3}$$

where SEC describes the measured energy efficiency of a process during a follow-up period, and $\mathrm{SEC}_{\text{ref}}$ the value of a reference production indicating

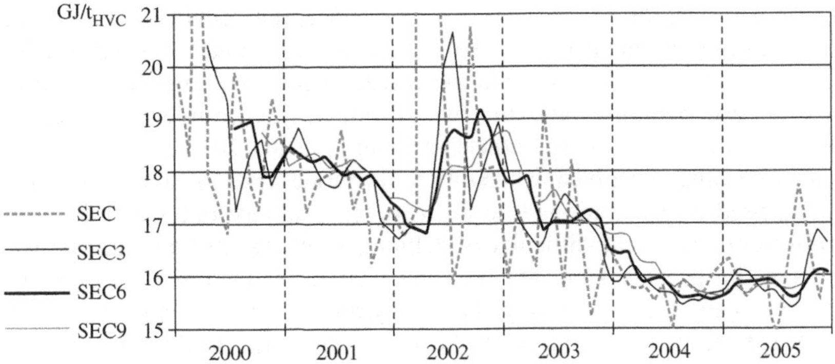

Fig. 1. Specific Energy Consumption, SEC [GJ/t_{HVC}] Based on Monthly Values of the Case Plant During 2000–2005 with the Floating Averages from the Previous 3 Months (SEC3), 6 Months (SEC6) and 9 Months (SEC9). HVC, High Value Chemicals, is the Sum of all Products Produced (Auvinen, 2008).

best performance. The EEI is only suitable for analyses of the process in question and not as a reference value for other processes (Sivill & Ahtila, 2009). The behaviour of the EEI is a "mirror image" of Fig. 1. Ideally, the value of EEI is one.

Energy-related data can be expressed in many ways. Patterson (1996) has categorized the existing indicators in four groups: physical–thermodynamic, economic–thermodynamic, thermodynamic and economic, of which the first two are mainly used in the process industry. In case of physical–thermodynamic indicators, the energy input is measured in thermodynamic units and the output in physical units, as in SEC (Eq. (2)). In case of economic–thermodynamic indicators, the energy input is measured in thermodynamic units and the output is measured, for example in product market prices.

Conceptual Challenges Posed by the Energy Efficiency Indicator

The energy efficiency literature lists the conceptual challenges posed by the energy efficiency indicator and the factors that affect it indirectly. In contrast, the performance measurement literature discusses criteria for good performance indicators in general and assesses the benefits and weaknesses of financial and non-financial indicators.

The technical literature discusses the conceptual challenges in energy efficiency measurement (Eqs. (1)–(3)). These are related to energy quality, balance boundary (i.e. calculation area) definition, energy allocation, denominator definition and the overall difficulty encountered in defining the factors affecting energy efficiency (Ahtila et al., 2010, p. 350). Energy quality becomes a problem when different energies (fuel, heat, electricity) are summarised in calculations (see Eq. (2)). The balance boundary definition problem relates to the difficulty of defining exact system boundaries; this is because the processes are always somewhat different and the boundaries between them may be vague. The energy allocation problem refers to the challenges in allocating energy to heat and power (in production sites with combined heat and power production, CHP) or to different production lines. The denominator definition can be a problem when energy efficiency data from different sources have to be summarised or the energy efficiency data from various processes are compared. Different products call for different manufacturing processes that in turn have different energy requirements.

An energy efficiency indicator is indirectly affected by several factors (see Fig. 1), most notably the capacity utilisation rate, including breaks and shutdowns. Ambient conditions, quality of materials and end products, operation of the control system, and operator actions affect energy use. Information on the relevant variables affecting the indicators is often unknown; it may not be available or is qualitative in nature (Sivill, Manninen, & Ahtila, 2009).

In other words, two other aspects affect the energy consumption of the process. These are: material efficiency and operational efficiency (Tuomaala, 2007, p. 108). For example, low raw material quality may result in an increase in energy consumption. Also, use of recycled materials generally decreases energy consumption (Siitonen, 2010, p. 29). Furthermore, an effective process and a high capacity utilisation rate mean fewer malfunctions and shutdowns, resulting in better performance with respect to energy use. In the petrochemical industry, failures in a process increase flaring, that is burning of hydrocarbons. This increases CO_2 emissions. Hence, looking solely at energy efficiency indicators in the evaluation of processes is not sufficient, since valuable information is left out. The challenge in evaluation lies in considering all relevant factors and their interconnections with energy efficiency.

Fig. 2 presents the hierarchical structure of a production plant schematically and depicts the alternative calculation areas (balance boundaries) for energy efficiency. Efficiency can be analysed at any of the levels in Fig. 2. Technical efficiency studies in equipment and unit processes

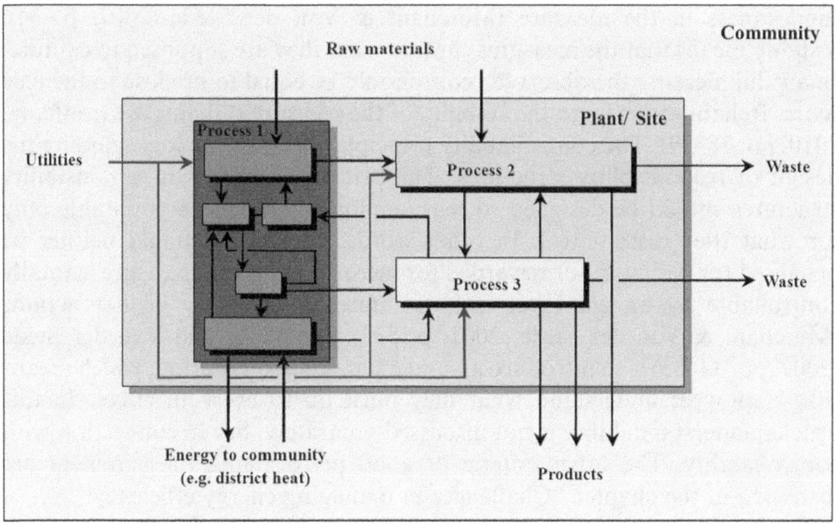

Fig. 2. The Hierarchical Structure of a Production Plant.

are mainly technical and numeric. In an analysis of the total site, the criteria become more strategic (raw material type, price, quality, etc.) and the interactions between the processes become complex.

The performance measurement literature discusses the criteria for determining good performance indicators. Various studies have emphasised different characteristics of good performance indicators (see e.g. Järvenpää, Partanen, & Tuomela, 2001, p. 186; Laitinen, 1998, p. 121; Merchant & Van der Stede, 2007; Partanen, 2007, p. 280). The characteristics suggested include, for example low measurement costs and the relevancy, timeliness, objectivity, precision, validity, reliability, controllability and understandability of the measurement results. In addition, the employees whose behaviours are being controlled should find the measure credible and the measure should assist them in identifying and solving problems. Furthermore, it should be possible to adapt the measure to the organisational level, set target values for the measure and integrate the measure with incentive systems.

Timeliness refers to the lag between the employee's performance and the measurement results (and the provision of rewards). Objectivity means freedom from bias. Measurement objectivity is low where either the choice of measurement rules or the actual measuring is conducted by the employees whose performances are being evaluated. Precision refers to the degree of randomness in the measure (Merchant & Van der Stede, 2007, p. 34). Validity means that the measures capture what they are supposed to capture. In a valid measure the observed score should be equal to or close to the true score. Reliability refers to the stability of the measure (Ghauri & Grønhaug, 2010, pp. 78–79). The controllability principle underlies the logic guiding the design of responsibility structures. The principle states that responsibility structures should be designed so that employees are held accountable only for what they can control. In other words, employees should neither be penalised for bad luck nor rewarded for mere good luck. A measure is totally controllable by an employee if it is influenced only by his/her actions (Merchant & Van der Stede, 2007, p. 533). Merchant and Van der Stede (2007, pp. 34–35) also introduce an understandability criterion, which means that employees understand what they must do to exert an effect. In this article, understandability is not discussed separately, but in connection with controllability. The other criteria of good performance measurement are discussed in the chapter "Challenges in managing energy efficiency".

Financial indicators are often used because they can be measured with relative precision, objectively, inexpensively and on a timely basis. Also, they are relatively congruent with the organisational goal of profit maximisation and they are usually controllable and understandable. However, there is also a well-documented list of shortcomings: myopia, focus on the past, risk aversion and sub-optimisation (Merchant & Van der Stede, 2007). Non-financial indicators have been found to have advantages in addressing these shortcomings; as they are more directly traceable to strategies, they are not affected by price fluctuations (Fisher, 1992) and are leading indicators of future financial performance (Merchant & Van der Stede, 2007). Instead of being limited to a diagnostic purpose that serves the implementation of an intended strategy, non-financial indicators may have a more active and constitutive role to play in the management process (Vaivio, 1999).

In summary, there are several conceptual challenges in applying the SEC as an energy efficiency indicator from the technical point of view. For this reason, it is difficult to generate an explicit measure for energy efficiency and the criteria for good performance measurement may not be met.

Challenges in Managing Energy Efficiency

While the technical literature focuses on how to construct an indicator for the complete site, the performance measurement literature discusses plant (company) management in more general terms.

Technically, development of energy efficiency indicators starts from the investigation of individual processes on the equipment level and proceeds to higher levels of aggregated information towards site and company level (Sivill, 2011, p. 31). Aggregate indicators are used on two levels: the process level and the plant/site level (see also Fig. 2). Since the output of each process (denominator in Eq. (1)) is different, the composition of a plant-level energy efficiency indicator is problematic. Tanaka (2008a) states that a comparison of energy use in different units and aggregate efficiency for the whole of manufacturing is effectively impossible without conversion of the value of the physical units to a common value. This is most often the same as the value of production.

In addition, designing indicators for different purposes and organisational levels is not simple when the production facilities are interlinked. Tanaka (2008b) states that in integrated facilities the material and energy streams are interconnected and there are interactions that make the cause-and-effect relationships difficult or even impossible to interpret.

In the performance measurement literature the performance indicators are used as management tools and they come from the organisation's strategy and goals. Epstein (2008) has proposed that corporate-level indicators should be a starting point to find indicators coherent with the corporate strategy down through the hierarchy. Performance indicators should be designed to fit the organisation's structure and the decision-making responsibility of individual employees.

The indicators used typically vary across organisational levels. At higher levels, most of the key indicators are defined in financial terms, for example stock price or return on equity. Lower-level managers, on the other hand, are typically evaluated in terms of operative data that are more controllable at local level (Merchant & Van der Stede, 2007, p. 30).

The use of integrated performance measurement frameworks, for example the Balanced Scorecard (BSC) (see Kaplan & Norton, 1992) in translating strategy into action has been increasingly emphasised in performance measurement literature. By translating mission and strategy into objectives and measures, the BSC provides a framework to communicate strategy to employees at all levels of the organisation. A well-designed set of measures makes it easier to understand how value can be created. For example, the causal linkages in a BSC provide managers with guidance as to what they must do to affect the financial performance measures, which are at the end of the causal chain (Merchant & Van der Stede, 2007, p. 474). In developing performance measurement systems it is also critical to consider both soft factors such as culture, behaviour and attitudes and harder factors such as reporting structures, responsibilities and the use of information technology (Bitici, Carrie, & McDevitt, 1997).

However, capturing the essence of strategy with a performance measurement system and implementing a successful strategy is a challenge (Ittner, Larcker, & Meyer, 2003; Neely et al., 2000). The first task is to choose measurable performance dimensions that reflect the organisation's objectives. The second step is to select the correct indicators (Merchant & Van der Stede, 2007). The third step is to set appropriate targets, that is the level of performance required for each of the measures is identified. Fourth, an appropriate package of incentive and reward mechanisms is created. The final step in the performance management process is the provision of information on the actual results. A further issue is the use of feedback information (Otley, 2005, pp. 88–93).

Target setting is a key aspect in implementing performance indicators. Lacking knowledge of appropriate performance levels may lead to too low or too high performance targets; either of these reduces motivation (Vaivio, 1995). In particular, the lack of measurement history makes it difficult to set targets for non-financial measures (Tuomela, 2005). In performance measurement literature, strict control is often considered to be most beneficial in the areas most critical to the organisation's success. However, targets for energy efficiency cannot always be expressed in a precise manner because the factors affecting them are partly unknown.

Rewards and compensation are an important factor affecting motivation. Motivation and productivity are higher when a compensation system is well designed (Van Herpen, Van Praag, & Cools, 2005). Motivation and attitude towards improving energy efficiency could be improved by focusing on training during the implementation of performance measurement systems and on clear instructions and ensuring dialogue in the organisation (see Malina & Selto, 2001; Ukko, Tenhunen, & Rantanen, 2008).

In summary, the technical literature focuses on the technical construction of an indicator value for a complete site. Energy efficiency indicators are traditionally used as monitoring tools (Sivill & Ahtila, 2009). With the increasing strategic importance of energy efficiency measurement, it would be important to combine the technical and performance measurement perspectives to improve understanding of the challenges related to the indicator and its use as a management tool.

RESEARCH METHOD AND THE CASE STUDY PLANT

The research took place in a single firm, representing the energy-intensive process industry. The group is major European manufacturer of polyolefin plastics in Europe and one of the ten largest in the world. The group has operations on three continents and employs 5,500 people. The firm was appropriate for the study thanks to the energy intensity of its manufacturing processes. In addition, it has actively invested in energy efficiency. The case site employs some 850 people. In 2008, it generated revenue of approximately 1.2 billion euros. There are 11 operational and support units at the case site, the Hydrocarbon and the Polyolefin operations being the most energy intensive (see Fig. 3).

The case study method was used to conduct the research, and theme interviews, that is semi-structured interviews, were conducted to obtain data on the energy efficiency indicators used in the plant and on their perceived challenges and weaknesses. The case study method was considered to be

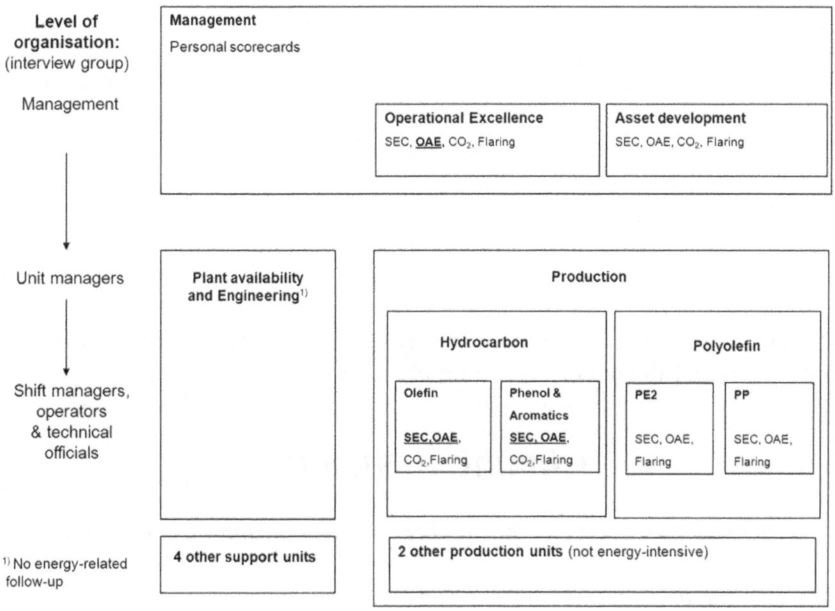

Fig. 3. Energy-Related Indicators Followed in the Scorecards of the Case Company. The Underlining Means the Indicator is a Part of the Bonus System.

most applicable to the circumstances of the study. The method is regarded as particularly useful when the concepts and variables are difficult to quantify and when the phenomenon under study is difficult to investigate outside its natural settings (Ghauri & Grønhaug, 2010, p. 109). In the study, the main interest was in the views of the personnel on energy-related issues, which supported the choice of qualitative research and interviews. In terms presented by Keating (1995), this study refers to theory discovery as energy efficiency and its measurement and management are a complex and cross-sectional phenomenon and the study seeks new theoretical perspectives.

The research began with a preliminary discussion with the Energy Specialist (later: Manager D) and the Controller (later: Manager F) of the case company to survey the topics that might come up in the interviews. The final theme interview forms were composed on the basis of the discussions and the literature review. The interviews were conducted at the turn of 2008–2009. Altogether 22 representatives from the case company participated in the research.

The first category of interviewees included eight representatives of top and middle management. Their interview questions focused on the strategic importance of energy efficiency and on measuring and rewarding energy efficiency in the company. A list of the participants is given in Appendix A.

The second category consisted of technical officials and floor-level employees, that is operators and shift managers from the company's Hydrocarbon operations. In total, eight technical officials and six operators and shift managers were interviewed in four group interviews using a similar interview form. A more detailed list of the interview groups is given in Appendix B. The group interviews included questions such as how employees can influence energy consumption or measurement of energy efficiency. The duration times of the interviews in both phases were from 45 minutes to 2 hours. All the interviews were recorded and transcribed in order to reveal the details as accurately as possible.

EMPIRICAL RESULTS

Energy Efficiency Performance Measurement in the Case Company

The Performance Measurement System

Performance was "measured in the spirit of the Balanced Scorecard (BSC)" (Manager B, Hydrocarbon). The system featured the four principle areas of the BSC, namely customers, finance, processes and learning.

There were 11 operational and support units at the case site (see Fig. 3). Four of the production units were energy intensive and they were situated in the Hydrocarbon and Polyolefins operations. Maintenance of the site was the responsibility of Plant Availability and Engineering. The unit scorecards were applied in all of the 11 units. The unit scorecards involved the floor-level employees, that is the shift managers and operators.

Middle and top management had business incentive plans of their own (see Fig. 3). Energy efficiency was followed in the Asset Development and Operational Excellence units. Asset Development took part in investment decision making and asset efficiency was the responsibility of Operational Excellence.

Energy Efficiency Indicator

The key energy efficiency indicator was SEC. There were two different types of SECs: (1) the SECs of the production units, later called SEC_{PU}s and (2) the plant-wide ("production location"-wide) SEC, later called SEC_{PW}. The calculation of SEC_{PU}s was based on Eq. (2). The plant-wide SEC_{PW} was the sum of all the SEC_{PU}s. The production tons (the denominator) of the SEC_{PU}s were replaced by a "weighting factor". After that, all of the SEC_{PU}s were summed up. The weighting factor for each unit was calculated by dividing the total tons of production at the case site by the production of the particular unit. The SEC_{PU}s were also followed by "product family" and thereby used as an internal benchmarking tool.

The other energy-related non-financial scorecard indicators included CO_2 emissions, flaring tons and OAE (overall asset effectiveness; see Fig. 3). CO_2 emissions originate from fuel burning, whereas OAE and flaring describe the functioning of the process. The latter two are indirect measures of energy efficiency. In addition, the company monitored financial energy indicators. These included the costs of fuels, steam and electricity per ton of product produced.

The Scorecard System

The SEC_{PU}s, CO_2 emissions, flaring tons and OAE were followed in the scorecards of the Olefin and the Phenol and Aromatics units of Hydrocarbon, in the PE2 and the PP units of Polyolefin (except for CO_2), in the Asset Development unit and in the Operational Excellence unit (see Fig. 3). The values were monthly averages. No energy-related information was monitored in the Plant Availability and Engineering units. Financial energy efficiency indicators were not included in any scorecards.

The scorecard indicators were monitored on a monthly basis in each production unit. The Leader Team at the case site reviewed energy performance and SEC_{PW} twice a year as part of the energy management system.

The rewarding of floor-level employees working in the Hydrocarbon operations was based on SEC (SEC_{PU}) and OAE (underlined in Fig. 3). No other employee groups were rewarded on the basis of SEC. The bonuses were paid at the turn of the year and were based on the average SEC_{PU} of the previous year. In addition to the scorecard-based rewards, the case firm had an initiative reward system that was built, among other things, to refine the energy efficiency ideas of individual employees.

Perceived Challenges with the Energy Efficiency Indicator

An important objective of the interviews was to obtain data on the characteristics of the energy efficiency indicators. Based on the interviews, opinions on SEC varied.

Technical Official A of Hydrocarbon stated that SEC is "just a number", but it is clear and everybody understands it directly. However, this view was not shared throughout the organisation. SEC suffered from several weaknesses, which affected the employees' attitudes. According to Manager A of Polyolefins, the indicators of energy efficiency were still fairly new:

> "These indicators are still searching for their form and they haven't been operationalized everywhere in the company." (Manager A, Polyolefins)

The accuracy of the measurement was also questioned by Operator 1 and Shift Manager B from Phenol and Aromatics, who mentioned their concern about a possible measuring error that had not yet been checked.

SEC was known to be affected by several factors. This was acknowledged by interviewees at various levels of the company:

> "Actually, it is not totally clear ... what issues affect it .... No-one has come at least directly to explain those matters to us." (Shift Manager A, Olefin)

> "If you have such an indicator as we have, Energy-KPI [SEC], which is kind of like an index number or a ratio, it won't help the operator greatly if you say that it has risen from last month by 10%, unless he or she knows whether it has been steam or electricity and in which part of the plant." (Manager A, Polyolefins)

The SEC was not perceived as being concrete enough for the operator level. The operators and shift managers requested more concrete aspects for measuring. Also, SEC was closely related to the capacity utilisation rate, which had declined because of the ongoing economic downturn. A more concrete aspect of measuring would have motivated the operators and shift managers to improve energy efficiency despite the economic situation:

> "At the operator level, it [energy efficiency measurement] should be connected to selected aspects of measuring instead of the indicator SEC. We could choose such places in the unit that we can truly affect, and then measure those." (Manager F, Polyolefins)

> "Now we have this big lump [current SEC], like gigajoules per ton, so why not these small, concrete things. Maybe we have tried to reach for the sky too much with this thing." (Technical Official A, Olefin)

The timing and punctuality of the indicator posed challenges. As the bonuses were always paid at the turn of the year, the motivation to improve energy efficiency was sometimes a bit low during the final months of the year. Monthly variations in SEC were not shown. According to Operator 2 from Olefin, real-time information on energy consumption would help operators in understanding how to influence it. Also, Shift Manager B at Phenol and Aromatics said that the scorecard came weeks later, when the results were no longer interesting. Operator 1 agreed:

> "It is counted monthly, but when it is done is another thing. Someone's on vacation, someone's ill – that kind of thing. But it is not that concrete. Everything should be like concrete for us. Here and now. That would be the best." (Operator 1, P&A)

The question of whether the indicator should also express energy consumption in euro divided opinions. Some thought that financial indicators would be easier to understand, but others doubted that they would provide any extra value.

Perceived Challenges in Managing Energy Efficiency

Plant-Wide Energy Efficiency Measurement

According to Manager D, energy consumption was followed monthly on the basis of production unit ("process" in Fig. 2) and product family but there was no real energy efficiency indicator for the total production location, that is the case site, because of the "volume differences". Summarising energy efficiency at production site level was not considered an effective practice.

Process integration posed a challenge to use of the indicator. According to Operator 1 from Phenol and Aromatics, although energy can be saved in one place, the gain may be lost elsewhere because the processes are so closely integrated.

Strategic Management of Energy Efficiency

Overall, energy issues were high on the management agenda of the case company and energy efficiency was regarded as a strategic choice and an important factor in the company's competitiveness. Manager A identifies rising energy prices as a reason for the increasing importance of energy-related issues:

> "Because the meaning of it [energy] has increased financially, it has become a big topic on the management agenda. As a consequence, new indicators are being developed and the indicators are more closely followed." (Manager A, Polyolefins)

Regarding the overall strategic guidelines, the case firm had an Energy and CO_2 Committee whose purpose was to provide strategic energy and CO_2 guidance to the energy and environment platform and procurement functions. The members of the committee were the CEO (chair) and all the relevant (executive) vice presidents.

In particular, employees who worked on energy efficiency projects felt that improving energy efficiency was important. Several managers stated that energy efficiency issues had not been sufficiently emphasised. According to Manager F, energy efficiency was not highlighted as much as it could have been, although the company had been able to improve its results slightly every year with regard to the goals. The main focus was more on the other areas, like quality and safety. But as Technical Official G from Phenol and Aromatics stated, if quality were compromised in order to save energy, it "might rebound in energy consumption in some other part of the process". Furthermore:

> "If it [energy efficiency] is an important matter, it has not been able to get a visible position in the company. Or maybe it has, but I have just missed it ..." (Technical Official A, Hydrocarbon)

> "... It [the role of energy efficiency in the company] has now been bothering me somewhat, of course. It has not been given as high a priority as it deserves." (Manager F, Polyolefins)

A common view on energy efficiency did not yet exist inside the company and energy efficiency issues did not concern all the relevant employee groups. For example, Manager E stated that the Plant Availability and Engineering employees felt that they were excluded from energy efficiency issues. However, it was expected that the energy management system, which was recently linked with the existing ISO 14001 EMS, would spread energy efficiency thinking to all the parts of the organisation:

> "… when the system [Energy Management System] starts to roll, then it should kind of force it [the energy efficiency] to all the units collectively and everyone will have to think from their own perspective what they can do for it." (Manager E)

As a result, the operators considered their abilities to affect the indicator to be fairly limited. It was obvious that the indicator had a minor role in the day-to-day activities:

> "If I'm honest, our possibilities to influence these are so small that what comes, comes. We just have this percentage that we can influence." (Operator 1, P&A)

> "Yes, we have the energy efficiency indicators, like megawatt-hour per ton. But there is still a problem with the cause-and-effect relationships: How can I contribute and how is it operationalized in the doings of everybody?" (Manager C)

The need to connect energy efficiency with daily life was also understood at managerial level. The challenge was that operators might have their own way of running the process: some wanted to try to reach the optimal level; this makes the process more sensitive. Others wanted to make certain that they had enough operating margin:

> "The challenge is, of course, also in making this kind of shift work function as a whole. They have their own assumed best ways of doing things. In this plant that I've mentioned they have a custom that when the shift changes-this is maybe a bit caricatured but it's true-then the on-coming shift changes the ratios or any values to what they consider correct." (Manager C)

Energy Efficiency Target Setting

Each unit set their own targets for the SEC_{PU}. However, the targets were not considered achievable and some matters beyond the control of the operators, such as the economic situation, apparently had the greatest impact on the indicator. When production is decreased, energy consumption does not decrease in a linear manner. The value of the SEC increases. The operators were disappointed that the ongoing economic downturn was not taken into account when setting the SEC target. This had a big effect on their motivation to improve energy efficiency:

> "… we saw already at an early stage that we had no chance to reach those targets. The target was way too high, and that isn't particularly motivating." (Operator 2, Olefin)

Achieving the targets during an economic downturn would have required more instructions from management level:

> "What do we need to do to reach those figures [targets] that we have. It seems so utopian. (Shift Manager A, Olefin)

> "Now that this recession came and products didn't sell, no-one came here to tell us that now we should save a lot of energy. More like, if you write in one of those weekly agendas that cut all unnecessary reuse and return and focus on energy efficiency, who gets motivated by that to do anything about it. It's read briefly that … oh, okay … there is that thing." (Shift Manager B, P&A)

The operators considered it rather difficult to reduce energy consumption when operating at or near full capacity. They pointed out that no systematic calculations on the energy saving potential of their work had been made previously. It also seemed clear that the operators in particular would have needed clearer instructions on how to improve energy efficiency and change the current operating method:

> "We operate as we have always run the process. If in the past it has been said to be fine, then we assume that it is fine unless someone shows us calculations that it is not fine. But if no-one shows that, then … . We don't know how to calculate it." (Operator 1, P&A)

In addition to simply keeping track of the indicator values, Manager C pointed out that everybody should be involved in thinking and improving energy efficiency in the spirit of continuous improvement in groups consisting of operators and supervisors. Collectively, the basic problem could be better understood and thus a solution reached. According to Manager C, sharing opinions about energy issues was important:

> "Very often it happens that we feel that we have a similar view of the problem. But most probably there are also differences in understanding." (Manager C).

DISCUSSION

Challenges Related to the Characteristics of the Energy Efficiency Indicator

Based on the interviews and the literature reviewed earlier, SEC, the energy efficiency indicator used in the case company, had several weaknesses related to its characteristics.

The challenges encountered in energy efficiency measurement are well documented in the technical literature. They include problems related to balance boundary, denominator definition, energy allocation, energy quality and the difficulty in defining the factors affecting energy efficiency. The issues being raised depend on the scale and scope of the study. For example, the balance boundary definition does not pose a challenge in cases where a certain production process (e.g. Hydrocarbon/Olefin) is measured and monitored on a continuous basis, but it is critical, for instance, when the objective is to compare manufacturing sites. The denominator definition is not regarded as problematic in cases where a certain production process (e.g. Hydrocarbon/Olefin) is measured and monitored on a continuous basis, but it is critical when the aim is to summarise the consumption figures of different units. Energy allocation is not critical in cases where energy efficiency is measured for a well-established mix of products (e.g. all products from Hydrocarbon/Olefin), but it becomes a problem when the energy needs to be allocated to each of the individual products or between the production units (see e.g. Siitonen, 2010, p. 30). This issue is rather more technical than managerial, and affects the ability to produce correct information.

Although the challenges related to energy quality are mostly technical, there is still a managerial side: as different energy vectors were summed up in the SEC it became more difficult to trace possible changes in the indicator value afterwards. The overall difficulty of defining the factors affecting energy efficiency measurement came up in several comments. The operators said that it was not clear how the indicator value was formed. It follows that the operators did not know how to influence the indicator and could not determine the potential for improvement. It was known that the energy efficiency is affected by other factors such as the capacity utilisation rate.

The characteristics of the energy efficiency indicator can be evaluated against the criteria for good performance indicators (e.g. Merchant & Van der Stede, 2007). Based on the interviews, the specific energy efficiency indicator seems to be relatively free of bias. On the other hand, concerns about possible measuring errors were expressed. Energy efficiency measurement in unit processes is technical. If a measurement is technically 'valid' (correct positioning, functioning, calibration, etc.), it presumably meets the objectivity, precision, validity and reliability criteria. In the case company, the measurement was quite new and lacked comparability. Hence, employees did not always find the measurement credible.

Neither was the requirement for timeliness met; the operators in particular commented on the lateness of the measurement results and on how this was demotivating for them. Real-time or fairly up-to-date information on energy consumption would show the reasons for changes in the indicator more clearly. Also, the controllability of the indicator seemed problematic. The employees were aware that energy efficiency measurement was influenced by various factors. The floor-level employees, that is the shift managers and operators, requested information on how to affect the measure. The floor-level employees consider their abilities to influence the indicator limited and they proposed that they should only be accountable for the functioning of production equipment. On the other hand, employees in Plant Availability and Engineering were not held accountable for energy efficiency. Furthermore, energy efficiency was not a basis for rewarding employees in Asset Development and Operational Excellence. It could therefore be worthwhile to reconsider the responsibilities related to energy efficiency improvement throughout the organisation.

An indicator should also assist in solving problems. As energy information is first summarised (Eq. (1)) and then aggregated to higher levels, information on energy quality is lost (energy quality problem). To be more exact, the source of information is lost. This was brought up by one of the operators requesting the original data.

Discussion of the advantages and disadvantages of financial vs. non-financial indicators did not end in consensus. While some operators believed a financial indicator would help them to understand the magnitude of the energy efficiency improvements, others did not expect any extra value. This may be explained by the fact that the financial energy efficiency information is obtained by converting the physical energy data [e.g. MWh/t] into financial information using energy costs [e.g. €/t]. This is also a way to overcome the denominator definition problem as the production tons [e.g. MWh/t] are expressed as a value of production [e.g. MWh/€]. Previous research describes non-financial indicators as more directly traceable to strategy than financial indicators (see e.g. Fisher, 1992). Going beyond financial information, non-financial indicators can connect managers regularly to operational issues of strategic significance (Vaivio, 1999). Thus, as a non-financial energy efficiency indicator, SEC could, as the leading indicator, deliver information on energy consumption faster than financial indicators. However, as the timeliness criterion was not met, this benefit was not exploited effectively in the case company.

To summarise, the challenges discovered regarding the characteristics of the energy efficiency indicator originating from both the energy efficiency and the performance literature constitute possible areas for attention when developing the indicator further. By using the energy efficiency index (EEI, see Eq. (3)) instead of the indicator SEC, target setting becomes somewhat more visible. Nevertheless, the other conceptual weaknesses will remain.

Challenges Related to the Management of the Energy Efficiency

Development of SEC starts from the investigation of unit processes (Eq. (2)). As management tries to create a plant-wide figure for energy efficiency, they face the problem of summarising the values of the indicator SEC from different processes (the denominator definition problem). These limitations in constructing a plant-wide energy efficiency indicator were well known at the case plant and they had adopted a special way of calculating plant-wide SEC_{PW}. From the technical perspective this can be considered a best practice. However, the problems associated with the complex interactions were not resolved. The consequences of interactions had been noticed by the employees: an operator commented that energy saved in one place might be lost in another place. This is a technical problem that cannot be solved until the calculation methods have been improved.

In the case company, energy efficiency was clearly considered a strategic objective and it was discussed in the management groups. However, the employees remarked that there could have been more emphasis on energy efficiency. Furthermore, energy efficiency was not included in the scorecards of Plant Availability and Engineering. Consequently, it would be important to clearly communicate the critical and strategic nature of energy efficiency from the corporate level to the lower levels of the company.

Moreover, the causal linkages between the measures should provide overall guidance as to what must be done to affect financial performance. One may argue that although the set of indicators (OAE, SEC, CO_2 and flaring) serves informative purposes, it does not provide information about causalities.

Each of the production units set their own targets for SEC. The target levels are perhaps the most accurate as the expected behaviour is modelled or the historically achieved best performance is used as a reference. But modelling or measuring may not always be possible or meaningful. Setting preferable targets is challenging since their effects on energy efficiency can only be defined through trial-and-error (Sivill et al., 2009). Coping with the variables affecting the measurements was perceived as problematic at various levels of the company.

The current rewarding system at the case company did not motivate the employees to improve energy efficiency. Determination of rewards on a yearly basis was problematic as the time period for observing performance was too long. In addition, some of the factors affecting energy efficiency were beyond the employees' control (e.g. capacity utilisation rate). If the target seemed unreachable, the employees did not feel motivated to continue their efforts to improve energy efficiency.

The employees of the case company requested instructions on how to influence the energy efficiency indicator. The floor-level workers also felt that the practical problems experienced at the lower levels were not fully understood at the corporate level. Thus, an internal dialogue emphasising the criticality of energy efficiency would be beneficial.

In addition to the *formal* performance measurement system, including the energy efficiency indicator and its target, instructions specifying how to perform the tasks and the rewards system, *informal* factors, that is organisational beliefs, values and traditions affect how control systems are used in practice (Riccaboni & Leone, 2009). Safety and production quality were in the corporate culture. Based on the empirical evidence, it seems to be difficult to integrate energy efficiency with day-to-day activities. This would require clear instructions. Integrating energy efficiency with the organisational way of thinking and operating would probably require inclusion of energy efficiency in both formal and informal controls.

CONCLUSIONS

This study examined the challenges imposed by current energy efficiency performance measurement. The case study method was used to conduct the research that took place in a single firm in an energy-intensive process industry. The paper combines two academic discussions, that is a technical discussion on measuring energy efficiency in the process industry and a discussion on performance measurement and management. It not only confirms the earlier documented conceptual challenges to the energy efficiency indicator, but also goes further by explaining their managerial consequences.

The results indicate that the general energy efficiency indicator, SEC, does not meet all of the criteria for good performance indicators (see e.g. Ghauri & Grønhaug, 2010; Merchant & Van der Stede, 2007). In particular, the indicator does not meet the timeliness and controllability criteria. Since measurements are not delivered on time, the timeliness criterion is not met. The controllability criterion seems problematic in several ways. First, the operators of the case plant did not know how to affect the indicator value. In addition, some of the factors affecting energy efficiency, such as the capacity utilisation rate, were beyond their control.

Second, energy efficiency was not a part of the management rewarding system. However, they are able to influence efficiency, for example through investments. The information needs and the ability to influence energy efficiency vary at different levels of an organisation. The organisation should consider how different employee groups can truly affect energy efficiency.

Based on the analysis, the energy efficiency target setting also seems problematic. This is due to the technical challenges, since many aspects influence the indicator and not all of them are known. Difficulties in target setting may cause difficulties in controlling the achievements. From the employee perspective, targets should be motivating at all organisational levels.

From the management perspective, attention should be paid to communication and instructions. In particular, the study underlines the different information needs at various organisational levels, and highlights the importance of learning through communication (see e.g. Epstein & Roy, 2001). Currently, the strategic importance of energy efficiency was not understood at the operational level in the case company, and as a consequence, SEC had a relatively minor role in day-to-day activities. To diffuse the strategic importance from the corporate level to all levels of the organisation and to share the practical ideas understood at the floor-level, this study proposes using the energy efficiency indicator increasingly as an interactive control tool (Simons, 1991, 1995). Improved dialogue would promote sharing of experiences and thoughts on energy efficiency and thus encourage learning throughout the company. Improved communication between the management and floor-level workers could also bring out new ideas for energy efficiency.

More research is needed to develop the energy efficiency indicator in order to communicate the energy efficiency improvement potential. In addition, it would be essential to consider management control systems as a whole, as many of them interact with energy efficiency. By examining all the elements in the control package (see Malmi & Brown, 2008) and the relationships between them, we would be more likely to obtain a better understanding of the effectiveness of the individual elements. This would foster a culture conducive to energy efficiency. Thus, tackling the conceptual problems alone may not improve energy efficiency if the formal and informal systems do not support improvement. Hence, in order to integrate energy efficiency effectively into the organisational way of thinking and operating, it should also be included in informal controls, for example culture, leadership and commitment.

POSTSCRIPT

Since the collection of data for this study at the turn of 2008–2009, the case company has been using a so called "corrected energy KPI" (corrected SEC). In this new version of the indicator, the tons of economic loss caused by the economic downturn are taken into account.

ACKNOWLEDGMENTS

This study has been financed by the Finnish Funding Agency for Technology and Innovation (Tekes) under the ClimBus programme. We would like to express our appreciation to the personnel of the case plant, for making themselves available for the interviews. The contribution of Mr. Matti Marttila during the process was significant. Also, our thanks go to Professors Teemu Malmi and Seppo Ikäheimo for their valuable comments during the writing period.

REFERENCES

Ahtila, P., Holmberg, H., Tuomaala, M., & Turunen, T. (2010). Energy management. In M. Karlsson (Ed.), *Papermaking part 2, drying*. Helsinki: Paperi ja Puu Oy.

Auvinen, H. (2008). *Measuring energy and CO_2 efficiency in petrochemical industry by using energy balance analysis*. M.Sc. thesis, Helsinki University of Technology, Espoo.

Bitici, U., Carrie, A., & McDevitt, L. (1997). Integrated performance measurement systems: A development guide. *International Journal of Operations & Production Management*, *17*, 522–534.

Caffall, C. (1995). *Learning from experiences with energy management in industry*. CADDET Analyses Series No. 17. Sittard: Centre for the Analysis and Dissemination of Demonstrated Energy Technologies.

EC. (2006). Directive 2006/32/EC of the European Parliament and of the Council of 5 April 2006 on energy end-use efficiency and energy services and repealing Council Directive 93/76/EEC.

EC. (2009). *European Commission, reference document on best available techniques for energy efficiency*. Retrieved from http://eippcb.jrc.es/reference/ene.html. Accessed on February 10, 2012.

EK (The Confederation of Finnish Industries). (2011, October 31). Retrieved from http://www.hpl.fi/www/fi/energia/energian_kaytto_ja_tehokkuus.php. Accessed on February 10, 2012.

Epstein, M. J. (2008). *Making sustainability work: Best practices in managing and measuring corporate social, environmental, and economic impacts*. Sheffield, England: Greenleaf Publishing.

Epstein, M. J., & Roy, M.-J. (2001). Sustainability in action: Identifying and measuring the key performance drivers. *Long Range Planning*, *34*, 585–604.

Fisher, J. (1992). Use of nonfinancial performance measures. *Cost Management*, *6*, 31–38.

Ghauri, P., & Grønhaug, K. (2010). *Research methods in business studies: A practical guide* (4th ed.). Harlow, England: Prentice Hall, Pearson Education Limited.

IEA (International Energy Agency). (2008a). *World energy outlook*. Retrieved from http://www.iea.org/textbase/nppdf/free/2008/weo2008.pdf. Accessed on February 10, 2012.

IEA (International Energy Agency). (2008b). *Worldwide trends in energy use and efficiency: Key insights from IEA indicator analysis*. Retrieved from http://www.iea.org/papers/2008/Indicators_2008.pdf. Accessed on February 10, 2012.

IEA (International Energy Agency). (2008c). Energy Technology Perspectives 2008, Scenarios & Strategies to 2050. OECD/IEA, Paris (pp. 471–517).

Ittner, C. D., Larcker, D. F., & Meyer, M. W. (2003). Subjectivity and the weighting of performance measures: Evidence from a balanced scorecard. *Accounting Review*, *78*, 725–758.

Järvenpää, M., Partanen, V., & Tuomela, T.-S. (2001). *Moderni taloushallinto – haasteet ja mahdollisuudet*. Helsinki: Edita Oyj.

Kaplan, R. S., & Norton, D. P. (1992). The balanced scorecard—Measures that drive performance. *Harvard Business Review*, *70*, 71–79.

Keating, P. J. (1995). A framework for classifying and evaluating the theoretical contributions of case research in management accounting. *Journal of Management Accounting Research*, *7*, 66–86.

Laitinen, E. K. (1998). *Yritystoiminnan uudet mittarit*. Helsinki: Kauppakaari Oyj.

Malina, M. A., & Selto, F. H. (2001). Communicating and controlling strategy: An empirical study of the effectiveness of the balanced scorecard. *Journal of Management Accounting Research*, *13*, 47–90.

Malmi, T., & Brown, D. A. (2008). Management control systems as a package – Opportunities, challenges and research directions. *Management Accounting Research*, *19*, 287–300.

Merchant, K. A., & Van der Stede, W. A. (2007). *Management control systems: Performance measurement, evaluation and incentives* (2nd ed.). Harlow, England: Prentice Hall, Pearson Education Limited.

Motiva. (2007). *Teollisuuden energiasäästösopimuksen vuosiraportti 2007* [Annual report of industrial energy conservation agreement 2007]. Retrieved from http://motiva.fi/files/2879/Teollisuuden_energiansaastosopimuksen_vuosiraportti_2007.pdf. Accessed on February 10, 2012.

Neely, A., Mills, J., Platts, K., Richards, H., Gregory, M., Bourne, M., & Kennerley, M. (2000). Performance measurement system design: Developing and testing a process-based approach. *International Journal of Operations & Production Management*, *20*, 1119–1145.

Otley, D. (2005). Performance management: A framework for analysis. In A. J. Berry, J. Broadbent & D. Otley (Eds.), *Management control: Theories, issues and performance* (2nd ed.). New York, NY: Palgrave Macmillan.

Partanen, V. (2007). *Talousviestintä johtamisen tukena*. Jyväskylä: Talentum.

Patterson, M. G. (1996). What is energy efficiency?: Concepts, indicators and methodological issues. *Energy Policy*, *24*, 377–390.

Riccaboni, A., & Leone, E. (2009). Implementing strategies through management control systems: The case of sustainability. *International Journal of Productivity and Performance Management*, *59*, 130–144.

Sardianou, E. (2008). Barriers to industrial energy efficiency investments in Greece. *Journal of Cleaner Production*, *16*, 1416–1423.

Siitonen, S. (2010). *Implications of energy efficiency improvement for CO_2 emissions in energy intensive industry*. Ph.D. (Tech) study, Aalto University School of Science and Technology, Espoo, Finland.

Simons, R. (1991). Strategic orientation and top management attention to control systems. *Strategic Management Journal, 12*, 49–62.

Simons, R. (1995). *Levers of control: How managers use innovative control systems to drive strategic renewal*. Boston, MA: Harvard Business School Press.

Sivill, L. (2011). *Success factors of energy management in energy-intensive industries: Energy performance measurement*. Ph.D. (Tech) study, Aalto University School of Engineering, Department of Energy Technology, Espoo, Finland.

Sivill, L., & Ahtila, P. (2009). Energy efficiency index as an energy efficiency indicator for integrated pulp and paper mills – a case study. In *Proceedings of international conference on applied energy (ICAE'09)*, Hong Kong (pp. 582–593).

Sivill, L., Manninen, J., & Ahtila, P. (2009). A combined approach to energy efficiency monitoring in the pulp and paper industry. In S. A. Nebra, Jr. S. de Oliveira & E. Bazzo (Eds.), *Proceedings of the 22nd international conference on efficiency, cost, optimization, simulation and environmental impact of energy systems (ECOS'09)*, Foz do Iguaçu, Brazil (pp. 1513–1521).

Sola, A. V. H., & Xavier, A. A. P. (2007). Organizational human factors as barriers to energy efficiency in electrical motors systems in industry. *Energy Policy, 35*, 5784–5794.

Statistics Finland. (2008). *Final energy consumption by sector 2008*. Retrieved from http://www.stat.fi/til/ehkh/2008/04/ehkh_2008_04_2009-03-24_kuv_017_en.html. Accessed on February 10, 2012.

Tanaka, K. (2008a). *Assessing measures of energy efficiency performance and their application in industry*. Retrieved from http://www.iea.org/papers/2008/cd_energy_efficiency_policy/6-Industry/6-%20JPRG_Info_Paper.pdf. Accessed on February 10, 2012.

Tanaka, K. (2008b). Assessment of energy efficiency performance measures in industry and their application for policy. *Energy Policy, 36*, 2887–2902.

Thollander, P., & Ottosson, M. (2008). An energy efficient Swedish pulp and paper industry – Exploring barriers to and driving forces for cost-effective energy efficiency investments. *Energy Efficiency, 1*, 21–34.

Tuomaala, M. (2007). *Conceptual approach to process integration efficiency*. Ph.D. thesis, Helsinki University of Technology, Helsinki.

Tuomela, T.-S. (2005). The interplay of different levers of control: A case study of introducing a new performance measurement system. *Management Accounting Research, 16*, 293–320.

Ukko, J., Tenhunen, J., & Rantanen, H. (2008). The impacts of performance measurement on the quality of working life. *International Journal of Business Performance Measurement, 10*, 86–98.

Vaivio, J. (1995). *The emergence of non-financial management accounting measures: A case study*. Licentiate thesis, Helsinki School of Economics and Business Administration.

Vaivio, J. (1999). Exploring a 'non-financial' management accounting change. *Management Accounting Research, 10*, 409–437.

Van Herpen, M., Van Praag, M., & Cools, K. (2005). The effects of performance measurement and compensation on motivation: An empirical study. *De Economist, 153*, 303–329.

APPENDIX A: INTERVIEWEES IN PHASE ONE (MANAGERS)

Interviews:

Technical official A, Hydrocarbon, 29.10.2008
Manager A, Decision Support, 29.10.2008, 11.11.2008 (and several informal discussions)
Manager B, Hydrocarbon, 12.11.2008
Manager C, Quality, 12.11.2008
Manager D, Health, Safety and environment, 10.10.2008 (and several informal discussions)
Manager E, Plant Availability and Engineering, 24.11.2008
Manager F, Polyolefins, 24.11.2008
Manager G, Human Resources, 26.2.2009 (by phone and e-mail)

APPENDIX B: INTERVIEWEES IN PHASE TWO (PLANT MANAGERS, TECHNICAL OFFICIALS AND OPERATORS)

Group 1: Technical officials, Olefin (3.2.2009)

Technical official A (whole Hydrocarbon)
Technical official B
Technical official C
Technical official D
Technical official E
Technical official F

Group 2: Operators, Olefin (3.2.2009)

Shift Manager A
Operator1
Operator2

Group 3: Technical officials, Phenol and Aromatics (5.2.2009)

Technical official G
Technical official H

Group 4: Operators, Phenol and Aromatics (5.2.2009)

Shift Manager B
Operator3
Operator4

PART III
MANAGEMENT CONTROL IN INTERORGANIZATIONAL SETTINGS

BUILDING OF A NETWORK AND THE CONSTRUCTION OF CONTROLS

Andrei Mineev

ABSTRACT

Purpose – *To describe the social process of building of an inter-organizational network and analyse the control practices implemented by its members.*

Methodology/approach – *The paper offers a holistic approach considering the institutional dynamics and management control practices in the network as a whole. This makes it possible to trace the development of practice variation, which is one of the most recent but up to now poorly studied problems in the field of institutional theory (Lounsbury, 2008). The paper draws on a case study of a newly established network of supply companies related to a petroleum project in North-West Russia.*

Findings – *Organizational arrangements in the network as well as corresponding controls are constructed as a result of interplay between several institutional logics such as Western network experience, international rules related to petroleum projects, the Soviet industrial background of the members, and their local community values. The controls demarcate organizationally members' behaviour, and thereby*

Performance Measurement and Management Control: Global Issues
Studies in Managerial and Financial Accounting, Volume 25, 169–193

ISSN: 1479-3512/doi:10.1108/S1479-3512(2012)0000025010

define the conditions for practice variations. On the other hand, use of such controls over time leads to unintended consequences and control dilemmas. Facing these dilemmas, the network members address them differently, and therefore more practice variations are being created continually.

Originality/value of paper – *The paper gives insight into an emerging business network, a setting normally neglected in the inter-organizational control literature, and establishes a dyadic view of management control, arguing that it both defines and creates conditions for practice variation.*

Practical implications – *Local supply chains play an important role in global petroleum projects. Decision-makers involved in petroleum projects will benefit from better understanding the cooperation processes prevalent in local industries.*

INTRODUCTION

This paper seeks to add to knowledge on control practices in inter-organizational business networks. Existing research texts on inter-organizational settings often address controls in already established inter-organizational relationships or network organizations functioning in a stabilized context of Western market economies (e.g. Castells, 2000; Dekker, 2003, 2004, 2008; Håkansson & Lind, 2004, 2007; Mouritsen & Thrane, 2005; Seal, Cullen, Dunlop, Berry, & Ahmed, 1999; Tomkins, 2001). Emerging inter-organizational settings related to new markets or the settings being developed in transitional economies, in particular Russia, are still neglected in the literature.

The Russian economy, which has undergone dramatic change during the last 20 years, represents a different and a more uncertain institutional context than Western economies. As shown in this paper, understanding the institutional conditions in which Russian organizations cooperate with one another appears to be important in order to understand inter-organizational controls in a network-based setting. Further, in the course of using the controls, new institutional conflicts arise in the network. Thus, the patterns of interplay between institutional and control elements in the network represent something new we can learn from the Russian context.

This paper offers an insight into a business network being established by supply companies in North-West Russia, The Murmansk Region. Taking upcoming oil and gas projects in the Russian part of The Barents Region into account, local supply companies are joining forces and resources in networks in order to become competitive when faced with competition from larger suppliers located in other parts of Russia and abroad. The purpose of the paper is *to describe the social process of building a network and analyse the control practices implemented by its members*. To do so, the empirical enquiry focuses on the development of a strategic consortium of construction companies (hereafter referred to as The Consortium). Members of The Consortium belong to a broader network association of suppliers (hereafter referred to as The Association) which was initiated by the Norwegian company Statoil in The Murmansk Region in 2006. The research question is threefold: *how has this supplier network and consortium been developed? What management control challenges have arisen? How have these challenges been handled?*

The remainder of the article is structured as follows. The next section, 'Analytical framework', specifies the theoretical assumptions chosen for the study. The following section presents the research method chosen. The following three empirical sections present the findings related respectively to the three research questions posed. The penultimate section offers a discussion of the empirical findings. The final section outlines the main results and limitations of the study and suggests prospects for future research work.

ANALYTICAL FRAMEWORK

To provide a background for analysing the research questions, this section develops an approach combining neo-institutional theory considerations with the theory of management control practice and management control in networks. In doing this, the section briefly reviews recent developments within the streams of literature, and then offers a framework synthesizing their respective strengths.

Early developments of neo-institutional theory (e.g. DiMaggio & Powell, 1983) were criticized as self-limiting due to the overemphasized role of isomorphic pressure in explaining organizational change (Lounsbury, 2008; Scott, 2008). This resulted in focusing on homogenization processes within organizational fields. The emergence of the concept of *institutional logic* has proved a key turning point in redirecting attention to the study of institutional sources of change towards organizational heterogeneity (Lounsbury, 2008). Lounsbury refers to Thornton's (2004) definition of logic as 'the socially constructed, historical pattern of material practices, assumptions, values, beliefs, and rules by which individuals produce and reproduce their material subsistence, organize time and space, and provide meaning to their social reality'.

Following Lounsbury, the focus on micro-actors and their interpretations of institutions is crucial to understanding how organizational fields are based on multiple logics. Micro-actors are very likely to be exposed to *practice variations* as they find themselves in contexts with multiple or even competing institutions. A possible approach to understanding the behaviour of micro-actors was offered by Scandinavian institutionalists (e.g. Czarniawska, 2008; Czarniawska & Joerges 1996; Czarniawska & Sevón, 1996; Sahlin-Andersson, 1996). One of the key concepts adopted by these scholars is *translation* (Latour, 1986, 1999). Following in Latour's footsteps, Czarniawska and Sevón (1996) argue that ideas involving change do not diffuse by themselves: people pass them on to one another, each one translating the ideas in question according to their own frame of reference.

Coinciding with criticism and shifts within institutional theory, the traditional contingency perspective was challenged by practice theorists in the field of accounting and control (e.g. Ahrens & Chapman 2004, 2007; Gendron, Cooper, & Townley, 2001; Mouritsen, Hansen, & Hansen, 2009). The contingency theory of management control (e.g. Otley, 1995, 1999) is strong in analysing the technical and economic conditions of the organizational environment, as well as structures of management control systems related to them. However, this theory offers only a limited view of controls as social practices carried out by people in organizations (e.g. Ahrens & Chapman, 2004). Recent practice-oriented studies highlight the indirect effects of accounting and control on company performance, and particularly with respect to their relations with other organizations (e.g. Håkansson & Lind, 2004; Hansen &Mouritsen, 2005; Mouritsen & Larsen, 2005; Mouritsen et al., 2009).

The aforementioned tendencies in neo-institutional and management control theory share common ground – the growing importance of attention to the processes involved and the social practices of micro-actors. In analysing the network-building process, this paper follows Lounsbury and Crumley (2007) and Lounsbury (2008), and attempts to apply an approach combining these two streams of research, namely focusing on both the institutional dynamics and the micro-processes of management control.

The analysis of control mechanisms is extended using the relatively new but growing body of research focusing on management control in inter-organizational settings (e.g. Dekker, 2003, 2004, 2008; Håkansson & Lind, 2004, 2007; Mouritsen & Thrane, 2005; Seal et al., 1999; Tomkins, 2001). This study follows up Dekker's (2004) work distinguishing between the formal and informal controls of network relationships. Formal control consists of contractual obligations (e.g. goal-setting, performance monitoring) and organizational mechanisms for cooperation (e.g. behavioural rules and procedures). Informal control, also referred to as social, trust-based control and relational governance, relates to informal cultures and systems influencing behaviour of cooperating counterparts (Dekker, 2004; Ouchi, 1979). Furthermore, the paper highlights how control practices affect the network and its interplay with the outside world. In doing so, I follow the approach of Mouritsen and Thrane (2005) who showed that control practices play an important role in visualizing network dilemmas in terms of conflict between individual and collective goals.

Summary of the Framework

To understand changes in the studied network, this study focuses on the practices of its members considered as micro-actors. Merging the perspectives of Czarniawska (2008) and Lounsbury (2008), the study sees micro-actors as translating agents influenced by various institutional logics. Furthermore, the framework suggests that focusing on controls can give a better understanding of the micro-actors' behaviour. Control mechanisms developed between the micro-actors can be classified as formal and social ones (Dekker, 2004). Furthermore, distinguishing between the individual and collective controls in the network (Mouritsen & Thrane, 2006) also appears important.

RESEARCH METHOD

The empirical enquiry of this study focuses on micro-actors (network members), and their organizational and control practices. The need to better understand the micro-actors in terms of their sociocultural context called for ethnographic approach (Alvesson & Skoldberg, 2000; Richardson, 2000). When it comes to organizational and control practices, reflexive methodology (Alvesson, 2003) was applied. Due to their interpretative character, both ethnographic and the reflexive parts of the inquiry were based on qualitative research method.

Data sources in the ethnographical part included the internal and public documents of the network, materials from about 30 oil and gas conferences and seminars in The Barents Region (in which I participated during the period 2006–2010) and informal conversations with the network members during these conference and seminars. However, as purposeful work with this particular paper started in the winter of 2010, the previously obtained information was re-visited in order to find how it could be used to inform my research questions.

The main method of data gathering for the reflexive part was reflexive interviewing (Alvesson, 2003; Kvale, 1997). Most of the informants were directors of supply companies – members of The Consortium, as well as the director of The Association coordination unit. To have a look from the outside, several additional interviews were carried out with Western experts working in the oil and gas industry in Russia, and with a technical director of a local company wanting to but not being able to join The Consortium.

Following the rules of reflexive interviewing, I did not address questions about challenges and corresponding control practices directly, rather using open-ended questions about cooperation in The Consortium and attempting to observe how the informants reflected on the challenges they face. As a result, several challenges and the corresponding actions to deal with them were articulated by the informants. Once identified, these were then discussed in more detail. Such reflexive discussions make the informants think and analyse during the interview instead of presenting things in a standardized (thought about in advance) way. This makes it possible to avoid potential representation biases (Czarniawska, 2001) and therefore increases the validity of the study. As a result, several control patterns applied in the network became more visible. In total, 15 interviews with 11 individuals were conducted during the spring of 2010.

The three following sections deal with the three research questions and describe respectively the network building process, faced challenges and corresponding controls.

BUILDING THE NETWORK

Mobilizing local companies in networks formed part of the Corporate Social Responsibility (CSR) programme of the Norwegian oil company Statoil on entering North-West Russia in 2005. Statoil was interested in contributing to the suppliers' preparation to development of the giant Shtokman gas field, located in the Russian part of the Barents Sea. As a result, a network association of suppliers (The Association) was established in 2006 in Murmansk in close cooperation with the local government and industry of this region. To develop the association Statoil used a learning example from

Table 1. Main Stages of the Network Development.

Network Development Phase	Starting up with Support from Norway 2006–2007	Transition to Self-Sufficiency 2008–2009	Early Self-Financing Phase 2010
Market conditions	Changing market prospects	Market optimism and growing recognition of the network	Returning market uncertainty
Organizing forms	Mobilization of local suppliers in regional network associations	Move to commercialization: Creating of the strategic Consortium	Further re-organizing: transition to joint venture
Members' interpretation	A chance to protect the local market from outside competitors	Facing internal and external challenges; development of corresponding controls	Pessimism, but ready for action?

Norway – The Petroarctic network association of suppliers related to offshore development in the Norwegian part of the Barents Sea. At the time the research was carried out, The Association had more than 200 members from different industries. The history of the association can logically be divided into three phases as presented in Table 1 and described below.

Starting up with Support from Norway (2005–2007)

When Statoil came to Murmansk, many local suppliers supported the initiative as they wanted to be involved in upcoming offshore projects such as Shtokman. But in addition to this, as explicitly mentioned in interviews of the directors of several local companies, they saw a challenge from larger competitors from outside the region, and therefore wanted to associate with one another, and find ways of protecting the local market. The network cooperation model offered by Statoil was seen as an opportunity.

The political situation surrounding the Shtokman project changed twice during this period. First up came Gazprom's 'No' to foreign bidders for shares in the Shtokman field in 2006. Statoil was then amongst the bidders. Gazprom's decision left a question mark hanging round the whole idea of The Association. However, Statoil decided not to break off support for the then nearly established association. After the merger of Statoil and Hydro in

2007, StatoilHydro introduced some new perspectives into the development of The Association pushing its focus towards increased operational sustainability via commercialising services for the members. The same year, soon after this, the second round of Shtokman negotiations was initiated and Gazprom chose Norwegian StatoilHydro and French Total as partners for development of the Shtokman field infrastructure.

Transition to Self-Sufficiency (2008–2009)

This phase can be characterized by growing market optimism due to more concrete plans about development of Shtokman by the project owners and Shtokman Development Company (SDAG).[1] As a result of negotiations between the management of The Association and SDAG, as well as promotion work done by StatoilHydro, and political support by the local authorities, SDAG mentioned in their Russian Content Policy that in international tenders, in case of equal conditions, priority should be given to Russian suppliers, with special attention to be paid to suppliers from Murmansk. Recognition of The Association has also increased at local government level. In 2009, even more highly ranking officials than previously received places on The Association board, thus securing solid political support.

The goal of Statoil, however, was not to run the Russian association, but help local actors develop it to become a self-sufficient (member-financed) organization by 2009. Addressing this issue, the members started thinking about commercialization of their cooperation. Some structural modifications took place in The Association. The members started to ally. Although the idea of creating alliances is not new, its implementation in Murmansk was initiated by local actors, not by the Norwegian partners. Thus, at the end of 2008, nine building companies – members of The Association – established a strategic consortium (The Consortium) with the purpose of pooling their resources, efforts and competence to meet the challenges of large on-land infrastructure projects related to development of the Shtokman field.

The Early Self-Financing Phase Since 2010

At the beginning of 2010, Statoil,[2] one year later than originally planned, stopped financing The Association. This resulted in raised fees for the members, but most of them did not drop out of the network, at least at the time of this research being conducted. The same year, referring to global uncertainty in the gas market, Shtokman investors decided to postpone the main investment decision on the project until 2011. Although some preliminary infrastructure tenders were constantly being announced, this created more contract uncertainty for the local suppliers. Altogether, these two factors, the beginning of the self-financing phase and the market uncertainty created new challenges for The Association. With regard to The Consortium, its board decided to reorganize to form a collectively owned joint venture coordinating the resources of the members. Member attitudes within The Consortium fluctuated considerably. On the one hand, members were pessimistic due to postponement of the Shtokman project, on the other hand, they understood that they had to be ready for action as on-site construction tenders are normally among the first ones to be announced in petroleum projects.

The Consortium

The Consortium was founded by the largest regional construction company (hereafter called the major company), with about 2,500 employees, together with 8 other key construction companies in December 2008. Altogether these nine companies (hereafter called The Board Companies) constituted The Consortium Board. The director of the major company was elected chairman. About 30 other regional Small and Medium-sized Enterprises (SMEs) which had joined The Consortium later became ordinary members. The board holds responsibility to decide the strategy of The Consortium but can involve ordinary members for consultation. Although The Consortium was established by local companies, it was done in close cooperation with The Association. The Consortium has been one of its strategic priorities, and only members of The Association could join The Consortium.

The Consortium is rather a network of persons (directors of member companies) than companies. Only the directors and not ordinary employees of member companies were involved in the coordination meetings. In an interview, the director of a board company (hereafter called Director B1) commented in this respect: 'If we (the directors) manage to come to an agreement with one another, then we would be able to secure cooperation between our respective company employees'.

As persons, the directors share several important commonalities. Most of them are energetic middle-aged men with professional backgrounds – some stretching back to the centralized Soviet system. Although today they are leaders of commercial companies, many of them are educated construction engineers or technicians with about 20–30 years' experience of working in the construction industry. Talking about the development of The Consortium, several members referred to their experience from the Soviet period. This was presented in a positive way with a tone of nostalgia, for instance, as in the following argument:

> Vertically-integrated companies such as in The Soviet Union are very important in the construction industry for combining resources and having responsibility for the whole process. Otherwise nobody is responsible for anything ..., at the same time, today, we (the firms) are separated and there are no perspectives for employees' development. In The Soviet Union a person could grow from a field specialist to a minister. But now growth is limited by the boundaries of the firm ... This consortium is an attempt to integrate the resources .... (Director B1)

The importance of professional background from the construction industry (also rooted in the Soviet period), and the responsibility for development of The Murmansk Region was also explicitly mentioned in the interviews, for example:

> Young businessmen think in different categories than we do: they want to conquer the market. But we want to develop our region ... we would cooperate with a young newcomer if, for example, he comes to us and says that he helps his father or uncle, who has long experience in the construction industry, to contribute to the region ... then we understand and know how to work together with such people. (Director of an ordinary member company – Director M1)

The directors know one another quite well, and some are on friendly terms. Informal talks via telephone and personal meetings are normal practice to update one another on the market situation.

FACED CHALLENGES

All the aforementioned uncertainties around Shtokman made the local industry feel quite sceptical about the development of project being 'on time'. Disbelief in big projects really arriving was one of the main concerns in the board meeting discussions. Anyway, the joint planning work started, although the members did not make any common investments in equipment or training by the time of study. The marketing and administration of the consortium was done through the association.

Some important challenges emerged when the consortium members started preparing for concrete tenders. The major company had to assert some pressure on the other members, trying to assure them that they would not withdraw before the project was over. This sort of claim and fear was quite natural as the company did not want to expose for risk of its reputation and name when dealing with the project owners on behalf of the consortium. In other words, there was a perceived risk that some smaller member companies would not be able to fulfil their obligations, as the expected projects were large and had a long lifespan. But, on the other hand, some members expressed their worries about the high dependence on the major company and the other board companies. For example, the major company could use its own capacity to cover at least a part of a project, but was challenged by the others who could take on this job as well. Since the major company could not take on the whole project alone, it had to cooperate with the other companies, but how much work it could retain or outsource had to be negotiated. However, the members also had to trust one another. For example, as Director M1 mentioned in an interview:

> we rely on one another because we have to do. We have to cooperate and look for compromises … otherwise we will not be able to enter the big projects.

Anyway, an issue receiving much attention and discussed quite emotionally by the members, was the distribution of tasks in relation to the announced infrastructure projects. Although one reason for this issue arising consisted purely of the difference in technical visions and opinions, another one, namely the overlapping capabilities of the members, could not be overestimated. The association director (hereafter called Director A) commented as follows:

> many of the members are competing fiercely with one another … only if a project is big enough will they cooperate … but even in such cooperation there will be place for competition … although competition and cooperation are antagonists, they cannot be separated.

In general, the attitude to internal competition in the consortium was of a twofold nature. The board members cultivated collegial decision-making and tried to avoid competition between their companies: 'we can always sit down and talk together … surely we will come to an agreement about how to share a project' (excerpt from an interview with Director B1). At the same time, the board members were not against competition-based selection of sub-contractors (e.g. based on internal tenders). The ordinary members were more exposed to competition with one another, as they rather depended on decisions made by the board companies and had overlapping capabilities.

Director M1 commented in an interview: 'we still do not know exactly if we would have to share or compete ... but we would like to compete in a civilized way'.

As already mentioned, social relations and trust were important in the consortium. The informants mentioned that the possible presence of non-transparent or unreliable partners presented a challenge. Directors of the board companies and those ordinary members who established the consortium at the beginning of 2009 were concerned about 'getting rid of non-transparent players' (quote from an interview with Director M1) at as an early stage as possible, and certainly before the real projects started. At the same time, for many regional construction companies, it was not easy to gain access to the consortium. For instance, an interviewed technical director from a local company (Director L1) mentioned:

> we and many other construction companies do not cooperate because there is nobody showing us how to do it, and we lack leaders ... as for the association and the consortium, it is like a club which is not open for everybody ... I do not even know what they do there.

Summing up this section we can see that the internal challenges emerged on the surface during preparation for project tenders. These evolved round a mutual distrust between buyers and suppliers, and internal competition within the consortium. The challenge of non-transparent or non-reliable players was also recognized by the members themselves. In the next section, I am going to present management control elements addressing these challenges.

MANAGEMENT CONTROL ELEMENTS

The observed control elements can be divided into two types, namely collective and individual. The collective controls are accepted by all members and directed at the achievement of the goals of the whole consortium, whilst the individual ones are applied by the members themselves. First, let us consider the key elements of collective control: the memorandum of the consortium, task distribution procedure and member controls. These elements were formally designed, but in practice their use has been extended by additional, rather informal ways of behaviour (Table 2).

Table 2. Collectively Accepted Control Elements.

Control Elements	Formal Design	Informal Extensions
Memorandum of The Consortium	Idea – 'Hunt big projects in a team' The board decides, the others contribute and follow	Member meeting negotiations; communication in social networks
Task distribution procedure	Matching technical assignment for a project with technical-economic capacities of the members	Primary board discussions Additional discussions at extended member meetings
Member requirements	Annual membership fees Commitment to the memorandums of the consortium and the association	'Reputation accounts' 'Trial contracts'

The Memorandum

The memorandum is a major document stating that members will jointly participate in bids for big infrastructure projects, sharing resources and responsibility. Each member on entering The Consortium signs this memorandum. However, organizational arrangements and responsibilities are only clarified in general terms. Numerical terms are not used in the memorandum to define the size of projects, the volume of investments, the responsibilities or the sanctions in case a member fails in his obligations or withdraws from a project. The director of another board company (Director B2) puts it like this with regard to this aspect: 'this memorandum is a document about responsibilities and rights, but it is like the basic law of a country (constitution document): it defines a general framework, but there is always room for interpretation'. The practical solutions were negotiated in rather an informal way at board meetings, at extended member meetings, as well as via social networks of directors and chief specialists of member companies. As Director A described it: 'they are going to hunt big projects in a team, but they have to elaborate mutual understanding'.

The memorandum differentiates between the decision-making rights of the board companies and the others. Commenting on the various roles in the consortium, Director B2 said: 'to understand the difference between the rights of the board companies and the others one may refer to the difference between majority and minority shareholders in an open joint stock company. All may contribute to decisions, but in accordance with their shares'. Thus, strategic decisions were taken by the board, although these could be questioned by others at extended member meetings.

Task Distribution

Task distribution was one of the most important procedures to be settled in the consortium. Preliminary estimates of the capacities and costs foreseen for the member companies in relation to the announced projects were done in the form of judgment decisions made by the directors and reported on at meetings. While assessing own capabilities was not regarded to be a difficult task, as soon as the members started talking about practical solutions, more difficulties arose. These were handled through negotiations. Although all members could participate in the discussions during extended member meetings, the final decision was made at a board meeting. Having understood the capabilities and standpoints of ordinary members, the board companies could decide on how to proceed with the tender. Board member Director B1 commented such in an interview: 'it is much better when we, about 8–9 people, meet and decide on the main things ... otherwise, if all 40 members meet together there will just be a bazaar, no decision-making ... we will never move forward'.

Member Requirements

The common idea behind the consortium was that only reliable, responsible and transparent, in other words trustworthy, companies could be members. Therefore, the leadership adopted a set of member requirements. A local construction company willing to join the consortium needed to have complied with the following formal criteria. First it had to be a member of the association (with member fees and commitments specified in the memorandum of the association). After this, the company had to be ready to take on the responsibilities outlined in the consortium memorandum. In addition, a trial contract for unknown companies was a sort of informal check to which any newcomer could be subjected. Some board companies asked a new company to be sub-contractor on a small project, and afterwards the results were evaluated by the 'old members' of the network.

Perhaps the strongest requirement for new members was a good reputation. Contract history and names of specialists in the company were named as important factors to be taken into consideration. As the founders of the consortium had known the regional industry for a long time, they did possess such information. For example, Director B1 mentioned in an interview: 'The Murmansk and Kola Region are a small place with only one road coming in ..., once anybody enters it, we know all about what happens to him'. The reputation of even established members of the consortium was continually paid attention to. This meant that their actions towards partners were considered by the others. Thus, everybody was visible and apparently knew about it: 'you can cheat only once ... people understand that it is much better to be a good player and take responsibility', as stressed by Director B1.

These member controls in effect established boundaries between the consortium and other construction companies (more than 300) in The Murmansk Region. The closer a company was to the leadership of the consortium, the more resources in terms of political support, market information and business contacts it could get from the association and the regional authorities. The Consortium consisting of 40 companies became a priority, a visible result of development of the regional construction industry. This meant that for others the way to get involved in organized preparation for the big infrastructure projects lay through the consortium.

Individual Safeguards

At individual level, some members applied various additional controls as safeguards in relation to their counterparts. Previously described challenges, such as perceived exposure to internal competition and relative supplier–buyer distrust were named as the main motives for doing this. These individual safeguards can be termed *informal communication, monitoring* and *negotiating*.

First, many ordinary members recognized the importance of continuous *informal communication* or 'keeping in touch' with the others. This was a 'cheap way' of picking up news firsthand, of not being forgotten by potential partners when there was a project, and just staying close to The Consortium. Thus, while waiting for big projects, such members could focus on their own businesses without using too much time in The Consortium.

Second, quite a few members carried out continuous *monitoring* of one another in a more detailed way. Taking into account the available information about a project (e.g. technical assignment, tender documentation), the members considered their possible positions in the supply chain, and tried to estimate who could be their possible counterparts (buyers, competitors or partners, and suppliers). The technical and economic capacities of their counterparts were also taken into consideration. This information was used to understand what to expect and what could be done in advance.

In their evaluation of potential partners, members used formal tools available from everyday business. For example, as described by Director B1, to develop project technical plans and budgets, most suppliers used 'Construction industry norms and rules' and 'Regional construction industry price-structure bulletins'. By referring to these documents the buyer could analyse the project budgets offered by the supplier and control the costs. The supplier normally insists on an 'open price' meaning that the final price is to be based on the budget execution sheet. This, in turn, requires from the buyer control of quality and cost variances during the whole project. According to Directors B1 and B2, such buyer–supplier control practice was also meant to be applied by members of the consortium.

The third type of individual controls, *negotiating own involvement* in concrete consortium projects was observed rather rarely, but several interviewed companies have applied it. For example, according to Director M1, directors of construction designing companies used to meet and negotiate possible cooperation strategies. They wished to avoid unnecessary competition in their segment of the market. As the major company had its own design department, they tried to develop a cooperative approach to cope with this possible challenge. In addition to such informal meetings within professional segments, quite a lot of members used to negotiate with potential suppliers and buyers and enter into preliminary agreements in the form of letters of intent. In this way, they tried to secure relations with potential partners preparing for the future.

Negotiating their own interests during formal member meetings in The Consortium was possible, but most of the members were not so active. As commented on by Director M1, among more than 40 persons participating in the extended meetings, there were only about 10–12 persons able to take the lead in proposing possible solutions. Only five or six persons were willing to take on the responsibility for collective decisions. The rest, nearly 30 people, normally remained silent, although they were not restricted by any rules to state their arguments and positions.

DISCUSSION

The first research question of this paper concerned the network development process. Based on the presented empirical findings, this development was influenced by various institutional logics. The most important of these can be called *Western network experience*, *international rules for petroleum project*, *'local community values'*, and *Soviet industry background*.

The Western network experience manifested by Statoil was communicated in The Association. There were at least two important reasons for local industry to wish to join the association. While one of these reasons related to business opportunities emerging in the Shtokman project, the second one was the perceived threat of competition with large companies outside the region. The petroleum project requirements corresponded with the new rules the members had to learn and relate to their businesses (i.e. the organization of bidding procedures by Shtokman Development AG, and positioning in the supply chain).

Due to the Shtokman delays, as often happens in uncertain situations, the activities of the consortium members were mostly directed at the development of social relations, and were less related to building competitive advantage in terms of quality and efficiency. Shared local community values, or local identity with the history and present situation of The Murmansk Region, constituted a sort of institutional logic, playing an important role in building relationships. This logic was strong because it was associated with common feelings, and even a common willingness to protect the local community from external competitors. Not least, this was also associated with the understanding that everybody knew, and could check on everybody else. In such transparent conditions it is rational to be cooperative.

In addition to adopting Western experience in Russia in dealing with the realities of business, members actively mobilized their own experience from Soviet Industry. Thus, Soviet industry background can also be understood as an important institutional logic within The Consortium. This resulted, in particular, in developing the above-described hierarchical arrangements within the network. Furthermore, belonging to the construction industry profession and understanding its traditions (also rooted in the Soviet system) were an important informal requirement for members.

Fig. 1 illustrates the interplay of institutional logics found in the process of network building. This interplay can be described using the translation model of institutional change (Czarniawska, 2008; Czarniawska & Sevón, 1996). The institutional logics classified as the 'new' and 'old' ones had a combined effect on the field of suppliers. The mingling of these logics, as well as increased exposure to competition and new business opportunities, led to a transformation of local thinking. This transformation happened as a result of the translation of all these ideas in the minds of the local suppliers and resulted in materialization of the network association, the consortium

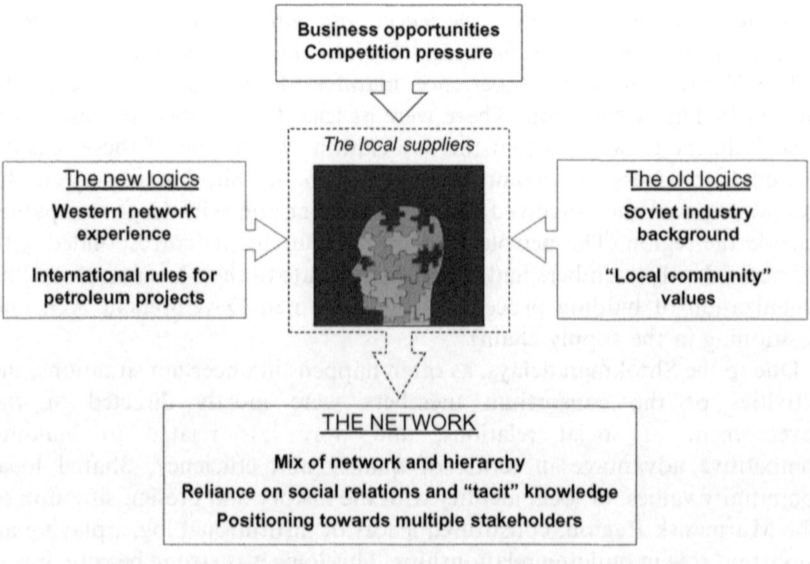

Fig. 1. Building of the Network as Institutional Interplay.

and its control elements. The consortium controls are characterized by a mix of network and hierarchy management forms, reliance on social relationships and 'tacit' knowledge, and strategic positioning towards multiple stakeholders.

The second and third research questions deal with the internal challenges for management control and the way these challenges were handled. Control elements in the network are summarized in Fig. 2. The controls at both network and member levels can be classified as either formal or social controls (Dekker, 2004). First, the formal control elements, collectively accepted by the members, in practice were extended by social ones (arrow 1 in Fig. 2). The development of the individual controls (safeguards) received a strong impetus as soon as the partners intensified their practical work in preparing the tenders. Some members still felt unsure about their own benefits, as they faced internal challenges associated with relative distrust between potential buyers and suppliers, internal competition and the fear of unreliable partners. Thus, contradictions between individual and collective goals in the network became more obvious. This supports Edwards and Jones (2008) who suggest that practical implementation of new ideas, in other words translation from words and plans to real actions (Czarniawska &

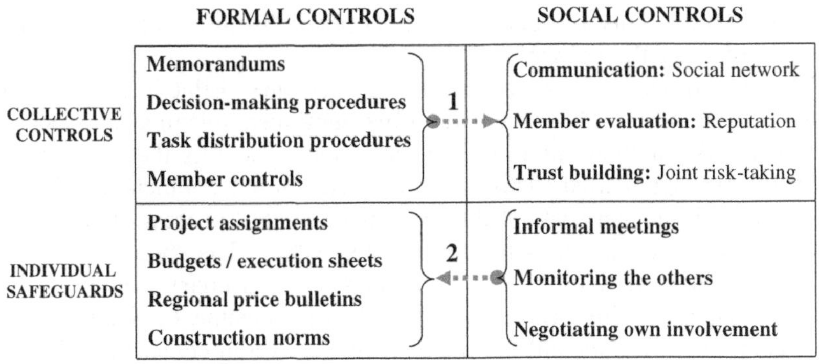

Fig. 2. Development of Management Controls in The Consortium Network.

Sevón, 1996) is one of the most difficult steps in cooperation, exposed to tensions and contradictions.

Using 'the safeguards' started with implementation of rather informal means of control, but later on it was extended by mobilization of formal control tools (arrow 2 in Fig. 2). These formal tools, borrowed by members from their ordinary businesses, were used as a sort of device for building arguments and carrying out negotiations. Interestingly, while the collective controls evolved from formal to social controls, the individual safeguards have moved in the opposite direction.

Management Control and Practice Variation

Focusing on both institutional prerequisites and further development of controls gives an opportunity to trace the origin of practice variations in the network. The first reason for practice variation is that controls demarcated the network, placing different groups of members in different conditions. For example, the above discussed hierarchical management forms, originating from Soviet industry experience, were supported by collectively accepted controls (the memorandum, the decision-making and the task distribution procedures). This led in practice to network fragmentation and the origin of two groups with different rights and opportunities – the board companies and the ordinary members. The member controls also defined a condition for practice variation. The intention of creating an atmosphere of trust and transparency and the reliance on strong social relationships resulted in the development of an entry barrier coming into place in front of the consortium. This automatically limits possibilities for other companies wishing to enter the network.

The second reason for practice variation becomes visible in the course of using the controls. Then the two aforementioned dilemmas, associated with network fragmentation and the high 'cost' of joining the network, are then further aggravated. This leads to institutional conflicts within and even outside the network. Table 3 illustrates this process. Let us first consider the decision-making procedure, which has two levels, and prioritizes the rights of board companies over ordinary members. For board companies, this procedure resulted in a dilemma between securing own company interests and the interests of the overall consortium. Taking advantage of their position they tended to avoid competing with one another.

For the ordinary members, this situation was also associated with a dilemma consisting of taking responsibility on the one hand, and transferring it to the board on the other hand. While some of them found ways to negotiate and influence strategic decisions, most preferred to transfer decision-making responsibility to the board. After a while, this power imbalance, being actually supported by both sides, has increased as the avoidance of competition at the level of board companies resulted in intensified competition at the level of ordinary members. This stipulated practice variation is seen as a mix of cartel- and market-based governance modes within the same network. This, in turn, underlined relative distrust and resulted in decoupling between collective and individual controls. Thus, a new kind of institutional conflict emerged – a conflict between the individual and

Table 3. Development of Control Dilemmas and Institutional Conflicts.

Key Controls	Activated Dilemmas	Outcomes	Institutional Conflicts
Two-level decision-making: Board and the others	*For board members*: How much advantage should we take? *For ordinary members*: taking responsibility or transferring it to the board	Mix of cartel- and market-based governance	Individual safeguards vs Collective controls
Member controls: Formal and social ones	Growing cost of joining the network	Differentiated political and informational support	Private enterprise vs regional network

the collective goals of the members. Paradoxically, interactions in the network became trust-based and distrust-driven at the same time.

The next dilemma arising concerned the growing cost of joining the network. The informal, reputation-based controls turned out to be even stronger than the formal ones related to membership fees and the memorandum. While the formal cost of joining the consortium was not so great, the informal cost, actually amounting to the cost of gaining trust, proved much tougher. This resulted in varying political and informational support from the association and regional authorities. Although the consortium was used as an important example of development of the regional supply chain, many regional enterprises were not actually included into it. In the worst case, further growing informal cost may even lead to imbalance in the development of the regional construction industry as a whole, and therefore to insufficient fulfillment of its potential. This creates institutional conflict between private and public interests in the geographic area – will The Consortium develop as a private enterprise or as a regional network?

Further aggravation of the outlined control dilemmas may cause even stronger variations because more pressure will be put on local industry. Being exposed to dilemmas, local construction companies have to make individual choices. Thus, the growing power exercised by the board may lead to too tough competition amongst the ordinary members. Too much reliance on personal relationships and professional background will close the network gates for younger businessmen, and they will have to consider alternative strategies for gaining access to new projects.

This study shows that by focusing on controls we can better understand the conditions for practice variations inside and also outside the network. The collectively designed controls, being influenced by various institutional logics, purely define such conditions. This fits in well with previous studies carried out concerning the role of institutions in the development of management control (e.g. Modell, 2002, 2005; Modell, Jacobs, & Wiesel, 2007; Wiesel, 2008). However, the collective controls are not only definers, but also activators of new variations. Their use over time leads to aggravation of dilemmas, development of individual controls and, finally, new institutional conflicts. This is consistent with practice-oriented studies (e.g. Mouritsen & Larsen, 2005; Mouritsen & Thrane, 2005; Mouritsen et al., 2009) arguing that the use of management controls by people leads to indirect effects and changes in organizational behaviour. Bridging these two streams of the literature, based on institutional and process-oriented analysis of control practices, this paper establishes a dyadic view with regard to management control, arguing that it both defines and creates conditions for practice variations.

CONCLUSIONS

The presented findings show that energetic institutional interplay already takes place at an early stage of the network-building process. This happens due to several institutional logics clashing at the same time when the new cooperation idea materializes in practice. These logics influenced the way the network members translated this idea and developed their organizational arrangements and corresponding controls.

Following up these controls enabled deeper understanding of the network. On the one hand, the controls, being influenced by the discussed logics, do place organizational demarcations on the members' behaviour. On the other hand, the use of controls leads to unintended consequences and the aggravation of control dilemmas. When facing such dilemmas the network members address them differently. Therefore, behaviour differences are created. Based on these findings, as a theoretical contribution, this paper offers insight into one of the problems of the institutional theory – the origin of practice variation among organizations within the same institutional context (Lounsbury, 2008; Lounsbury & Crumley, 2007). The paper establishes a dyadic view of management control, arguing that it both defines and creates conditions for practice variation.

In the studied network, collective controls were combined with individual ones – the members' safeguards. The study leaves, however, for further research on the feedback loop between them. Do the individual safeguards represent resistance to top-down pressure, and are they therefore directed at disengaging from collective controls and growing distrust? Or, alternatively, could they be implemented to repair things (Ahrens & Chapman, 2004), and therefore make the existing collective control system simultaneously more efficient and trust-based? Probably a detailed study of behaviour of the network members over a longer period of time will be needed to shed more light on these questions.

The paper argues that paying attention to emerging networks would seem important in that this allows for understanding of their underlying institutional logics. At the same time, this is a relatively under-researched presented topic in management control literature which mostly focuses on developed networks. In this particular study, one does run, however, the risk of considering incompletely formed social practices as already having been institutionalized. In this respect, longitudinal studies of networks can again prove valuable. Focusing on changes in control practices over time then could help towards gaining a deeper understanding of and theorizing about parts of a life cycle in network life. In this respect, research can benefit from comparative studies of various networks.

NOTES

1. Shtokman Development AG (SDAG) is operator of the Shtokman Gas field in the Russian part of the Barents Sea. SDAG is owned by Gazprom (51%), Total (25%) and Statoil (24%).
2. After merger of Statoil and the oil and gas division of Norsk Hydro in 2007, StatoilHydro became the name of the joint company until the fall of 2009. The name was then changed to Statoil (source: www.statoil.com).

REFERENCES

Ahrens, T., & Chapman, C. S. (2004). Accounting for flexibility and efficiency: A field study of management control systems in a restaurant chain. *Contemporary Accounting Research, 21*(2), 271–301. doi:10.1506/VJR6-RP75-7GUX-XH0X

Ahrens, T., & Chapman, C. S. (2007). Management accounting as practice. *Accounting, Organizations and Society, 32*(1–2), 1–27. doi:10.1016/j.aos.2006.09.013

Alvesson, M. (2003). Beyond neopositivists, romantics, and localists: A reflexive approach to interviews in organizational research. *Academy of Management Review, 28*(1), 13–33. doi:10.2307/30040687

Alvesson, M., & Skoldberg, K. (2000). *Reflexive methodology – New vistas for qualitative research.* London: Sage Publications.

Castells, M. (2000). *The rise of the network society.* Oxford, UK: Blackwell Publishers.

Czarniawska, B. (2001). Is it possible to be a constructionist consultant? *Management Learning, 32*(2), 253–266. doi:10.1177/1350507601322006

Czarniawska, B. (2008). *A theory of organizing.* Cheltenham, UK: Edward Elgar Publishing Ltd.

Czarniawska, B., & Joerges, B. (1996). Travels of ideas. In B. Czarniawska & G. Sevón (Eds.), *Translating organizational change* (pp. 13–48). Berlin: De Gruyter.

Czarniawska, B., & Sevón, G. (1996). *Translating organizational change.* Berlin: De Gruyter.

Dekker, H. C. (2003). Value chain analysis in interfirm relationships: A field study. *Management Accounting Research, 14*(1), 1–23.

Dekker, H. C. (2004). Control of inter-organizational relationships: Evidence on appropriation concerns and coordination requirements. *Accounting, Organizations and Society, 29*, 27–49.

Dekker, H. C. (2008). Partner selection and governance design in interfirm relationships. *Accounting, Organizations and Society, 33*(7–8), 915–941.

DiMaggio, P. J., & Powell, W. W. (1983). The iron cage revisited: Institutional isomorphism and collective rationality in organizational fields. *American Sociological Review, 48*(2), 147–160.

Edwards, T., & Jones, O. (2008). Failed institution building: Understanding the interplay between agency, social skill and context. *Scandinavian Journal of Management*, *24*(1), 44–54.

Gendron, Y., Cooper, D., & Townley, B. (2001). In the name of accountability – State auditing, independence and new public management. *Accounting, Auditing & Accountability Journal*, *14*(3), 278–310.

Håkansson, H., & Lind, J. (2004). Accounting and network coordination. *Accounting, Organizations and Society*, *29*(1), 51–72.

Håkansson, H., & Lind, J. (2007). Accounting in an interorganizational setting. In C. S. Chapman, A. G. Hopwood & M. D. Shields (Eds.), *Handbook of management accounting research* (Vol. 2, pp. 885–902). Oxford, UK: Elsevier. doi:10.1016/S1751-3243(06)02017-7

Hansen, A., & Mouritsen, J. (2005). Strategies and organisational problems: Constructing corporate value and coherence in balanced scorecard processes. In C. S. Chapman (Ed.), *Controlling strategy: Management, accounting, and performance measurement* (pp. 125–150). Oxford, UK: Oxford University Press.

Kvale, S. (1997). *Interviews: An introduction to qualitative research interviewing*. Oslo: Ad Notam Gyldendal.

Latour, B. (1986). The powers of association. In J. Law (Ed.), *Power, action and belief* (pp. 261–277). London: Routlege and Kegan Paul.

Latour, B. (1999). *Pandora's hope: Essays on the reality of science studies*. Cambridge, MA: Harvard University Press.

Lounsbury, M. (2008). Institutional rationality and practice variation: New directions in the institutional analysis of practice. *Accounting, Organizations and Society*, *33*(4–5), 349–361. doi:10.1016/j.aos.2007.04.001

Lounsbury, M., & Crumley, E. T. (2007). New practice creation: An institutional perspective on innovation. *Organization Studies*, *28*(7), 993–1012. doi:10.1177/0170840607078111

Modell, S. (2002). Institutional perspectives on cost allocations: Integration and extension. *European Accounting Review*, *11*(4), 653–679. doi:10.1080/0963818022000000993

Modell, S. (2005). Students as consumers? An institutional field-level analysis of the construction of performance measurement practices. *Accounting, Auditing & Accountability Journal*, *18*(4), 537–564. doi:10.1108/09513570510609351

Modell, S., Jacobs, K., & Wiesel, F. (2007). A process (re)turn? Path dependencies, institutions, and performance measurement in Swedish central Government. *Management Accounting Research*, *18*(4), 453–475.

Mouritsen, J. (2005). Beyond accounting change: Design and mobilisation of management control systems. *Journal of Accounting & Organizational Change*, *1*(1), 97–113. doi:10.1108/18325910510635317

Mouritsen, J., Hansen, A., & Hansen, C.Ø. (2009). Short and long translations: Management accounting calculations and innovation management. *Accounting Organizations and Society*, *34*(6–7), 738–754.

Mouritsen, J., & Larsen, H. T. (2005). The 2nd wave of knowledge management: The management control of knowledge resources through intellectual capital information. *Management Accounting Research*, *16*, 371–394.

Mouritsen, J., & Thrane, S. (2005). Accounting, network complementarities and the development of inter-organisational relations. *Accounting, Organizations and Society*, *31*, 241–275.

Otley, D. (1995). Management control, organizational design and accounting information systems. In D. Ashton, T. Hopper & R. W. Scapens (Eds.), *Issues in management accounting*. London: Prentice-Hall Europe.
Otley, D. (1999). Performance management: A framework for management control systems. *Management Accounting Research*, *10*, 363–382.
Ouchi, W. (1979). A conceptual framework for the design of organizational control mechanisms. *Management Science*, *25*, 833–848.
Richardson, L. (2000). Evaluating ethnography. *Qualitative Inquiry*, *6*(2), 253–255.
Sahlin-Andersson, K. (1996). Imitating by editing success: The construction of organizational fields. In B. Czarniawska & G. Sevón (Eds.), *Translating organizational change*. Berlin: De Gruyter.
Scott, W. R. (2008). *Institutions and organizations* (3rd ed.). London: Sage Publications.
Seal, W., Cullen, J., Dunlop, A., Berry, T., & Ahmed, M. (1999). Enacting a European supply chain: A case study of the role of management accounting. *Management Accounting Research*, *10*(3), 303–322. doi:10.1006/mare.1999.0105
Thornton, P. H. (2004). *Markets from culture: Institutional logics and organizational decisions in higher education publishing*. Stanford, CA: Stanford University Press.
Tomkins, C. (2001). Interdependencies, trust and information in relationships, alliances and networks. *Accounting, Organizations and Society*, *26*(2), 161–191. doi:10.1016/S0361-3682(00)00018-0
Wiesel, F. (2008). Kundorientering och ekonomistyrning i offentlig sector. Doktoravhandling i företagsokonomi vid Stockholms universitet 2008. Stockholm University, School of Business.

THE ROLE OF MANAGEMENT CONTROL SYSTEMS ON INTER-ORGANISATIONAL EFFICIENCY: AN ANALYSIS OF EXPORT PERFORMANCE

Raquel Florez, Juan M. Ramon, Maria Velez, Maria Concepcion Alvarez-Dardet, Pedro Araujo and Jose M. Sanchez

ABSTRACT

Purpose – *Resource-based literature argues that firms' export performances are influenced by a proper combination of their own resources and capabilities, allowing for sustainable competitive advantages. Because export activities are usually based on relationships between firms and foreign intermediaries, the adequate management of inter-organisational activities should also be analysed as a key determinant of export performance.*

Originality – *Our research adds to the existing literature by examining the role that management control systems (MCS) play in exploiting firms' exporting capabilities and resources to optimise export performance.*

Performance Measurement and Management Control: Global Issues
Studies in Managerial and Financial Accounting, Volume 25, 195–222

ISSN: 1479-3512/doi:10.1108/S1479-3512(2012)0000025011

Methodology – *Utilising empirical data from Spanish exporters, an initial analysis of export efficiency was performed based on DEA and segmentation techniques. From efficiency scores, we tested causal relationships between MCS design and use and the optimisation of resources and capabilities when performing export activities.*

Findings and practical implications – *The main conclusions are that any type of control system was found to have a positive influence in export performance, but only social control establishes a difference in terms of efficiency. The results show that strong social controls increase firms' efficiencies when managing export channels, allowing firms to achieve outputs superior to competitors with similar resources and capabilities. In addition, an efficiency 'lifecycle' was identified for relationships between exporters and intermediaries. A pro-cyclical effect was found among MCS design, use, and export efficiency; intense MCS were established in the first stages of a relationship but were relaxed after a certain level of efficiency was achieved, leading to a reduction of efficiency in the long term that compromised the continuity of relationships.*

INTRODUCTION

A growing number of firms maintain collaborative relations with foreign intermediaries to access foreign markets (Leonidou, Katsikeas, & Samiee, 2002). However, in spite of their importance, many export ventures perform poorly over time. Research efforts have focused on the identification of the factors that influence export performance (Zou & Stan, 1998). Recently, Kaleka (2002) approached the issue from a resource-based view (RBV) and presented a comprehensive classification framework of sources, resources and capabilities in terms of competitive advantage in export markets.

However, business strategists have recognised that export performance may not be based solely on the acquisition of resources. Because the success of export agreements is negatively affected by possible incongruences between partners' objectives due to the separation of ownership (Gordon, 1993), the inadequate management of relationships with overseas intermediaries has been identified as a key contributor to poor export performance (Styles & Ambler, 2000). In an increasingly competitive global marketplace, the manufacturer's ability to manage inter-firm relationships with foreign intermediaries is a major determinant of export success

(Aulakh, Kotabe, & Sahay, 1996; Cavusgil, Deligonul, & Zhang, 2004). In the broad sense, management control systems (MCS) appear to be a source of performance (Kimura & Mourdoukoutas, 2000).

The aim of this research is to explore the role that MCS play in improving export performance from a resource-based view. This preliminary study presents an improved approach to evaluating the performance of exporting firms that manage externalised export channels and is interesting for several reasons. First, departing from Kaleka (2002), survey data from a questionnaire of Spanish exporters is used to define inputs and outputs. We then analyse the efficiency of exporting firms through a DEA model to obtain non-parametric scores of performance efficiency. Previous scores are used to check whether the design and use of specific MCS impacts the realised performance and efficiency considering the characteristics of the exporting firms. Finally, a 'lifecycle' of the exporter–intermediary relationship is identified.

The remainder of the work is organised as follows. The second section focuses on export performance from the resource-based perspective. The third section argues for the role of control systems in the setting of independent export channel management and considers the types of control system and dimensions of utilisation. In the fourth section, methodological issues are considered and efficiency is measured through DEA. Finally, empirical results are discussed, and the main preliminary conclusions are presented.

EXPORT PERFORMANCE: A RESOURCE-BASED VIEW PERSPECTIVE

The literature has underlined the need for a better understanding of export performance (Leonidas & Katsikeas, 1996; Zou, Fang, & Zhao, 2003; Zou & Stan, 1998), and existing knowledge is far from perfect. Two main problems can be highlighted in the literature: the lack of a uniform metric of export performance (Cavusgil & Zou, 1994) and inconsistent findings regarding the determinants of export performance (Fraering, 1996).

Export performance is broadly defined as the outcome of a firm's activities in export markets (Shoham, 1998) and is of central interest in the international management literature. However, there is currently no universal measure of export performance (Cavusgil & Zou, 1994; Leonidou et al., 2002). As a result, alternative operationalisations make comparing the results of different studies difficult, and more empirical research is needed to capture the multi-dimensionality of export performance (Zou, Taylor, & Osland, 1998).

Conceptual definitions of export performance include export effectiveness, efficiency, and continuous engagement in exporting (Shoham, 1991). Measures of export performance have included a myriad of indicators (Table 1), and export intensity, export sales growth, export profit level, export sales volume, export market share, and export profit contribution have been the most frequently used measures of export performance (Dhanaraj & Beamish, 2003; Leonidou et al., 2002; Zou et al., 1998). Ling-Yee and Ogunmokun (2001) add that management's perceptions of export advantages are an important determinant of export performance.

Because individual measures are insufficient to capture the rich complexity of the export performance construct, multi-dimensional measures emerge as useful alternatives. Following this view, Zou et al. (1998) developed a broad export performance measure, the EXPERF scale, which integrates the major perspectives presented in previous studies. This scale is considered to be one of the most comprehensive export performance scales, and it gathers insights from the literature in strategy, management, marketing, and exporting. The EXPERF scale uses three dimensions to measure export performance:

- *Financial performance* is the common approach to performance based on financial measures. Prior research has assessed financial export

Table 1. Export Performance Measures Used in Previous Research.

Export intensity variability	Change in return on assets	Perception of dynamic success
Change in export intensity	Return on investment	Perception of static success
Total export sales	Change in return of investment	Entering new difficult markets
Export intensity relative to industry	Gross export profits	Number of new export markets
Change in export sales	Export gross profit margin	Number of export markets
Export market share	Export sales operating profit margin	Six years' export survival
Change in market share	Variability of export sales ratio	Perceived sales relative to industry
Satisfaction with change in market share	Change in net profits	
Six years' export survival	Satisfaction with change in net profits	

Source: Adapted from Shoham (1991, 1998).

performance through indicators such as export sales, export sales growth, export profits and export intensity (Zou et al., 1998).

- *Strategic performance* considers the strategic goals that firms aim to meet through exporting, including improved competitiveness, increased market share or strengthened strategic position (Cavusgil & Zou, 1994). Prior research has measured strategic performance through indicators such as market share, strategic presence and competitive position (Zou et al., 1998).
- *Perceived performance* considers perceptual and attitudinal measures of export performance because being satisfied with exporting operations is a strong indicator of export success. Prior studies have measured the perceived performance either directly, through perceived export success and satisfaction, or indirectly, by considering the firm's attitudinal changes towards exporting and towards barriers to exporting (Zou et al., 1998).

Finally, all of these dimensions can be aggregated into a global measure of performance to overcome the difficulties of performance measurement (Katsikeas, Leonidou, & Morgan, 2000).

Three major theoretical approaches to studying the determinants of export performance have emerged (Zou et al., 2003): (i) industrial organisation theory, which ascribes the firm's export performance to its external market position; (ii) transaction cost analysis, which uses market imperfections to explain a firm's international strategy choice and efficiency in international marketing; and (iii) the RBV, which focuses on internal resources to identify the determinants of a firm's export performance. It has been suggested that including the RBV in studies is a fruitful direction that would enrich the export management literature (Zou & Stan, 1998). The contribution of the RBV to strategic management is its emphasis on firms' specific resources as the real source of sustainable competitive advantage and above-average performance. This approach is important to managers involved in strategy formulation, as it examines factors, elements and processes that they control, change or manage (Kaleka, 2002). From the RBV, exporting firms consist of assortments of assets and individuals that possess skills. Not all resources have the potential to lead to competitive advantage, and those that do must have four attributes: they must (1) be valuable, (2) be rare, (3) be difficult to imitate, and (4) have no strategically equivalent substitutes (Barney, 1991). The synergistic effects generated by combinations of these factors are more important than the simple accumulation of these attributes in establishing competitive advantage (Amit & Schoemaker, 1993). Thus, firms are idiosyncratic in terms of the bundle of resources that they accumulate over time, and organisational resources are considered to be the ultimate sources of competitive advantage. Departing from these arguments, Kaleka (2002) has identified two related classes of sources that are necessary to create competitive advantages in export markets: resources and capabilities.

Resources represent assets controlled by the exporting firm that are used as inputs to export processes. Kaleka (2002) identifies four types of competitive resources for exporters: (a) physical assets (e.g. the use of modern technology, preferential access to supply sources, geographical proximity to the market, and production capacity availability); (b) the scale of operation (e.g. annual turnover, full-time staff and the percentage of employees engaged in exporting); (c) financial assets (e.g. the available funds to be devoted to export ventures); and (d) the firm's experience in export market operations (e.g. the length of time exporting, the number of on-going export ventures, export market knowledge and past performance).

Capabilities concern the firm's ability to combine, develop and use its resources to create competitive advantage. Capabilities are a firm's complex bundles of skills and accumulated knowledge exercised through organisational processes that enable the firm to coordinate activities and make the best use of its resources. Kaleka (2002) identifies four main capabilities relevant to export markets: (a) informational (e.g. the acquisition of market information, the ability to make contacts, the identification of prospective customers overseas, the wareness of competing products and the ability to capture foreign market information); (b) customer relationship building (e.g. the exporter's ability to understand the requirements of overseas customers and establish and maintain close export business relationships); (c) product development (e.g. the development of new products, the improvement and modification of existing products, and the adoption of new methods and ideas in the production process); and (d) supply chain (e.g. the identification of attractive sources of supply and the establishment, development and maintenance of strong supplier relationships).

MANAGEMENT CONTROL SYSTEMS: EFFECTS ON PERFORMANCE AND EFFICIENCY

Performance can suffer in export channels as the separate ownership and profit claims of the exporter and intermediary often lead to arms-length trading arrangements (Cavusgil & Zou, 1994; Gordon, 1993). As a result, relationships with intermediaries are hard to coordinate, and high performance is difficult to achieve (Rosson & Ford, 1982; Styles & Ambler, 2000). Geographical distances and cultural differences also increase the cost of monitoring and coordinating geographically dispersed operations (Aulakh & Gençtürk, 2000). To face these difficulties, different non-market forms of governance can be employed to effectively organise and regulate exchange conduct (Bello & Gilliland, 1997; Gundlach, 1994).

MCS have been recognised as an important aspect of managing inter-firm relationships (Dekker, 2004; Vélez, Sánchez, & Álvarez-Dardet, 2008) and as a source of firm performance: an effective MCS allows a company to improve operational effectiveness, employee creativity, and company competitiveness (Kimura & Mourdoukoutas, 2000). Although control considerations are important in any organisation, they are especially relevant to managing international exchanges (Aulakh & Gençtürk, 2000). In an exporter–foreign intermediary relationship, MCS are designed and used both to create conditions that motivate the intermediary to achieve predetermined outcomes and to contribute to the definition and implementation of the strategy (Pahud de Mortanges & Vossen, 1999; Sachdev, Bello, & Pilling, 1994). In this context, the term MCS refers to specific, exporter-initiated directives, policies and procedures aimed to influence foreign intermediaries to perform marketing actions in ways that support the exporter's objectives and strategies (Bello & Gilliland, 1997; Coletti, Sedatole, & Towry, 2005).

However, while the literature has focused on understanding the control exercised as well as its antecedents and consequences, it provides very little guidance about the real effectiveness of MCS in exchange relationships. According to the RBV, we consider MCS to be collections of several mechanisms that orient the combinations of physical and organisational resources (Baraldi & Strösten, 2009). The implementation and use of MCS can help export managers to identify interdependencies between the resources and capabilities assigned to the exporter–intermediary relation and to exploit them. To create value, partners have to pool resources, determine tasks to be performed, and decide on a division of labour, which demands coordination efforts, a function that can be performed by MCS. However, MCS are not necessarily desirable per se but are subject to efficiency considerations and deliberate choice (Aulakh & Gençtürk, 2000): when export managers implement and use MCS to exercise control over foreign intermediaries, they do so because the intermediaries are instrumental in achieving their objectives. Therefore, the performance consequences of MCS should be focused on the positive influences of MCS on exporters' performance efficiencies.

Consequently, the challenge facing export managers is to implement and use MCS that induce the desired performance while inhibiting dysfunctional behaviours within the particular governance structure of a cross-border exchange relationship (Aulakh & Gençtürk, 2000). The efficient management of interdependent foreign activities necessitates sophisticated MCS that can respond to local differences while taking advantage of global opportunities (Gençtürk & Aulakh, 1995). To this end, managers need to address two fundamental questions when choosing an MCS: (1) what type of control should be implemented and (2) what use of control should be exercised.

Export managers can influence intermediaries' behaviour through different types of control; these have been identified in the literature (Aulakh & Genç türk, 2000; Solberg, 2006) as (i) output control, (ii) process control and (iii) social control. Output and process controls represent formal MCS, while social control is an informal control design. *Output control* refers to the degree to which the exporter monitors the results or outcomes produced by intermediaries. Output control is achieved in three steps: defining the dimensions and standards of performance, measuring the performance and comparing with targets and rewarding results that reach these goals. *Process control* refers to the extent to which the exporter monitors the intermediaries' behaviours or means used to achieve the desired outputs. Process control ensures that desirable actions are performed while also preventing undesirable actions. *Social control* represents an informal control mechanism based on prevailing social perspectives and patterns of interpersonal interactions.

Differences in the focus of MCS designs may infer different but positive effects on exporters' efficiencies, which motives the first set of research hypotheses:

H_{11}. Efficiency in transforming resources and capabilities into export performance is stronger when output controls are developed.

H_{12}. Efficiency in transforming resources and capabilities into export performance is stronger when process controls are developed.

H_{13}. Efficiency in transforming resources and capabilities into export performance is stronger when social controls are developed.

The management control literature also suggests that the effectiveness of MCS is dependent on the utilisation of the information that they generate (e.g. Bisbe & Otley, 2004). Consequently, the three MCS designs can support different uses with different focuses in the management of the inter-firm relationships. A well-established taxonomy argues that control systems serve two main purposes: (i) diagnostic use and (ii) interactive use (Simons, 1995). This typology has achieved great relevance in the intra-organisational arena (Bisbe & Otley, 2004; Tuomela, 2005; Vaivio, 2004; Widener, 2007), and we propose that it is especially pertinent in the context of export firm–foreign intermediary relations due to strategic uncertainties arising from both physical and cultural distances.

Diagnostic use implies the traditional role of monitoring performance and correcting deviations from target performance. Diagnostic use assumes the establishment and negotiation of goals and objectives, the alignment of performance measures with strategic goals and priorities, the establishment of incentives based on performance, and the monitoring of significant deviations or exceptions. *Interactive use* represents a more proactive style of management than diagnostic use and is based on monitoring strategic uncertainties and sharing existing MCS information between export managers and intermediaries (Simons, 1995). Likewise, interactive use implies identifying key strategic areas, constructively discussing data with intermediaries, challenging or debating data, and establishing assumptions and plans of action (Ferreira & Otley, 2006). Interactive use encourages learning regarding strategic uncertainty, uncovering the need for strategic changes for the entire organisation (Tuomela, 2005), and acting as an active element of strategy formulation, emergence and implantation (Simons, 1990).

Because MCS design and use are two sides of the same coin, to examine one dimension while omitting the other can result in an unnecessarily limited perspective of the operation of the MCS. Thus, differences in MCS use might have additional effects on exporters' efficiencies, leading to a second set of hypotheses:

H_{21}. The positive effect of MCS designs on export performance efficiency is stronger when MCS are used in a diagnostic way.

H_{22}. The positive effect of MCS designs on export performance efficiency is stronger when MCS are used in an interactive way.

MEASURING EFFICIENCY: THE DATA ENVELOPMENT ANALYSIS (DEA)

Efficiency is a significant topic in business research that compares disposable inputs and produced outputs (Lovell & Knox, 1993): a decision making unit (DMU) is said to be efficient if it is able to produce maximal outputs given a certain level of inputs (output orientation) or if it uses minimal inputs to produce specific outputs (input orientation). Different parametric and non-parametric approaches have been proposed to measure the efficiencies of DMUs, and Data Envelopment Analysis (DEA) is one of the most useful in practice.

DEA is a non-parametric model that evaluates the efficiencies of comparable DMUs in transforming inputs into outputs (Charnes, Cooper, Lewin, & Seiford, 1994). DEA is a frontier production method that separates efficient units (B, E, F, G in Fig. 1) and inefficient units (A, C, D). An efficiency score that indicates distance to the frontier (e.g. A–A′ in Fig. 1) is obtained for each DMU: efficient units receive scores of 1, while inefficient units receive scores between 0 and 1. Each inefficient DMU is also compared to an efficient set of reference DMUs to identify weak points that can be improved.

The first DEA model was proposed by Charnes, Cooper, and Rhodes (1978) (CCR model). Let Y_{rk} be the amount of output r produced by DMU 'k', and let X_{ik} represent the amount of input i used by k. DEA estimates the best positive set of input multipliers $\{v_{ik}\}_{i=1}^{I}$ and output multipliers $\{u_{rk}\}_{r=1}^{R}$

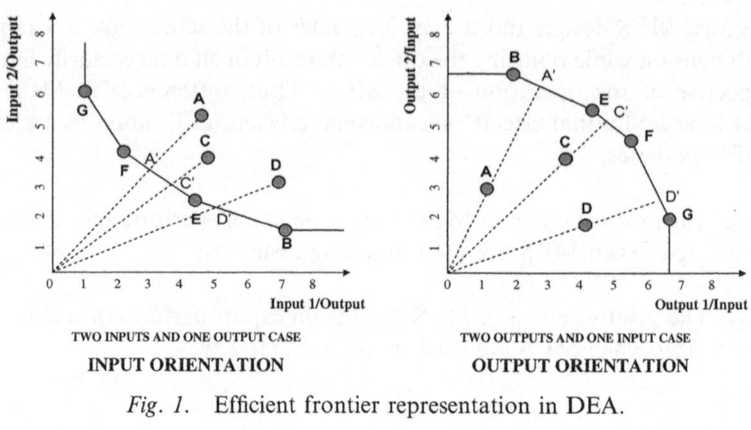

Fig. 1. Efficient frontier representation in DEA.

such that the ratio of the total weighted outputs to the total weighted inputs is maximised but does not exceed 1 (100%),

$$e_k = \max \frac{\sum_r u_{rk} Y_{rk}}{\sum_i v_{ik} X_{ik}}, \quad \text{s.t.} \quad \frac{\sum_r u_{rk} Y_{rk}}{\sum_i v_{ik} X_{ik}} \leq 1 \quad \text{for all} \quad k \tag{1}$$

$$u_{rk} \geq \varepsilon, v_{ik} \geq \varepsilon, \quad \text{for all} \quad r, i$$

This fractional problem can be reduced to a linear programming problem (vector notation),

$$\begin{aligned} & e_k = \max \mu Y_k \\ & \text{s.t.} \;\; vX_k = 1 \\ & \mu Y_k - vX_k \leq 0, \quad \text{for all} \quad k \\ & \mu_r \geq \varepsilon, v_i \geq \varepsilon, \quad \text{for all} \quad r, i \end{aligned} \tag{2}$$

The CCR model assumes constant returns to scale (CRS), so that if (X, Y) is feasible, then (tX, tY) is also feasible $(t>0)$. Alternatively, Banker, Charnes, and Cooper (1984) (BCC model) considered variable returns to scale (VRS), but the CCR model is usually assumed in practice because it avoids elements that are external to managers' control and produces comprehensible and robust results in the presence of scarce datasets. An interesting variant was introduced by Anderson and Petersen (1993), who produced a ranking among efficient units by comparing each efficient DMU to a linear combination of the rest of the efficient units (so that the efficiency scores ≥ 1).

Because DEA focuses on non-parametric frontiers rather than central tendencies, it is particularly useful for discovering relationships that remain hidden under linear hypotheses such as those used by regression models and structural equations. The main limitation of DEA is that it does not provide specific measures of model fit in terms of the parameters defining the model.

EMPIRICAL ANALYSIS

Research Setting and Data Collection

The previously defined hypotheses were tested through a cross-sectional sample of Spanish exporters (primary, manufacturing and service industries). The sample was limited to regular exporters who made on-going use of externalised export channels. An online questionnaire was administered to export managers. Differences between managerial levels were controlled by delivering the questionnaire to mid-level managers, and the questions focused on an intermediary of secondary importance to avoid bias towards the most successful relationship (Anderson & Narus, 1990; Bello & Gilliland, 1997; Skarmeas, 2006; Zhang, Cavusgil, & Roath, 2003). The answers referred to a specific product exported to a particular market (Lages, Lages, & Lages, 2005) to favour consistency in the answers by keeping respondents focused on the export relationship and related management systems (Cavusgil & Zou, 1994).

Dillman's (2000) procedure of using two methodological stages was followed to guarantee data quality: (i) a literature review, interviews with stakeholders (six export managers, one foreign employee from a trade office, EXTENDA[1] and one international marketing consultant), and expert forums (two export management meetings); and (ii) survey design, pre-test (eight researchers, three EXTENDA employees and six export managers) and survey administration (online platform and confidential questionnaires). The aim of the questionnaire was to assess the resources and capabilities used in the export venture and the realised export performance. The questionnaire also enquired about the designs and uses of MCS developed to manage intermediary relationships. Supplementary control data were also included (e.g. gender, experience and industry).

An initial population of 2,355 exporting firms was identified, and three successive filters were applied to this set[2]: (i) they must be active and localised at the time of the survey (302 deleted firms); (ii) they must have regular exporting activities (657 deleted firms); and (iii) they must use external export channels on a regular basis (740 deleted firms). The first criterion prevents time-bias, while the second and third filters guaranteed that only sufficiently experienced firms were included to ensure that MCS were implemented. Only 656 companies passed all filters, 224 of which declined to participate, so the target sample included 432 firms. Export managers received an e-mail that included a specific password to access the online survey, which was administered over five weeks, as well as several reminder calls and an incentive of €5 charitable donation. Finally, 193 usable questionnaires were returned, for a response rate of 29.42% (over population), which is similar to response rates in the existing literature (Aulakh & Gençtürk, 2000; Van der Stede, Young, & Chen, 2005). Respondents were primarily male (78.57%), had an average of 9.76 years of experience in exporting, had worked for their current employer for an average of 8.32 years and had been employed as export managers in the current firm for an average of 6.92 years.

Variable Measurement

Resources and Capabilities

Resources and capabilities were measured using Kaleka (2002) instrument. A list of 11 items was defined to capture competitive resources (experiential resources, the scale of operation, financial resources, and physical resources). A list of 9 items was employed to capture export capabilities[3] (informational, customer relationship building, and product development capabilities). Export managers were asked to rank the position of the firm compared to its primary direct competitors using a 7-point Likert scale ranging from '1 – much worse' to '7 – much better'.

Export Performance

The EXPERF scale (Zou et al., 1998) was used to define alternative performance dimensions: financial export performance (3 items), strategic export performance (3 items), and perceived export success (3 items). A 7-point Likert scale was utilised to define respondent's statements, which ranged from '1 – I disagree very much' to '7 – I agree very much'.

MCS Design and Use

The definitions of control types were based on items adapted from Aulakh and Gençtürk (2000): output control (five items), process control (four items) and social control (five items). Export managers were asked to assess their agreement on a 7-point Likert scale ranging from '1 – I disagree very much' to '7 – I agree very much'.

MCS uses were defined by considering the traditional distinction between diagnostic use (five items) and interactive use (four items) (Ferreira & Otley, 2006; Simons, 1990), adapted to the inter-organisational context. Responses were measured on a 7-point Likert scale that ranged from '1 – I disagree very much' to '7 – I agree very much'.

Control Variables

Several variables were introduced to control specification error and model bias. From a review of the literature on exporters' performance (Leonidou et al., 2002; Sousa, 2004; Zou & Stan, 1998), three groups of variables were identified as the antecedents of export performance: managerial issues (characteristics of managers), organisational issues (characteristics of the exporting firm) and environmental issues (macro factors). From these groups, five control variables were defined: (i) *manager export experience* (years), which has been positively associated to management's international orientation (Zou & Stan, 1998), export propensity (Ibeh, 2003) and export performance (Suárez-Ortega & Alamo-Vera, 2005); (ii) *global export intensity* or volume of sales from export activities (as % of total sales), a proxy of export commitment that is positively related to export performance (Leonidou et al., 2002); (iii) *intermediary export intensity* or volume of sales through this intermediary (as % of total foreign sales), a proxy of the importance of the channel that has a positive impact on export performance (Babakus & Yavas, 2006; Cavusgil & Zou, 1994); (iv) *length of relationship with intermediary* (years), related to the establishment of a stable foreign network (Babakus & Yavas, 2006); and (v) *industry*, which includes four categories to control for sector effects (final food goods, other final goods, industrial goods, other goods).

Factor Analysis and Descriptive Statistics

The reliability and validity of each variable was assessed. Response bias was tested for by analysing the average differences in any variable in the responses of the early and late respondents (Abernethy & Vagnoni, 2004); no evidence of response bias was found ($\alpha = 0.05$).

The validity of the measurement instruments was tested by factor analysis. The uni-dimensionality of the constructs was analysed in terms of the average variance explained (AVE), Cronbach's α coefficient for scale consistency (expected > 0.7), Kaiser–Meyer–Olkin statistics (expected > 0.5), Bartlett test for item correlation (expected $p < 0.05$) and factor loadings (expected > 0.30) (Hair, Anderson, Tatham, & Black, 1999).

Seven factors surfaced regarding resources and capabilities (factor loadings > 0.73, AVE > 0.72). Correlation analysis revealed reliable constructs in terms of the KMO (> 0.50) and Bartlett test ($p = 0.000$). Consistency was confirmed by Cronbach's alpha (> 0.7). Export performance was considered through three different factors (financial, strategic and perceived success) with AVE values over 0.77 and Cronbach's alpha values over 0.86. The reliability of these items was confirmed (factor loadings > 0.65), and correlation analysis confirmed that the constructs were robust (KMO > 0.68, Bartlett test $p = 0.000$). A second-order factor that explained 81% of the variance (KMO > 0.6, Bartlett test $p = 0.000$) was also identified and confirmed the reliability of an overall performance variable.

Regarding MCS design, two factors emerged for both process control and social control (factor loadings > 0.78, AVE > 0.63, KMO > 0.77, Bartlett test

$p = 0.000$). Output control was redefined to exclude two non-explanative items (R3, R4) to obtain a consistent (Cronbach's alpha value: 0.732, factor loadings > 0.67, AVE > 0.69) and reliable (KMO > 0.6, Bartlett test $p = 0.000$) construct. Two separate items emerged regarding MCS uses (diagnostic and interactive) and exhibited internal consistency (Cronbach's alpha > 0.89), convergence validity (factor loadings > 0.8, AVE > 71) and factor reliability (KMO > 0.8, Bartlett test $p = 0.000$)

Table 2 provides descriptive statistics and correlations among exogenous variables. It must be highlighted that any type of control and use positively impacts every dimension of performance ($p < 0.1$). Output control is highly correlated with the overall, financial and strategic performance ($p < 0.05$). Process control is also correlated with all performance types but most strongly with financial performance, so controls on the execution of tasks have direct effects on the exporter's profit. On the contrary, perceived success is not significantly affected by formal controls but is strongly influenced by social controls, which play a critical role in all dimensions of export performance. Finally, both diagnostic and interactive uses of MCS information have positive effects on the performance of export relationships.

Table 2. Exogenous Variables (Descriptive Statistics and Correlation Analysis).

Performance	Mean	S.D	MCS Design			MCS Use	
			Output control	Process control	Social control	Diagnostic use	Interactive use
Overall performance	4.592	1.141	0.184**	0.160*	0.341**	0.288**	0.282**
Financial performance	4.491	1.208	0.211**	0.179**	0.328**	0.298**	0.264**
Strategic performance	4.667	1.220	0.192**	0.160*	0.324**	0.272**	0.266**
Perceived success	4.610	1.237	0.114†	0.109†	0.303**	0.230**	0.260**
Output control	4.842	1.170	1.000				
Process control	4.525	1.357	0.696**	1.000			
Social control	4.516	1.303	0.457**	0.528**	1.000		
Diagnostic use	4.983	1.010	0.614**	0.499**	0.436**	1.000	
Interactive use	5.028	1.077	0.607**	0.508**	0.515**	0.819**	1.000

Note (one-tailed Pearson): $^{*}p < 0.05$; $^{**}p < 0.01$; $^{\dagger}p < 0.1$.

Results

Data Envelopment Analysis

Four DEA models were defined to test the export efficiency. Inputs included resources (experiential, scale of operation, financial and physical resources) and capabilities (information, customer relationship building and product development), while outputs included alternative measures of performance (overall, financial, strategic and perceived success). The DEA models focused on maximising the export outcomes obtained from the current level of inputs (output orientation). Because variables were measured using Likert scales, models were reduced to the standard CCR format. The super-efficiency variant was also included to obtain a global ranking of DMUs. Table 3 summarises the DEA scores, which provide disaggregated information on the nature of export efficiency.

The results reveal a limited number of efficient DMUs, ranging between 3.11% (strategic model) and 6.74% (overall model). Efficient units get high super-efficiency scores, with the best unit ranging from 2.100 to 2.615 points. Despite the large number of non-efficient DMUs, their minimum (0.250) and mean (0.558–0.616) scores are far above the minimum, so no extremely inefficient behaviours are observed. Because the distances between better and worse units are not large, non-efficient units could improve their position in the short term by implementing corrective measures, as DEA provides a target level of inputs/outputs for non-efficient DMUs that could orient improvement policies.[4]

Table 3. DEA Scores.

	Global Efficiency (Second-Order Factor)		Financial Efficiency		Strategic Efficiency		Perceived Success Efficiency	
	(*Super*) efficient	Not efficient	(*Super*) efficient	Not efficient	(*Super*) efficient	Not efficient	(*Super*) efficient	Not efficient
No. of DMUs	13	180	9	184	6	187	12	181
Min. DEA score	1.015	0.250	1.053	0.250	1.027	0.250	1.015	0.250
Max. DEA score	2.615	0.947	2.500	0.977	2.615	0.943	2.100	0.994
Average DEA score (SD)	1.350 (0.450)	0.616 (0.132)	1.368 (0.499)	0.558 (0.138)	1.489 (0.693)	0.561 (0.145)	1.291 (0.366)	0.559 (0.135)

Source: EMS© software.

Hierarchical Multiple Regression
Using the DEA scores, hierarchical multiple regressions (HMR) were used to test the hypotheses on the influence of MCS design and uses on performance efficiency. HMR provides information regarding the impact of predictors in a sequential way, testing whether adding one or more features will increase the predictability of the endogenous variable (Aron & Aron, 1999; Cohen, 2001; Jaccard & Turrisi, 2003). Three groups of variables were successively introduced into the model: (i) control variables (base model); (ii) MCS design (Model 1); and (iii) MCS uses (Model 2). Interactions between types of control and uses were also tested, but no significant relationships were found. Four HMR were developed, one for each of the categories of efficiency (overall, financial, strategic, and perceived success) (Appendix A).

Overall Efficiency. The base model does not explain a significant share of the data variance ($R^2 = 0.091$) even though the model's reliability is confirmed ($p < 0.05$). Both the intermediary export intensity and the length of the relationship appear to be significant ($p < 0.01$) but have different effects on the overall performance; while the first positively affects efficiency, the second has a negative impact on efficiency; thus, longer relationships appear to be less efficient than shorter agreements. Model 1 makes a significant contribution above the base model ($\Delta R^2 = 0.033$, G statistics $p < 0.05$). Social control appears to have a significant positive effect on the overall export efficiency, so hypothesis H_{13} is supported. In contrast, output control and process control do not have significant effects on efficiency (rejecting H_{11} and H_{12}), even though they were positively correlated with overall performance (Table 2). As a consequence, while any type of management control impacts performance in a positive way, only social control establishes a difference between efficient and inefficient exporters in terms of the optimal exploitation of resources and capabilities. Model 2 introduces MCS uses but does not make a significant contribution beyond Model 1 ($\Delta R^2 = 0.002$, G statistics $p = 0.997$, $\Delta F = -0.425$), and hypotheses H_{21} and H_{22} are thus rejected. Again, MCS uses have significant impacts on the overall performance (Table 2) but do not differentiate between more and less efficient exporters.

Financial Efficiency. The base model indicates that the intermediary export intensity and the length of relationship are significant predictors of financial efficiency ($p < 0.05$, $R^2 = 0.070$), with a positive and a negative impact, respectively. Model 1 slightly improves the explained variance ($\Delta R^2 = 0.024$, G statistics $p < 0.15$), and social control again appears to be the most explanative factor ($p < 0.05$), while no other types of control provide statistically significant contributions ($p > 0.4$). Model 2 includes MCS uses in the equation, and reveals a small positive effect of diagnostic use on financial efficiency ($p = 0.1$). While the final model has a reduced statistical significance ($\Delta R^2 = 0.014$, G statistics $p = 0.170$, $\Delta F = -0.068$), the inclusion of diagnostic use and social control partially confirms hypotheses H_{13} and H_{21}. Comparing the previous results with those from Table 2 shows that any MCS design and use positively impacts financial efficiency, but only social control and diagnostic use establish a difference between efficient exporters and non-efficient units.

Strategic Efficiency. The base model reveals significant effects from intermediary export intensity (positive) and length of relationship (negative) ($R^2 = 0.096$, p value $= 0.009$), as did the previous models. Model 1 makes a moderate contribution above the base model ($R^2 = 0.122$, $p < 0.05$, G statistics $p = 0.072$), and reveals that social control has a strong positive effect ($p < 0.05$, H_{13} confirmed). Finally, Model 2 shows that including MCS uses does not add significant information that differentiates between efficient and non-efficient firms.

Perceived Success Efficiency. The results regarding perceived success reveal a reduced effect of the control variables on efficiency (only the length of relationship is statistically significant at $p < 0.05$). The variance explained by the base model is strongly reduced ($R^2 = 0.062$), and the explanatory power is limited ($p < 0.1$). However, Model 1 provides the highest contribution compared to the previous results: social control has a strong impact on perceived success efficiency ($p < 0.05$), producing a significant model over and above the base approach ($\Delta R^2 = 0.046$, G statistics $p < 0.01$, $\Delta F = 0.428$) and confirming H_{13}. Model 2 includes MCS uses but does not add additional information affecting perceived success efficiency; even though Table 2 shows a significant relationship between diagnostic and interactive uses and perceived success, these variables do not establish a distinction in terms of optimising the utilisation of exporters' resources and capabilities.

THE 'LIFECYCLE' OF THE EXPORTER–INTERMEDIARY RELATIONSHIP

The models presented above revealed that the length of relationship has a significant negative effect on exporters' efficiencies; thus, a particular analysis was performed on this issue. Time analyses of the overall, financial, strategic, and perceived success efficiencies were performed to study the 'lifecycle' of relationships between exporters and intermediaries. An inverse 'U' function was expected and would indicate a positive learning effect in the earlier stages of relationships and a reversal in the late stages due to relationship maturity and relaxed control.

The results confirm the expected 'lifecycle' (Appendix B). Increased efficiency scores are observed in the first four years of relationships (growth stage), followed by a maturity period that lasts from the fifth and ninth year. During the maturity period, efficiency scores are stable, and no significant variations are observed along any efficiency dimension. Finally, a decline stage arises at the tenth year of the relationship, and efficiency scores continuously decrease at longer times.

To reveal the causes of this efficiency lifecycle, a particular analysis of control types and uses was performed at each stage. During the growth period, very strong formal controls are progressively replaced by social controls as the relationship advances. This increase in social control seems to assist the learning effects of relationships, leading to increased efficiency. The use of MCS also decreases after the second year of the relationship.[5]

During the maturity stage, both the diagnostic and interactive MCS uses are reduced. Formal and informal controls get relaxed, though there is some increase in control at the end of the stage, when the first symptoms of decline are observed. This decrease is particularly significant for social controls in the fifth year of the relationship and leads to a significant decrease in exporters' efficiencies.

The decline stage is characterised by moderated implementation and use of MCS: both are higher than during the mature stage, but lower than during the growth period. An increase in process control is observed in particular, as the firms attempt to more strongly control the way tasks are developed. These stable MCS are unable to impede the progressive efficiency decline observed in the longest relationships.

These results suggest that export managers who face relationships with foreign intermediaries exhibit 'pro-cyclical' behaviour: superior results in the growth stage produce a relaxation in MCS design and uses, which lead to a decline of efficiency that cannot be reversed through later attempts to increase managerial controls.

DISCUSSION AND CONCLUSIONS

This study applies the RBV and contributes to the literature by offering empirical evidence about the influence of MCS, design and use on how exporting firms' exploit their capabilities and resources to optimise their export performance. This research focused on externalised export channel settings, which are frequent vertical relationships managed by dominant exporters.

Descriptive statistics of the variables employed revealed that any type of MCS control and use has a positive influence on every dimension of performance. Output control mainly impacts the overall, financial and strategic performances, while process control primarily affects financial performance. Social control plays a key role in any dimension of performance but particularly on the success perceived by managers. Both diagnostic and interactive uses of MCS have significant and positive effects on exporters' performance levels.

DEA models are developed to study the overall, financial, strategic and perceived success performances and provide additional information about exporters' efficiencies. Interesting managerial implications are highlighted. An optimistic view of this setting is obtained: although a limited number of exporting firms behave efficiently in the DEA sense, the average distances between better and worse firms are not large.

Hierarchical multiple regression models are utilised to better understand the predictors of performance efficiency. The results partially confirm our research hypotheses. MCS design is considered, and the effects of output, process and social controls on the nature of efficiency scores are analysed. The hierarchical multiple regression models reveal that only informal controls, such as social controls, appear to have a significant positive effect on the export efficiency. As a result, even though any type of control has a positive impact on export performance, only social control differentiates firms in terms of efficiency: the increased use of social controls leads to more optimal transformation of resources and capabilities into performance. The results also demonstrate that the uses of MCS do not significantly affect the export efficiency; a reduced but positive effect of diagnostic use is observed only on the financial efficiency. As for controls, MCS uses do play a direct role on exporter performance, but they do not differentiate firms in terms of efficiency in the use of resources and capabilities.

Our results offer other attractive managerial implications. Analysing the export intensity reveals that more intense relationships produce higher efficiencies. However, the length of the relationship has a negative impact on efficiency, that is, longer relationships are less efficient than shorter agreements. A more detailed analysis of this observation has highlighted a 'lifecycle' of export relationships that includes three stages: growth, maturity and decline. Pro-cyclical behaviour is observed: good results in the first stage lead to relaxed management controls in the maturity period, which lead to the decline stage. This result has significant managerial implications, revealing that export managers need to maintain vigilance in consolidated relationships with intermediaries to prevent efficiency losses that reduce performance and can remove a firm from the market.

The study suffers the normal limitations of survey data collection. Future studies could investigate the role of foreign intermediaries in the process of efficiency generation in exporter firms.

NOTES

1. EXTENDA is a public agency that promotes external trade between firms and offers grants and consultation services to encourage internationalisation.
2. The population was filtered through a pre-notified letter followed by phone calls which confirmed the firm's validity, the export manager's identity, and his/her disposition to survey participation.
3. The 'supplier relationship building' capability was not included, as the questionnaire was focused on the downstream relationship between the exporter and the foreign intermediary.
4. With the aim of identifying different competitive strategies, a cluster analysis was done using the DEA scores; seven groups were identified (two non-efficient, five efficient groups). The results revealed different mixes of resources and capabilities for each group and find more intense MCS in the efficient clusters, particularly in terms of social control and diagnostic use.
5. A higher use is observed in the second year than the first one, probably due to full implementation of MCS.

ACKNOWLEDGMENTS

This research has been partially funded by research projects ECO2008-05550, SEJ-5067 (with funds from the European Union PO FEDER-FSE 2007-2013) and PO6-SEJ-01799 and a research agreement signed with EXTENDA.

REFERENCES

Abernethy, M., & Vagnoni, E. (2004). Power, organization design and managerial behaviour. *Accounting, Organizations and Society*, *29*(3/4), 207–225.

Amit, R., & Schoemaker, P. (1993). Strategic assets and organizational rent. *Strategic Management Journal*, *14*(1), 33–46.

Anderson, J., & Narus, J. (1990). A model of distributor firm and manufacturer firm working partnerships. *Journal of Marketing*, *54*, 42–58.

Anderson, P., & Petersen, N. (1993). A procedure for ranking efficient units in data envelopment analysis. *Management Science*, *39*(10), 1261–1264.

Aron, A., & Aron, E. N. (1999). *Statistics for psychology*. Upper Saddle River, NJ: Prentice Hall.

Aulakh, P. S., & Gençtürk, E. F. (2000). International principal-agent relationships. Control, governance and performance. *Industrial Marketing Management*, *29*, 521–538.

Aulakh, P. S., Kotabe, M., & Sahay, A. (1996). Trust and performance in cross-border marketing partnerships: A behavioral approach. *Journal of International Business Studies*, *27*(5), 1005–1032.

Babakus, E., & Yavas, U. (2006). Perceived uncertainty, networking and export performance. *European Business Review*, *18*(1), 4–13.

Banker, R. D., Charnes, A., & Cooper, W. W. (1984). Some models for estimating technical and scale inefficiencies in Data Envelopment Analysis. *Management Science*, *30*(9), 1078–1092.

Baraldi, E., & Strömsten, T. (2009). Controlling and combining resources in networks – form Uppsala to Stanford, and back again: The case of a biotech innovation. *Industrial Marketing Management*, *38*, 541–552.

Barney, J. (1991). Firm resources and sustained competitive advantage. *Journal of Management*, *17*(1), 99–120.

Bello, D. C., & Gilliland, D. I. (1997). The effect of output controls, process controls and flexibility on export channel performance. *Journal of Marketing*, *61*, 22–38.

Bisbe, J., & Otley, D. R. (2004). The effects of the interactive use of management control systems on product innovation. *Accounting, Organizations and Society*, *29*, 709–737.

Cavusgil, S. T., Deligonul, S., & Zhang, C. (2004). Curbing foreign distributor opportunism: An examination of trust, contracts, and the legal environment in international channel relationships. *Journal of International Marketing*, *12*(2), 7–27.

Cavusgil, S. T., & Zou, S. (1994). Marketing strategy-performance relationship: An investigation of the empirical link in export market ventures. *Journal of Marketing*, *58*, 1–21.

Charnes, A., Cooper, W. W., Lewin, A. Y., & Seiford, L. M. (1994). *Data envelopment analysis: Theory, methodology, and applications*. Boston, MA: Kluwer.

Charnes, A., Cooper, W. W., & Rhodes, E. (1978). Measuring the efficiency of decision making units. *European Journal of Operational Research*, *2*(6), 429–444.

Cohen, B. H. (2001). *Explaining psychological statistics*. New York, NY: Wiley.

Coletti, A. L., Sedatole, K. L., & Towry, K. L. (2005). The effect of control systems on trust and cooperation in collaborative environments. *The Accounting Review*, *80*(2), 477–500.

Dekker, H. C. (2004). Control of inter-organizational relationships: Evidence on appropriation concerns and coordination requirements. *Accounting, Organizations and Society*, *29*, 27–49.

Dhanaraj, C., & Beamish, P. W. (2003). A resource-based approach to the study of export performance. *Journal of Small Business Management*, *41*(3), 242–261.

Dillman, D. A. (2000). *Mail and Internet surveys: The tailored design method*. New York, NY: John Wiley & Sons.

Ferreira, A., & Otley, D. (2006). *Exploring inter and intra-relationships between the design and use of management control systems*. Working Paper, Monash University. Retrieved from http://papers.ssrn.com/sol3/papers.cfm?abstract_id=896228

Fraering, M. (1996). Explanatory and predictive indicators of export performance: A synthesis of findings. In: *Enhancing knowledge development in marketing* (Vol. 7, pp. 104–105). Chicago, IL: American Marketing Association.

Gençtürk, E. F., & Aulakh, P. S. (1995). The use of process and output controls in foreign markets. *Journal of International Business Studies*, *26*(4), 755–786.

Gordon, J. S. (1993). *Profitable exporting*. New York, NY: Wiley.

Gundlach, G. T. (1994). Exchange governance: The role of legal and nonlegal approaches across the exchange process. *Journal of Public Policy and Marketing*, *13*(2), 246–258.

Hair, J. F., Jr., Anderson, R. E., Tatham, R. L., & Black, W. C. (1999). *Multivariate data analysis: With readings*. Upper Saddle River, NJ: Prentice Hall.

Ibeh, K. I. N. (2003). On the internal drivers of export performance among Nigerian firms: Empirical findings and implications. *Management Decision*, *41*(3), 217–225.

Jaccard, J., & Turrisi, R. (2003). *Interaction effects in multiple regression*. Newbury Park, CA: Sage.

Kaleka, A. (2002). Resources and capabilities driving competitive advantage in export markets: Guidelines for industrial exporters. *Industrial Marketing Management*, *31*, 273–283.

Katsikeas, C. S., Leonidou, L. C., & Morgan, N. A. (2000). Firm-level export performance assessment: review, evaluation, and development. *Journal of the Academy of Marketing Science*, *28*(4), 493–511.

Kimura, S., & Mourdoukoutas, P. (2000). Effective integration of management control systems for competing in global industries. *European Business Review*, *12*(1), 41–45.

Lages, C., Lages, C., & Lages, L. F. (2005). The RELQUAL Scale: A measure of relationship quality in export market ventures. *Journal of Business Research*, *58*(8), 1040–1048.

Leonidas, L., & Katsikeas, C. S. (1996). The export development process: An integrative review of empirical models. *Journal of International Business Studies*, *27*(3), 517–551.

Leonidou, L. C., Katsikeas, C. S., & Samiee, S. (2002). Marketing strategy determinants of export performance: A meta-analysis. *Journal of Business Research*, *55*(1), 51–67.

Ling-yee, L., & Ogunmokun, G. (2001). Effect of export financing resources and supply chain skills on export competitive advantages: Implications for superior export performance. *Journal of World Business*, *36*(3), 260–279.

Lovell, C. A. K., & Knox, A. (1993). Production frontier and productive efficiency. In H.O. Fried, C. A. K. Lovell & S. S. Schmidt (Eds.), *The measurement of productive efficiency: Techniques and applications* (pp. 3–67). New York, NY: Oxford University Press.

Pahud de Mortanges, C., & Vossen, J. (1999). Mechanisms to control the marketing activities of foreign distributors. *International Business Review*, *8*, 75–97.

Rosson, P. J., & Ford, I. D. (1982). Manufacturer-overseas distributor relations and export performance. *Journal of International Business Studies*, *13*(2), 57–72.

Sachdev, H. J., Bello, D. C., & Pilling, B. K. (1994). Control mechanisms within export channels of distribution. *Journal of Global Marketing*, *8*(2), 31–49.

Shoham, A. (1991). Performance in exporting: A state of the art literature review and synthesis and directions for future research. *Conference of the Academy of International Business*, Miami, FL.

Shoham, A. (1998). Export performance: A conceptualization and empirical assessment. *Journal of International Marketing*, *6*(3), 59–81.

Simons, R. (1990). The role of management control systems in creating competitive advantage: New perspectives. *Accounting, Organizations and Society*, *15*, 127–143.

Simons, R. (1995). *Levers of control: How managers use innovative control systems to drive strategic renewal.* Harvard Business School Press.

Skarmeas, D. (2006). The role of functional conflict in international buyer-seller relationships: Implications for industrial exporters. *Industrial Marketing Management, 35*(5), 567–575.

Solberg, C. A. (2006). Relational drivers, controls and relationship quality in exporter–foreign middleman relations. *Advances in International Marketing, 16*, 81–105.

Sousa, C. (2004). Export performance measurement: An evaluation of the empirical research in the literature. *Academy of Marketing Science Review, 9*. Retrieved from http://amsreview.org/articles/sousa09-2004.pdf. Accessed on February 10, 2012.

Styles, C., & Ambler, T. (2000). The impact of relational variables on export performance: An empirical investigation in Australia and the UK. *Australian Journal of Management, 25*(3), 261–281.

Suárez-Ortega, S. M., & Alamo-Vera, F. R. (2005). SMES' internationalization: Firms and managerial factors. *International Journal of Entrepreneurial Behaviour and Research, 11*(4), 258–279.

Tuomela, T.-S. (2005). The interplay of different levers of control: A case study of introducing a new performance measurement system. *Management Accounting Research, 16*(3), 293–320.

Vaivio, J. (2004). Mobilizing local knowledge with 'provocative' non-financial measures. *European Accounting Review, 13*(1), 39–71.

Van der Stede, W., Young, S. M., & Chen, C. X. (2005). Assessing the quality of evidence in empirical management accounting research: The case of survey studies. *Accounting, Organizations and Society, 30*(7–8), 655–684.

Vélez, M. L., Sánchez, J. M., & Álvarez-Dardet, C. (2008). Management control systems as inter-organizational trust builders in evolving relationships: Evidence from a longitudinal case study. *Accounting, Organizations and Society, 33*, 968–994.

Widener, S. K. (2007). An empirical analysis of the levers of control framework. *Accounting, Organizations and Society, 32*(7/8), 757–788.

Zhang, C., Cavusgil, S. T., & Roath, A. S. (2003). Manufacturer governance of foreign distributor relationships: Do relational norms enhance competitiveness in the export channel? *Journal of International Business Studies, 34*, 550–566.

Zou, S., Fang, E., & Zhao, S. (2003). The effect of export marketing capabilities on export performance: An investigation of Chinese exporters. *Journal of International Marketing, 11*(4), 32–55.

Zou, S., & Stan, S. (1998). The determinants of export performance: A review of the empirical literature between 1987 and 1997. *International Marketing Review, 15*(5), 333–356.

Zou, S., Taylor, C. R., & Osland, G. E. (1998). The EXPERF scale: A cross-national generalized export performance measure. *Journal of International Marketing, 6*(3), 37–58.

APPENDIX A: HIERARCHICAL MULTIPLE REGRESSIONS

	Overall Efficiency						Financial Efficiency					
	Base		Model 1		Model 2		Base		Model 1		Model 2	
	Coefficient	*p*	Coefficient	*p*	Coefficient	*p*	Coefficient	*p*	Coefficient	*p*	Coefficient	*p*
Control variables												
Manager export experience	0.080	0.288	0.088	0.238	0.088	0.243	0.024	0.752	0.036	0.638	0.031	0.679
Global export intensity	0.048	0.537	0.027	0.725	0.023	0.767	0.139	0.077	0.115	0.148	0.104	0.093
Intermediary export intensity	0.196	0.006**	0.189	0.009**	0.189	0.009**	0.158	0.029*	0.151	0.037*	0.153	0.036*
Length of relationship	−0.241	0.002**	−0.259	0.001**	−0.259	0.001**	−0.203	0.011*	−0.219	0.006**	−0.220	0.006**
Industry (final, not food)	−0.106	0.163	−0.128	0.094	−0.124	0.106	−0.061	0.427	−0.078	0.310	−0.073	0.350
Industry (industrial)	−0.092	0.216	−0.099	0.182	−0.101	0.176	−0.094	0.212	−0.101	0.180	−0.107	0.157
Industry (services)	−0.036	0.616	−0.020	0.778	−0.020	0.783	−0.015	0.841	−0.009	0.907	−0.008	0.910
Independent variables												
Output control			−0.130	0.199	−0.143	0.200			−0.073	0.476	−0.089	0.431
Process control			−0.096	0.361	−0.099	0.348			−0.065	0.541	−0.072	0.499
Social control			0.170	0.047*	0.171	0.058†			0.174	0.047*	0.193	0.035*
Diagnostic use					0.067	0.599					0.209	0.106
Interactive use					−0.043	0.741					−0.192	0.146
Reliability and validity												
R^2	0.091		0.124		0.126		0.070		0.094		0.108	
F statistics (p value)	2.638 (0.013)		2.572 (0.006)		2.147 (0.016)		2.058 (0.050)		1.880 (0.050)		1.812 (0.049)	
ΔR^2			0.033		0.002				0.024		0.014	
ΔF			−0.066		−0.425				−0.178		−0.068	
G statistics (p)			2.285 (0.019)*		0.206 (0.997)				1.406 (0.115)		1.413 (0.170)	

APPENDIX A. (*Continued*)

	Strategic Efficiency						Perceived Success Efficiency					
	Base		Model 1		Model 2		Base		Model 1		Model 2	
	Coefficient	*p*	Coefficient	*p*	Coefficient	*p*	Coefficient	*p*	Coefficient	*p*	Coefficient	*p*
Control variables												
Manager export experience	0.071	0.345	0.083	0.270	0.080	0.290	0.080	0.296	0.089	0.239	0.091	0.229
Global export intensity	0.072	0.352	0.048	0.541	0.042	0.596	0.034	0.670	0.011	0.890	0.005	0.946
Intermediary export intensity	0.196	0.006**	0.190	0.008**	0.191	0.008**	0.134	0.065†	0.127	0.079†	0.129	0.075†
Length of relationship	−0.255	0.001**	−0.274	0.001**	−0.274	0.001**	−0.185	0.021*	−0.207	0.009**	−0.207	0.009**
Industry (final, not food)	−0.066	0.383	−0.087	0.255	−0.084	0.274	−0.096	0.212	−0.122	0.111	−0.115	0.137
Industry (industrial)	−0.108	0.148	−0.115	0.121	−0.118	0.114	−0.113	0.139	−0.120	0.110	−0.123	0.102
Industry (services)	0.010	0.886	0.019	0.795	0.019	0.796	−0.101	0.171	−0.082	0.265	−0.081	0.273
Independent variables												
Output control			−0.099	0.328	−0.104	0.353			−0.163	0.112	−0.198	0.079†
Process control			−0.057	0.586	−0.060	0.567			−0.099	0.349	−0.105	0.322
Social control			0.185	0.031*	0.199	0.029*			0.196	0.024*	0.182	0.047*
Diagnostic use					0.113	0.376					0.070	0.585
Interactive use					−0.113	0.387					0.007	0.960
Reliability and validity												
R^2	0.096		0.122		0.126		0.062		0.108		0.111	
F statistics (p value)	2.787 (0.009)		2.505 (0.008)		2.151 (0.016)		1.752 (0.099)		2.180 (0.021)		1.858 (0.042)	
ΔR^2			0.026		0.004				0.046		0.003	
ΔF			−0.202		−0.354				0.428		−0.322	
G statistics (p)			1.796 (0.072)†		0.412 (0.950)				3.128 (0.002)**		0.304 (0.984)	

Note: Standardised regression coefficients are provided. $^{*}p<0.05$; $^{**}p<0.01$; $^{***}p<0.001$; $^{\dagger}p<0.1$.

APPENDIX B: 'LIFECYCLE' OF THE EXPORT–INTERMEDIARY RELATIONSHIP

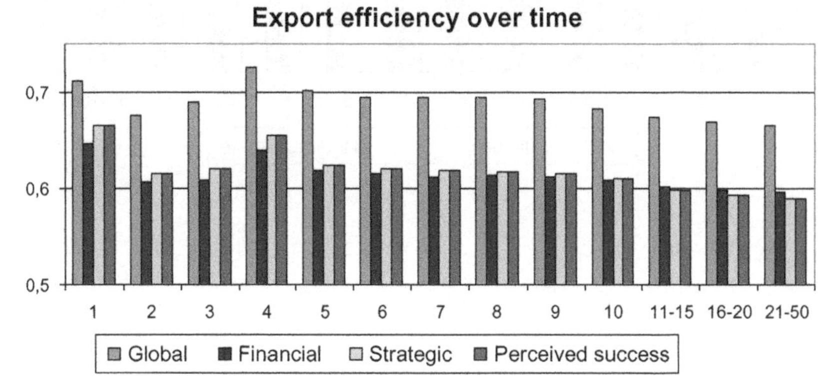

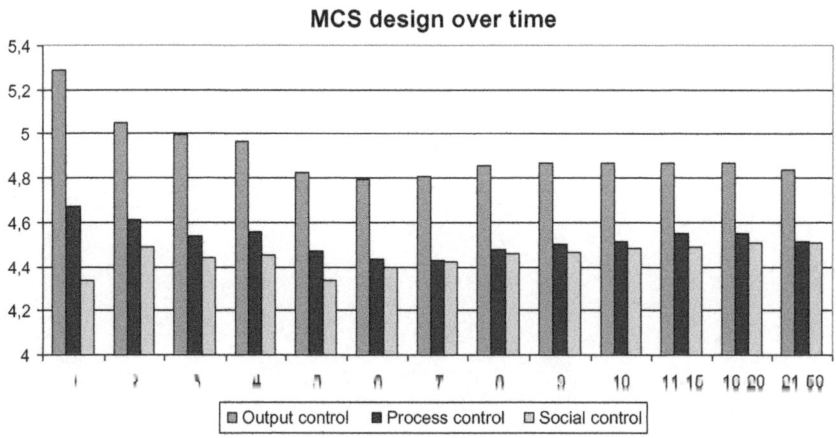

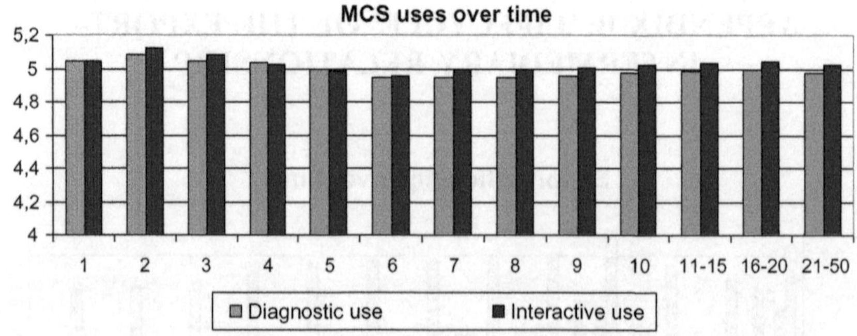
MCS uses over time
5,2
5
4,8
4,6
4,4
4,2
4
1
2
3
4
5
6
7
8
9
10
11-15
16-20
21-50
Diagnostic use
Interactive use

DETERMINANTS OF THE IMPLEMENTATION AND LONG-TERM USE OF INTERORGANIZATIONAL COST MANAGEMENT TOOLS – AN EXPERIMENTAL INVESTIGATION

Jan O. Piontkowski, Andreas Hoffjan,
Maik Lachmann and Lukas D. Schuchardt

ABSTRACT

Purpose – *Interorganizational cost management among companies can lead to significant cost reductions. However, the determinants of the implementation and long-term use of open book accounting as a tool in interorganizational cost management still remain unclear. We contribute to the academic literature by examining the influence of different determinants on the propensity to use open book accounting.*

Design/methodology/approach – *We conduct an experiment and use a covariance-based structural equation model to analyze the influence of the amount of the initially offered cost information, the offer of a relation-specific asset, and the relative power structure. The model introduced in*

Performance Measurement and Management Control: Global Issues
Studies in Managerial and Financial Accounting, Volume 25, 223–246

ISSN: 1479-3512/doi:10.1108/S1479-3512(2012)0000025012

this paper also integrates aspects of user acceptance that are derived from the Technology Acceptance Model.

Findings – *The results demonstrate that both groups of variables have a significant effect on the willingness to use open book accounting. We also show that users of a management device are influenced in their choice by the perceived ease of use of the instrument; yet the extent to which open book accounting can help them achieve their goals (perceived usefulness) has an even stronger influence.*

Research limitations/implications – *Our findings contribute to a better understanding of the determinants that lead to the successful implementation of open book accounting as an interorganizational cost management tool, and help companies to avoid pitfalls during the implementation process.*

Originality/value – *This is the first study to analyze the simultaneous influence of different situational and attitudinal determinants on the propensity to engage in interorganizational cost information exchange.*

INTRODUCTION

The ever-growing importance of interfirm relationships is omnipresent; firms are, with increasing frequency, cooperating with one another to successfully navigate a more intense and competitive business environment. To effectively cope with this environment, firms can no longer rely simply on their own resources but, rather, must develop close interfirm collaborations, thereby blurring organizational boundaries and contributing to the emergence of new firm networks (Das & Teng, 1996, 2000; Langfield-Smith, 2005; Mouritsen & Thrane, 2006; Otley, 1994; Thrane & Hald, 2006; Tomkins, 2001; Van der Meer-Kooistra & Vosselman, 2006). For example, the automotive industry is steadily increasing its number of outsourced components while the number of suppliers is simultaneously being reduced (Baiman, Fischer, & Rajan, 2001). Increased outsourcing activities lead to the problem of information asymmetry between the buyer and the supplier in a transaction (Cooper & Slagmulder, 2004), which can lead to a cost increase on either side of the supply chain. This situation cannot be completely governed by contracts because no matter how specific a contractual agreement between the firms may be, it will always be incomplete (Coad & Cullen, 2006).

We define open book accounting (OBA) as the mutual exchange of cost information between supply chain partners. The literature has identified the different functions of OBA in an interfirm relationship. OBA can be used to check the integrity of a quote (Hoffjan & Kruse, 2006) or serve as a tool to legitimate and justify prices and price changes between the supplier and the buyer (Carr & Ng, 1995; Seal, Cullen, Dunlop, Berry, & Ahmed, 1999). It also helps to reduce the information asymmetry that evolves between the buyer and the supplier as more and more significant items of the production process are outsourced (Cooper & Slagmulder, 2004). Kajüter and Kulmala (2005) argue a similar case, proposing that OBA could be used to identify the over-specifications of a product that unnecessarily drive up costs. OBA can be used to analyze the effects that structural changes to the supply chain have on cost efficiency (Cooper & Slagmulder, 2004). Thus, the sharing of cost information can serve as a basis for value chain analysis, which helps to discover potential cost reductions across all of the firms involved (Dekker, 2003). OBA can have a positive effect on intrafirm costs as well. Mouritsen, Hansen, and Hansen (2001) argue that, before the interorganizational cost level can be optimized, the individual firm must be transparent about its own cost structure.

The collaborative exchange of proprietary cost information, if conducted properly and in a fair manner, have been shown to effectively foster interfirm relationships (for a review, see Hoffjan & Kruse, 2006). As a consequence, recent research activities in the field of interorganizational management accounting have started to tackle the question of which factors can positively influence and enhance the collaborative exchange of information, and which situational factors cause resistance or, worse, a misuse of revealed cost information (e.g., Drake & Haka, 2008; Kajüter & Kulmala, 2005; Piontkowski & Hoffjan, 2009; Van den Abbeele, 2006). In this context, we contribute to the academic literature by experimentally examining the *simultaneous* influence of different situational and attitudinal determinants on the propensity to use OBA by developing a structural equation model. Determinants derived from the Technology Acceptance Model are integrated into our model and analyzed. In addition, we provide insight into possible barriers and promoters during the implementation of OBA.

The remainder of this paper is structured as follows. Our hypotheses are developed in the second section. The experimental setting is described in the third section. The fourth section presents the main results, and the fifth section summarizes our conclusions and points out important implications drawn from the model.

LITERATURE REVIEW AND HYPOTHESIS DEVELOPMENT

OBA, as an interfirm cost management tool, requires the disclosure of cost information between the involved firms (Kajüter & Kulmala, 2005). The *exchange of cost information* has been analyzed in some depth. Formal analytic approaches show, based on rational considerations, that the denial of information exchange occurs even within symmetrical power relations (Baiman & Rajan, 2002; Miraglia & Leotta, 2010). However, the assumption of completely rational behavior eliminates the effects of soft influencing factors, such as trust. This simplification is somewhat questionable because the social context in which interactions take place should not be dismissed (Ghoshal & Moran, 1996; Van den Abbeele, 2006). For example, Drake and Haka (2008) show that individuals' concerns about inequity may negatively influence their willingness to share necessary information. However, the initial offer of relevant cost information seems to be related to a firm's likelihood of using OBA (Piontkowski & Hoffjan, 2009). So far, initial offers have mainly been investigated in the context of purchasing and selling negotiations. For instance, Buelens and Van Poucke (2004) investigated factors that determine the initial offer in a negotiation process. Galinsky and Mussweiler (2001) examined the role of first offers as anchors. The experimental variable *initially offered quantity of cost information* is proposed to have an effect on two variables: the perceived partner's willingness to engage in OBA and the actual amount of cost information that is reciprocated by the participants. For both variables, a positive effect is expected.

> **H1.** If the participants are faced with an extensive initial offer of cost information, they will (a) perceive the partner firm to be more willing to truly cooperate and (b) actually reciprocate with more cost information items.

The role of *commitment devices* to foster collaboration among partner firms has been thoroughly analyzed in the literature (Andaleeb, 1996; Anderson & Weitz, 1992; Heide & John, 1990; Rokkan, Heide, & Wathne, 2003). Mutual investments can significantly increase the extent of joint action in interfirm relationships. These positive effects have been found in the long term (Kwon & Suh, 2005); for the short term, Gundlach, Achrol, and Mentzer (1995) report mixed results. Moreover, Clee and Wicklund (1980) argue that the offering of a specific investment may reduce the partner's freedom of choice, provoking a move in the opposite direction. However, most of the cited studies find that commitment devices have a positive impact on collaboration among partner firms. The second experimental variable, the offer of a *relation-specific asset*, is therefore proposed to directly affect a partner's willingness to further reveal cost information. As reasoned before, the offer of a relation-specific asset is intended to show commitment to the interfirm relationship and convince the partner of one's own good intentions. However, the idiosyncratic investment is not expected to have a direct effect on the amount of cost information reciprocated by individuals. The reason for this assumption is that individuals will tend to return in kind (as in a tit-for-tat strategy), and it is questionable whether they will transform the perceived commitment by the other party into a greater amount of reciprocated information.

H2. The offer of a specific asset has a positive impact on the willingness to engage in interorganizational cost information exchange.

The *role of power* has been widely discussed in recent literature. Van den Abbeele (2006) points out that the disadvantages of less powerful buyers are less pronounced when they are able to make detailed cost information available in the exchange process to create a cooperative and constructive environment. Further, if the less powerful buyer offers information that helps to reach a prominent solution, the target firm, even though more powerful, will more likely reciprocate the effort (Campbell, 1963; Gouldner, 1960). For the third experimental variable, the *relative power structure*, three effects are expected. Following the results reported by Van den Abbeele (2006), it is expected that participants who were in the less powerful position and hence, more dependent on the relationship with the partner firm, will show a greater willingness to engage in further cost information exchange. This is anticipated because engaging in interorganizational cost management provides the less powerful party with a means to engage the more powerful party in a cooperative environment and to improve its own relative position (Bonoma, 1976). However, power should be positively related to the perceived partner's willingness. It is expected that a buying firm will perceive the partner to be more willing to earnestly exchange cost information when it is in a more powerful position than when it is in a less powerful position. The third dependent variable that is expected to be influenced by the relative power structure is the general long-term intention to use OBA as a management tool. In this particular case, it is anticipated that the same considerations apply for this variable as those that hold for the factor of willingness. It is anticipated that participants in the powerful position oppose using OBA. By contrast, participants in the less powerful position will have greater intentions to use OBA because they may perceive it as a chance to improve their relative position compared to the more powerful partner firm.

H3. Power is (a) negatively related to the own willingness to engage in interorganizational cost information exchange and the long-term intention to use OBA (c), but (b) positively related to the perceived partner's willingness to engage in interorganizational cost information exchange.

The next group of variables is derived from the technology acceptance model (TAM). The TAM (introduced by Davis, 1989; Davis, Bagozzi, & Warshaw, 1989) constitutes the theoretical starting point for aspects of user acceptance; it has gained substantial empirical support over the last two decades (Venkatesh & Bala, 2008). User acceptance comprises the concepts of *perceived ease of use* and *perceived usefulness*. Supplementing the theoretical aspects covered by the dependent variables in the experimental investigations, the proposed path model incorporates variables that cover the user's acceptance of a management device and the attitude toward OBA. These aspects are not expected to be directly affected by the experimental variables. Nonetheless, they are expected to have an effect on the general propensity to use OBA. Following the principles of the TAM and its extension by Venkatesh and Davis (2000), it is suggested that perceived ease of use will have a positive influence on perceived usefulness. This is proposed following the findings by Venkatesh and Davis (2000) who indicate that degree of effort necessary for tool use is inversely correlated with job performance.

H4. Perceived ease of use has a positive influence on perceived usefulness.

The *perceived usefulness* variable evaluates the extent to which OBA is, in general, perceived to be an applicable instrument to achieve individual as well as corporate objectives. Venkatesh and Davis (2000) mention this aspect as the job relevance of a target system, which has a positive influence on the perceived usefulness of a system. In the present model, it is assumed that perceived usefulness will positively influence both a firm's willingness to exchange cost information within this particular interfirm relation and the general (long-term) intention to use OBA in an interfirm context. Both propositions conform to the effects implied by the TAM, which suggests that the higher the perceived usefulness, the higher the tendency to use a device or a system. Further, perceived usefulness is proposed to positively influence perceived own benefit.

H5. Perceived usefulness positively influences (a) willingness to use OBA, (b) long-term intention to use OBA and (c) perceived benefits for one's own firm.

In addition to the three experimental variables and two variables that were derived from the TAM, the path model consists of five endogenous variables. The relationships between those variables are stated in Hypotheses 6–8.

One of the main endogenous variables is *perceived own benefit*, which can be regarded as an evaluation of the ability of OBA to increase own performance as well as the performance of the own firm in the present interfirm relationship. Perceived usefulness, in turn, represents a general evaluation of the instrument OBA and its ability to foster the achievement of individual and corporate objectives. In the proposed model, perceived own benefit is considered to be a variable that has one main influencing factor (perceived usefulness), but has several effects, which are explicitly mapped in the model and which are proposed to cause change in other variables. We assume that those target firms that perceive a higher own benefit of OBA, are more likely to assess the initiating partner firm as truly willing to share information because they recognize that the exchange situation will not only yield a positive effect for the initiator but also for themselves. Therefore, perceived own benefit is expected to positively influence the perceived partner's willingness. Further, it can be argued that, if OBA is perceived to be own-benefit increasing in the management of interfirm relations, the intention for the long-term application also increases. Different from perceived usefulness, perceived own benefit evaluates an aspect of the relationship with the actual partner in the present scenario. Hence, participants who experience a high-perceived own benefit through the collaborative exchange of cost information with the partner firm are assumed to actually show a positive reaction toward the partner firm. They are expected to do this by reciprocating more cost information than participants with a lower perceived own benefit. Thus, the amount of cost information reciprocated by the participants is expected to be positively influenced by perceived own benefit.

H6. Perceived own benefit positively influences (a) the own willingness to engage in an OBA relationship, (b) the perceived partner's willingness to engage in an OBA relationship, (c) the long-term intention to use OBA and (d) the amount of cost information reciprocated.

Own willingness to further exchange cost information with the partner firm has thus far been proposed to be influenced by perceived usefulness, and the two experimental variables offer a relation-specific asset and the relative power structure. Willingness to engage in further information exchange is presumed to influence the amount of cost information reciprocated as a reaction to the initial offer by the supplier. Because both variables evaluate the present interfirm exchange situation, the amount of cost information reciprocated is likely to tend in the same direction as the willingness. Hence, own willingness to engage in cost information exchange is posited to positively influence the amount of cost information actually returned in response to the initial offer.

H7. Own willingness to use OBA positively influences the amount of cost information reciprocated.

The perception of the *partner's willingness* is proposed to be influenced by the amount of offered cost information, relative power position and perceived own benefit. The perceived partner's willingness, in turn, is expected to influence own willingness. Fig. 1 summarizes our hypotheses.

H8. The perceived partner's willingness positively influences own willingness to use OBA.

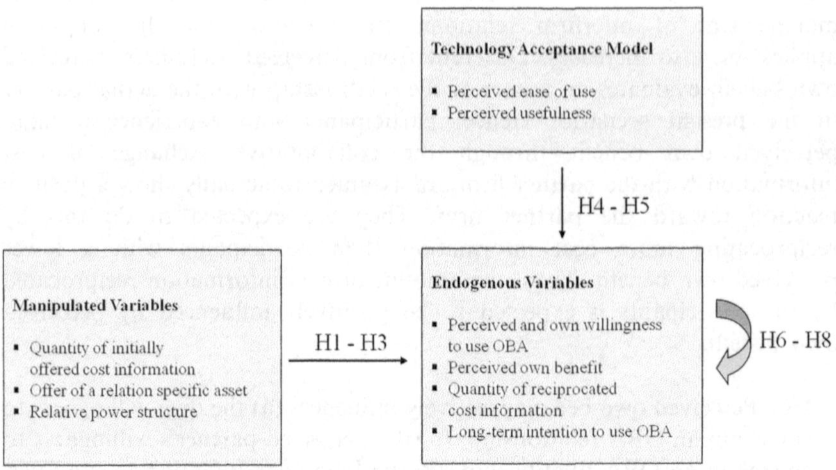

Fig. 1. Hypotheses on Determinants of the Propensity to Use OBA.

EXPERIMENTAL SETTING

General Design

An experimental approach was applied to examine the determinants of the propensity to use OBA. The sample consisted of 150 participants recruited from Master and Ph.D. classes at a European University. The reactions of the Master's- and Ph.D.-level students were compared with each other. The results did not show any prevailing difference in behavior. All participants had a sound knowledge on the principles of OBA and interorganizational cost management, which had been acquired in management accounting courses. Participants were randomly assigned to different experimental conditions. Student participation was encouraged by the opportunity to receive credits for the course. Subjects were told that they took part in an investigation about cost accounting in interfirm relationships. No specific reference was made to the phrase open book accounting. The participants were told that the outcome of the investigation would not be graded and anonymity was guaranteed. On average, it took the participants 25–35 minutes to read through the material and complete the required tasks.

Participants were provided a scenario dealing with the supplier–buyer relationship of two companies (Piontkowski, 2008). The material was designed by adopting the approach presented by Drake and Haka (2008). The authors provide two sets of cost data that are interdependent in their structure, thereby making the exploitation of joint cost-saving potentials possible, which is one of the main goals of OBA.

All subjects received the same background information about the relationship between the supplier (Framing.Inc) and the buyer (nameless) whose perspective the participants were to take. In the experimental scenario, the participants face an initial offer of cost information and/or a relation-specific asset from Framing.Inc, the "phantom" supplier (meaning that no real person acts as the supplier because the supplier only exists within the scenario). Participants were told that their firm produces electronic devices for which casings, such as those produced by Framing.Inc, were needed.

The cost structure of the two firms was interdependent; thus, both parties could jointly search for cost-cutting opportunities by negotiating attributes and properties of the casings. For example, if the casings were delivered double-bagged, the cost incurred by the supplier would be higher; yet the buyer would incur lower quality assurance costs. The participants' attention was drawn to the interdependent cost structure; they were asked to calculate the total cost of different combinations of casing properties. A detailed illustration of the used variables is given in the next section. The experimental material was pretested on several Ph.D.-students, to clarify potential misunderstandings and to ensure successful experimental manipulation.

Operationalization of Variables

Overall, the model consists of three groups of variables that stem from different theoretical backgrounds. The first group of variables consists of the three experimental variables: *amount of cost information initially offered, offer of a relation-specific asset* and *relative power structure*, which were manipulated across the different experimental conditions. All variables in the first group serve as exogenous variables.

Information quantity was manipulated by the number of cost information items that Framing.Inc initially disclosed. In the conditions with a small quantity of initially offered information, only one cost information item (which represented about 8% of the total cost information that could be shared) was provided. In the condition with a large quantity of initially offered cost information, nine cost information items (which represented about 80% of the cost information) were revealed. The participants were asked to indicate which of their own cost information (at maximum 12) items they were willing to share in response to the initial move of the supplier. All information that was at the participants' disposal was relevant and accurate and could not be manipulated. Hence, if exchanged with the partner, these cost data could be used to achieve a lower overall cost level.

The offering of a *specific asset* was manipulated such that Framing.Inc either suggested the construction of a conveyor belt between the two adjacent production sites as a joint investment or not. The specificity of the investment was stressed by pointing out that the conveyor belt could actually only be used by the two companies. The joint investment could reduce quality assurance costs, which was one of the cost categories, by 10–20%. The cost of the investment was to be shared equally between the two firms, and the investment cost related to this specific conveyor belt would account for approximately 2% of each of the firms' annual revenue. This sharing of the cost ensured the mutual nature of the investment. The reciprocity of the investment was a crucial requirement that confirmed a genuine commitment to the relationship on the part of both parties. The operationalisation of the specific investment was implemented following an example described by Dyer and Singh (1998).

Last but not the least, *relative power structure* was operationalised by manipulating the relative dependence of the supply chain partners on each other. Power is modeled dichotomously to analyze the most interesting contrast of low relative power vs. equal or high relative power. Those participants who were either in an equal power setting or held at the high power position were in one group whereas those who were in the low power position formed the second group. The procedure was chosen to differentiate between participants operating at least on a level of parity (i.e., equal or higher power position) and those occupying a supposedly subordinate position (i.e., lower power position).

The second group of variables is derived from concepts of user acceptance. User acceptance is constituted by the concepts of ease of use and usefulness. These aspects are not expected to be directly affected by the experimental variables. Nonetheless, they are expected to have an effect on the general propensity to use OBA. Derived from the Technology Acceptance Model (Davis, 1989), two variables play a role in the proposed research model: *perceived ease of use* and *perceived usefulness*. Perceived ease of use will also be treated as an exogenous variable because reasons why participants perceive OBA to be (or to not be) easy to use are not further analyzed herein (for further information, see the study by Venkatesh and Davis (1996), who tackle the question of what factors lead to perceived ease of use). Perceived usefulness is treated as an endogenous variable that is influenced only by perceived ease of use.

Participants were asked to evaluate OBA as a cost management tool in terms of perceived usefulness. The following four items were used: (a) using OBA improves my performance as a manager; (b) using OBA increases the productivity of my firm; (c) using OBA enhances the establishment of an effective interfirm relationship; and (d) I find OBA to be useful in the relationship between Framing.Inc and my firm (all items adapted from Davis, 1989; Davis et al., 1989). Concerning perceived ease of use, participants evaluated the effort expectancy by answering the following two items: (a) OBA requires a lot of effort and (b) I perceive the principles of OBA as easy to use. These items were also adapted from Davis (1989) and Davis et al. (1989).

The third group consists of endogenous variables, which differ from exogenous variables in that their causes are explicitly represented in the model; however, they can still be causes of other variables. This means that endogenous variables can take a dual role: both as an independent and as a dependent (criterion) variable (Kline, 2005, p. 68–69). In the present model, five endogenous variables belong to the third group of variables that are essential to the concept of OBA: the *own* and the *perceived partner's willingness* to engage in cost information exchange and the amount of *cost information actually reciprocated*. These endogenous variables are the central indicators for OBA used in the study. The model aims to extend the scope of investigation beyond the situational assessment by predicting the *long-term propensity* to use OBA. Therefore, the long-term intention to use OBA in interorganizational cost management is added to the set of dependent variables. The long-term intention was assessed by two items. The first item indicated whether individuals would use OBA in further interactions with Framing.Inc. The second item asked whether participants would use OBA when faced with a similar issue in another interfirm relation.

The important position of *perceived own benefit* for the application of interorganizational cost management has been identified in prior research (Piontkowski & Hoffjan, 2009). The variable served as a moderating influence that revealed further information on the scope of validity for certain effects. Consequently, the model will contain this variable as one of the central endogenous variables.

RESULTS

The analysis of the path model will be conducted as follows. In the first step, the overall fit of the model will be evaluated and, if necessary, the model will be adjusted. In the second step, based on the then-optimized model, the proposed hypotheses will be tested.

Table 1 displays the results for the proposed model. The different goodness-of-fit indicators of the model show acceptable values. The normed fit index (NFI) of the proposed model is 0.92 and the comparative fit index (CFI) is 1.00. Because of several weaknesses – for example, the negligence of

Table 1. Goodness-of-Fit Indicators Model 1.

	p	χ^2/df	NFI	RMSEA	GFI	AGFI	CFI	TLI	SRMSR
Model value	0.51	0.97	0.92	0.00	0.96	0.93	1.00	1.00	0.06
Proposed cutoff		<2		<0.06	>0.90	>0.90	>0.95	>0.95	<0.08

Notes: NFI, normed fit index; RMSEA, root mean square error of approximation; GFI, goodness of fit index; AGFI, adjusted goodness of fit index; CFI, comparative fit index; TLI, Tucker–Lewis index; SRMSR, standardized root mean square residual.

the degrees of freedom or its dependence on the sample size – researchers have suggested to substitute the NFI by other, more powerful indicators of the model fit (Hu & Bentler, 1998; La Du & Tanaka, 1989). For completeness, the NFI will be reported among other, more sophisticated goodness-of-fit measures, but no recommendation for a cutoff is proposed.

The Tucker–Lewis index (TLI) (Tucker & Lewis, 1973) is 1.00. This value is acceptable when adopting a cutoff value of 0.95 for the CFI and the TLI (Bentler, 1990; Hu & Bentler, 1999). The goodness of fit index (GFI) is 0.96 and the adjusted GFI (AGFI) is 0.93. Because these indicators range from 0 to 1, both values are acceptable. For the next indicator, the standardized root mean square residual (SRMSR), a value of 0.06 is obtained. Because low values indicate a better model fit and the SRMSR value is below the cutoff value of 0.08 for poor model fit (Hu & Bentler, 1999), the assumption of a good approximation by the model is not rejected. The root mean square error of approximation (RMSEA) for the model is 0.00, and the lower bound of the 90% confidence interval is 0.00 (Browne & Cudeck, 1992). This represents the optimal case for the lower bound (Kline, 2005, p. 139). The upper bound of the confidence interval is 0.06. These three values indicate that not only does the RMSEA have an acceptable value but also the value of the lower bound or of the upper bound of the confidence interval does not exceed the cutoff value. In summary, the different goodness-of-fit indicators demonstrate a very good fit of the proposed research model.

After the overall model fit has been evaluated, the different paths – which represent the hypothesized effects in the model – are analyzed with the critical ratio (cr) that describes a z-value as an indicator of significance. Table 2 shows the statistics of the hypotheses tests for the research model. It is apparent that one of the seventeen paths is not significant. The critical relation is the assumed positive relation between perceived own benefit and own willingness to engage further in cost information exchange ($cr = 1.22$, $p = 0.221$). Thus, Hypothesis 6a is *not* supported. Therefore, we remove the casual connection from our model in order to improve the overall model fit because new causal paths can lead to higher goodness-of-fit values. The proposed model already shows an acceptable goodness-of-fit to the empirical data. For this reason, the primary objective of the model adaptation is improving the parsimony of the model by reducing the number of causal paths. Self evidently, an anticipated causal connection that turned out to be nonsignificant is likely to be scrutinized. Hence, the model is recalculated without the insignificant path. Fig. 2 shows the modified model including the correlations of constructs in the path model. The goodness-of-fit indices

Table 2. Statistics of Structural Equation Model 1.

Hypothesis	Path	Critical Ratio	*p*-Value
1a	Info. quantity → Perceived partner's willingness	4.12	0.000***
1b	Info. quantity → Info. reciprocated	5.49	0.000***
2	Spec. Asset → Own willingness	2.29	0.022*
3a	Power → Own willingness	−2.145	0.032*
3b	Power → Perceived partner's willingness	2.26	0.024*
3c	Power → Long-term intention to use OBA	−2.22	0.026*
4	Ease of use → Usefulness	1.74	0.082•
5a	Usefulness → Own willingness	2.10	0.035*
5b	Usefulness → Long-term intention to use OBA	4.18	0.000***
5c	Usefulness → Own benefit	9.79	0.000***
6a	Own benefit → Own willingness	1.22	0.22
6b	Own benefit → Perceived partner's willingness	3.70	0.000***
6c	Own benefit → Long-term intention to use OBA	4.85	0.000***
6d	Own benefit → Info. reciprocated	1.97	0.049*
7	Own willingness → Info. reciprocated	2.98	0.003**
8	Perceived partner's willingness → Own willingness	8.71	0.000***

*Significant at the 0.05 level.
**Significant at the 0.01 level.
***Significant at the 0.001 level.
•Significant at the 0.1 level.

for the modified model are displayed in Table 3, which shows that the fit indicators remain on a high level.

With the exception of one variable, the proposed hypotheses were supported. Finally, based on the adapted model, the proposed hypotheses are analyzed and the main results are stated. Hypothesis 1a, which states that the amount of initially offered cost information influences the perceived partner's willingness to use OBA, is supported (cr = 4.12, $p = 0.000$). The anticipated effect of the initially offered information quantity on the amount of reciprocated cost information was significant (cr = 5.486, $p = 0.000$). Therefore, Hypothesis 1b is supported. As expected, there is a significant effect of the offer of a relation-specific asset on the own willingness to engage in further cost information exchange (cr = 2.30, $p = 0.021$). Thus, Hypothesis 2 is supported.

Hypothesis 3a proposed that the party in the less powerful position would be more willing to engage in OBA activities. That is, power is negatively related to own willingness to use OBA. This assumption is supported by the results (cr = −2.24, $p = 0.025$). The expected effect that a

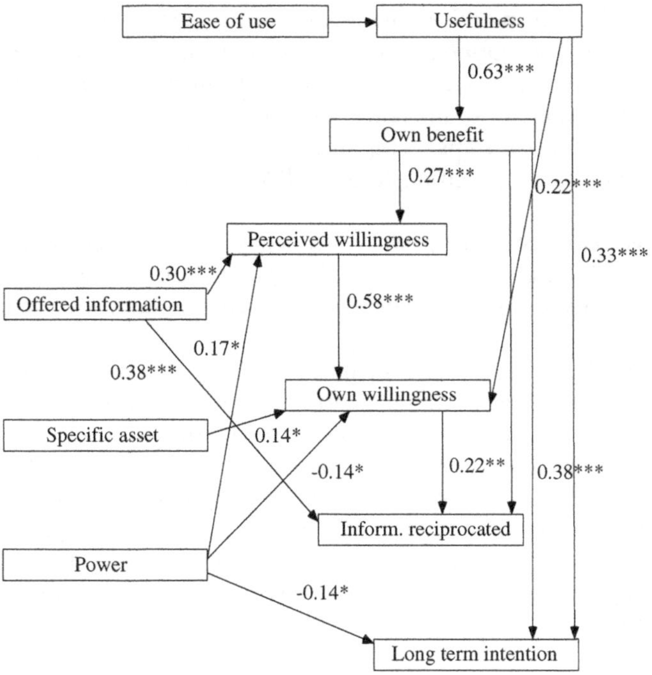

*** Significant at the 0.001 level

** Significant at the 0.01 level

* Significant at the 0.05 level

. Significant at the 0.1 level

Fig. 2. Results of the Modified Path Model (Amos Output).

Table 3. Goodness-of-Fit Indicators Model 2.

	p	χ^2/df	NFI	RMSEA	GFI	AGFI	CFI	TLI	SRMSR
Model value	0.48	0.99	0.92	0.00	0.96	0.93	1.000	1.00	0.06
Proposed cutoff		<2		<0.06	>0.90	>0.90	>0.95	>0.95	<0.08

Notes: NFI, normed fit index; RMSEA, root mean square error of approximation; GFI, goodness of fit index; AGFI, adjusted goodness of fit index; CFI, comparative fit index; TLI, Tucker–Lewis index; SRMSR, standardized root mean square residual.

more powerful firm perceives the less powerful partner as more willing to engage in interorganizational cost management was found, and Hypothesis 3b was supported (cr = 2.26, $p = 0.024$). Last but not the least, the power structure is assumed to influence long-term intention to use OBA in (similar) interfirm set-ups. Hypothesis 3c, which states that power correlates with lower intention to use OBA, is also supported (cr = −2.22, $p = 0.026$).

Perceived ease of use shows a tendency to positively influence the perceived usefulness of OBA (cr = 1.740, $p = 0.082$). Hence, Hypothesis 4, which was derived from considerations based on the TAM, tends to be supported.

As expected, the perceived usefulness of OBA as a management device positively influences own willingness to engage in cost information exchange and supports Hypothesis 5a (cr = 3.60, $p = 0.000$). The direct and positive effect of the perceived usefulness on the intention to use OBA as a management device (cr = 4.18, $p = 0.000$) supports Hypothesis 5b. The assumption that the perceived usefulness positively influences perceived own benefit (Hypothesis 5c) is also supported (cr = 9.79, $p = 0.000$).

Regarding the expected effect of perceived own benefit, three anticipated effects are found:

(1) Perceived own benefit positively influences the perceived partner's willingness to further exchange information (cr = 3.70, $p = 0.000$), which supports Hypothesis 6b.
(2) There is a significant effect of perceived own benefit on the long-term intention to use OBA (cr = 4.85, $p = 0.000$). This supports Hypothesis 6c.
(3) Perceived own benefit positively influences the number of cost information items that the participants revealed as a reaction to the initial offer. This was anticipated in Hypothesis 6d, which is therefore supported (cr = 1.97, $p = 0.049$).

Hypothesis 7 suggests that own willingness to further engage in cost information exchange has a positive effect on the amount of reciprocated cost information by the participant. A significant effect was found, and Hypothesis 7 was supported (cr = 2.98, $p = 0.003$). There was also a positive effect of the perceived partner's willingness on own willingness, which supports Hypothesis 8 (cr = 9.16, $p = 0.000$). Table 4 shows a summary of the statistics of the improved path model.

Table 4. Statistics of Structural Equation Model 2.

Hypothesis	Path	Critical Ratio	*p*-Value
1a	Information quantity → Perceived partner's willingness	4.12	0.000***
1b	Information quantity → Information reciprocated	5.49	0.000***
2	Specific asset → Own willingness	2.30	0.021*
3a	Power → Own willingness	−2.24	0.025*
3b	Power → Perceived partner's willingness	2.26	0.024*
3c	Power → Long-term intention to use OBA	−2.22	0.026*
4	Ease of use → Usefulness	1.74	0.082•
5a	Usefulness → Own willingness	3.60	0.000***
5b	Usefulness → Long-term intention to use OBA	4.18	0.000***
5c	Usefulness → Own benefit	9.79	0.000***
6b	Own benefit → Perceived partner's willingness	3.70	0.000***
6c	Own benefit → Long-term intention to use OBA	4.85	0.000***
6d	Own benefit → Information reciprocated	1.97	0.049*
7	Own willingness → Information reciprocated	2.98	0.003**
8	Perceived partner's willingness → Own willingness	9.16	0.000***

*Significant at the 0.05 level.
**Significant at the 0.01 level.
***Significant at the 0.001 level.
•Significant at the 0.1 level.

CONCLUSIONS AND MANAGERIAL IMPLICATIONS

The path model extends the scope of the investigation from situational, manipulable variables to attitudinal aspects. The latter were derived from the TAM (Davis et al., 1989). Even though these aspects, at least in this specific research design, were not expected to be – and actually are not – influenced through the experimental variables, they play an important role in the propensity to engage in cost information exchange. These attitudinal aspects cannot be influenced in an ad hoc experimental investigation, which manipulates the situational conditions. Rather, these attitudes are based on long-term considerations by the respective person or institution. Participants (and therefore the users of a management device) are influenced in their choice by the perceived ease of use of the instrument; yet the extent to which OBA can help them achieve their goals (perceived usefulness) has an even stronger influence.

Further, the important role of perceived own benefit as a determining influence was confirmed. Directly influenced by perceived usefulness, it plays an important role in the development of the propensity to use OBA.

Perceived own benefit exerts a positive influence on key (dependent) variables of the path model. If a target firm anticipates that exchanging cost information to jointly manage the cost structure of the supply chain will have beneficial consequences for both parties, it does not look for the catch in the proposal but, rather, perceives the partner as having proper and earnest intentions. Increased own benefit not only has a positive effect on the cognitive process (perception of willingness) but also fosters the reciprocal sharing of cost information as a reaction to the initial offer by the partner firm. Own benefit as a key determinant is even more consequential because it enhances the disposedness and intention to use OBA as a management device not only in the present interfirm scenario but also in other current or future buyer–supplier relationships.

In general, aspects of user acceptance prove to be important determinants of propensity. Perceived ease of use shows the expected positive influence on perceived usefulness; the effect was highly significant. However, the effect is not as strong as, for example, that of perceived usefulness on the general intention to use OBA. This result is in line with findings by Adams, Nelson, and Todd (1992), Venkatesh and Davis (2000) and Venkatesh, Morris, Davis, and Davis (2003), who all indicate that perceived usefulness is the stronger determinant of usage intention and actual usage of a system.

The effects of the experimental variables were confirmed. First, the positive effect of the offer of a relation-specific asset on own willingness to engage further in collaborative cost information exchange is confirmed. Second, the amount of cost information plays a crucial role in determining the participants' behavior; a firm reacts positively to a larger initially offered quantity of cost information by reciprocating more information themselves. Thirdly, the results for power (less powerful vs. equal or more powerful position) indicate that the power position is negatively related to the intention to use OBA. Participants show a lower inclination to use OBA in an interfirm relation when they are in a powerful position than when they are in a less powerful position. One possible reason for this behavior is that the less powerful firm tries to attenuate the disadvantages of its position by accentuating its collaborative efforts and, by extension, its genuine commitment. This confirms results found by Van den Abbeele (2006). More precisely, because a less powerful party perceives itself to be more dependent on the interfirm relation than its partner firm, it utilizes OBA as a means of convincing the partner firm of the benefit of the interfirm relation. Another reason – that, however, causes an effect in the same direction – may be that participants perceive engaging in interorganizational cost management as a chance to improve their own situation and shift the exchange relation toward a mixed power system in which both parties have some value at stake in the relationship (Bonoma, 1976).

Prior studies have identified pitfalls and inhibitors of interorganizational cost management (Kajüter & Kulmala, 2005; Kulmala, 2004). Additional conclusions related to successful initiation of OBA as understood from a managerial point of view are discussed in the next section. These serve as guidelines for avoiding pitfalls and inhibitors of prosperous interfirm cost management.

Evaluate the Power Relation

Bonoma (1976) indicated that a bilateral power set-up provides the most suitable conditions for a prosperous interfirm relation. This hypothesis is supported by the findings of this study; that is, the relative power structure was found to change behavior. Whereas, in the context of equal power, a feeling of mutuality and unity is very likely to evolve, under power asymmetry, perceived risk for the more and the less powerful drift apart, indicating a divergent evaluation of the particular position. Further, under asymmetric power conditions, the application of commitment devices needs to be thoroughly reconsidered. Rokkan et al. (2003) identified that relation-specific assets can have ambiguous effects that depend on the relational context in which they are applied.

Combining these results with those of the present study, it is rather questionable whether conclusions of the effects of specific assets in interfirm set-ups can be generalized. However, it is obvious that special care must be taken when applying idiosyncratic assets to interorganizational relationships because the consequences are far from unambiguously determinable. If at all, relation-specific assets in the initiation phase of an interfirm partnership need to be carefully applied. Relation-specific assets can lower the fear of opportunistic behavior by the partner if, and only if, offered by the firm in the more powerful position. In an interfirm relationship, very frequently, one of the involved parties will perceive itself to be more/less powerful than the partner. If this is the case, the application of a specific asset becomes rather critical, as has been discussed in detail.

Anticipate Possible Concerns

Proactively anticipating possible concerns that a partner might have during the establishment of an interfirm relationship is a relevant aspect of interorganizational alliance forming. Thus, providing the necessary information and, if applicable, the required management capacity, are important issues for the formation of interfirm alliances (Kelly, Schaan, & Joncas, 2002). This will not only increase perceived willingness to make the interfirm cooperation work but also will put the manager in a position of exerting a stronger influence on the interfirm alliance (Mjoen & Tallman, 1997; Muthusamy & White, 2006). This is especially important in situations similar to the one depicted in the scenario of the present study in which one partner is persuading the other to engage in OBA. In such situations, there is increased danger that either the persuaded party grows resistant or, even though agreeing on the principles of OBA, the targeted organization lacks firm commitment to the project. This means that the situational pressure results in a pseudo-accordance, which is not supported by true conviction. Therefore, pushy persuasion strategies are to be avoided because they will only evoke reluctance. Instead, the managers should count on long-term commitment and compliance (for further empirical findings see Gundlach et al., 1995).

Increase Partner's User Acceptance

No management tool has the chance to be applied in the long run if it is not accepted by the user. The general evaluation of OBA as a cost management tool was very positive across all groups because intention to use OBA in general was very high. This indicates that, in general, OBA is seen as an instrument that can optimize supply-chain-wide cost and foster interfirm relationships.

Therefore, it becomes the manager's responsibility to transform this general propensity to use OBA into actual user acceptance. The results of the present study together with findings from prior research identify starting points for persuading others to use OBA. First, attitudinal aspects should be targeted. In general, for the short term, decisions to adopt a new management technique will most likely produce resistance. Persuasion strategies at this stage should be prevention oriented and focus on weakening resistance by reducing the fear of misconduct.

During the later stages, a promotion-oriented persuasive strategy that concentrates on possible gains through the collaboration should be chosen, as recommended by the Temporal Construal Theory by Trope and Liberman (Sherman, Crawford, & McConnell, 2004; Trope & Liberman, 2000, 2003). As shown in the study, perceived usefulness and – for a specific situation, perceived own benefit – are crucial for the acceptance of a management device. A firm that initiates interorganizational cost management activities with a target firm should consult domain-specific persuasion strategies and abstain from non-domain specific strategies. For instance, it should stress that joint cost management does not aim at eroding the partner's profit margin. Rather, it aims at eroding the partner's cost. Perceived own benefit could be pinpointed by presenting prior successful examples of interorganizational cost management activities and/or by presenting possible scenarios for the intended relationship with the potential partner firm. Further, OBA aims at optimizing the cost structure in a supply chain. Hence, arguments presented to the target firm should concentrate on the benefits obtainable through the engagement in interorganizational cost management.

Because we answer our research question with the help of a laboratory experiment, our findings have several limitations. Our paper investigates the behavior of the participants in a one-move experiment. That is, the initial move and the reaction to it are analyzed. However, the analysis neglects possible effects that can emanate from multiple steps of interaction. For example, a history of (other, non-OBA-specific) collaborative and reliable interactions between the partner firms can pave the way for the introduction of joint cost management.

For any empirical investigation, the aspect of a possible answering bias (due to the presence of social desirability) must be taken into account (Pavlou & Fygenson, 2006; Podsakoff & Organ, 1986). In the case of an investigation on interorganizational cooperation and collaboration, it might be possible that the participants indicate a higher propensity to share information and to collaborate than they actually have. The reason for this sort of behavior lays in the fact that people might be influenced by thoughts about what may be socially desirable: that is, which conduct is expected of them in a certain situation.

The proposed model exhibited rather high goodness-of-fit indicators. However, this does not mean that a claim for completeness could be made. The proposed relations, which integrate attitudinal variables in the theoretic framework, constitute one possible model for explaining the evolvement of OBA. Other approaches based on other theoretical constructs – for example, the influence of different cultural aspects, which could also provide a well founded explanation – cannot be ruled out. Considering these limitations, a significant amount of work remains to be done in this field.

REFERENCES

Adams, D. A., Nelson, R. R., & Todd, P. A. (1992). Perceived usefulness, ease of use, and usage of information technology: A replication. *MIS Quarterly*, *16*(2), 227–247.

Andaleeb, S. S. (1996). An experimental investigation of satisfaction and commitment in marketing channels: The role of trust and dependence. *Journal of Retailing*, *72*(1), 77–93.

Anderson, E., & Weitz, B. (1992). The use of pledges to build and sustain commitment in distribution channels. *Journal of Marketing Research*, *29*(1), 18–34.

Baiman, S., Fischer, P. E., & Rajan, M. V. (2001). Performance measurement and design in supply chains. *Management Science*, *47*(1), 174–188.

Baiman, S., & Rajan, M. V. (2002). Incentive issues in inter-firm relationships. *Accounting, Organizations and Society*, *2*(3), 213–238.

Bentler, P. M. (1990). Comparative fit indexes in structural models. *Psychological Bulletin*, *107*(2), 238–246.

Bonoma, T. V. (1976). Conflict, cooperation and trust in three power systems. *Behavioral Science*, *21*(6), 499–514.

Browne, M. W., & Cudeck, R. (1992). Alternative ways of assessing model fit. *Sociological Methods & Research*, *21*(2), 230–258.

Buelens, M., & Van Poucke, D. (2004). Determinants of a negotiator's initial opening offer. *Journal of Business and Psychology*, *19*(1), 23–35.

Campbell, D. T. (1963). *Social attitudes and other acquired behavioral dispositions*. New York, NY: McGraw-Hill.

Carr, C., & Ng, J. (1995). Total cost control: Nissan and its UK supplier partnerships. *Management Accounting Research*, *6*(4), 347–365.

Clee, M. A., & Wicklund, R. A. (1980). Consumer behavior and psychological reactance. *The Journal of Consumer Research*, *6*(4), 389–405.

Coad, A. F., & Cullen, J. (2006). Inter-organisational cost management: Towards an evolutionary perspective. *Management Accounting Research*, *17*(4), 342–369.

Cooper, R., & Slagmulder, R. (2004). Interorganizational cost management and relational context. *Accounting, Organizations and Society*, *29*(1), 1–26.

Das, T. K., & Teng, B.-S. (1996). Risk types and interfirm alliance structures. *Journal of Management Studies*, *33*(6), 827–843.

Das, T. K., & Teng, B.-S. (2000). A resource-based theory of strategic alliances. *Journal of Management*, *26*(1), 31–61.

Davis, F. D. (1989). Perceived usefulness, perceived ease of use, and user acceptance of information technology. *MIS Quarterly*, *13*(3), 318–340.

Davis, F. D., Bagozzi, R. P., & Warshaw, P. R. (1989). User acceptance of computer technology: A comparison of two theoretical models. *Management Science*, *35*(8), 982–1003.

Dekker, H. C. (2003). Value chain analysis in interfirm relationships: A field study. *Management Accounting Research*, *14*(1), 1–23.

Drake, A. R., & Haka, S. F. (2008). Does ABC information exacerbate hold-up problems in buyer-supplier negotiations? *The Accounting Review*, *83*(1), 29–60.

Dyer, J. H., & Singh, H. (1998). The relational view: Cooperative strategy and sources of inter-organizational competitive advantage. *Academy of Management Review*, *23*(4), 660–679.

Galinsky, A. D., & Mussweiler, T. (2001). First offers as anchors: The role of perspective-taking and negotiator focus. *Journal of Personality and Social Psychology*, *81*(4), 657–669.

Ghoshal, S., & Moran, P. (1996). Bad for practice: A critique of the transaction cost theory. *The Academy of Management Review*, *21*(1), 13–47.

Gouldner, A. W. (1960). The norm of reciprocity: A preliminary statement. *American Sociological Review*, *25*(2), 161–178.

Gundlach, G. T., Achrol, R. S., & Mentzer, J. T. (1995). The structure of commitment in exchange. *Journal of Marketing*, *59*(1), 78–92.

Heide, J. B., & John, G. (1990). Alliances in industrial purchasing: The determinants of joint action in buyer-supplier relationships. *Journal of Marketing Research*, *27*(1), 24–36.

Hoffjan, A., & Kruse, H. (2006). Open book accounting in supply chains – When and how is it used in practice? *Cost Management*, *20*(6), 40–47.

Hu, L., & Bentler, P. M. (1998). Fit indices in covariance structure modeling: Sensitivity to underparameterized model misspecification. *Psychological Methods*, *3*(4), 424–453.

Hu, L., & Bentler, P. M. (1999). Cutoff criteria for fit indexes in covariance structure analysis: Conventional criteria versus new alternatives. *Structural Equation Modeling*, *6*(1), 1–55.

Kajüter, P., & Kulmala, H. I. (2005). Open-book accounting in networks: Potential achievements and reasons for failures. *Management Accounting Research*, *16*(2), 179–204.

Kelly, M. J., Schaan, J.-L., & Joncas, H. (2002). Managing alliance relationships: Key challenges in the early stages of collaboration. *R&D Management*, *32*(1), 11–22.

Kline, R. B. (2005). *Principles and practice of structural equation modeling*. New York, NY: Guilford.

Kulmala, H. I. (2004). Developing cost management in customer-supplier relationships: Three case studies. *Journal of Purchasing & Supply Management*, *10*(2), 65–77.

Kwon, I.-W. G., & Suh, T. (2005). Trust, commitment and relationships in supply chain management: A path analysis. *Supply Chain Management: An International Journal*, *10*(1), 26–33.

La Du, T. J., & Tanaka, J. S. (1989). Influence of sample size, estimation method, and model specification on goodness-of-fit assessments in structural equation models. *Journal of Applied Psychology*, *74*(4), 625–635.

Langfield-Smith, K. (2005). *Trust, risk and control in strategic alliances: A case study in the construction industry*. Retrieved from http://www.mbs.ac.uk/-research/accounting-finance/documents/imLangfieldSmith_000.pdf. Accessed on March 23, 2010.

Miraglia, R. A., & Leotta, A. (2010). The interaction between information and trust in the control of transactional relationships: Theoretical perspectives and empirical support. *Studies in Managerial and Financial Accounting, Performance Measurement and Management Control: Innovative Concepts and Practices*, *10*(1), 143–175.

Mjoen, H., & Tallman, S. (1997). Control and performance in international joint ventures. *Organization Science*, *8*(3), 257–274.

Mouritsen, J., Hansen, A., & Hansen, C.Ø. (2001). Inter-organizational controls and organizational competencies: Episodes around target cost management/functional analysis and open book accounting. *Management Accounting Research*, *12*(2), 221–244.

Mouritsen, J., & Thrane, S. (2006). Accounting, network complementarities and the development of inter-organisational relations. *Accounting, Organizations and Society*, *31*(3), 241–275.

Muthusamy, S. K., & White, M. A. (2006). Does power sharing matter? The role of power and influence in alliance performance. *Journal of Business Research*, *59*(7), 811–819.

Otley, D. (1994). Management control in contemporary organizations: Towards a wider framework. *Management Accounting Research*, *5*(3–4), 289–299.

Pavlou, P. A., & Fygenson, M. (2006). Understanding and predicting electronic commerce adoption: An extension of the theory of planned behavior. *MIS Quarterly*, *30*(1), 115–143.

Piontkowski, J. O. (2008). *Determinants of the propensity to use open book accounting*. Retrieved from http://www.scribd.com/doc/15696361/Determinants-of-the-propensity-to-use-open-book-accounting

Piontkowski, J. O., & Hoffjan, A. (2009). Less is sometimes more: The role of information quantity and specific assets in the propensity to engage in cost data exchange processes. *Journal of Purchasing and Supply Management*, *15*(2), 71–78.

Podsakoff, P. M., & Organ, D. W. (1986). Self-reports in organizational research: Problems and prospects. *Journal of Management*, *12*(4), 531–544.

Rokkan, A. I., Heide, J. B., & Wathne, K. H. (2003). Specific investments in marketing relationships: Expropriation and bonding effects. *Journal of Marketing Research*, *40*(2), 210–224.

Seal, W., Cullen, J., Dunlop, A., Berry, T., & Ahmed, M. (1999). Enacting a European supply chain: A case study on the role of management accounting. *Management Accounting Research*, *10*(3), 303–322.

Sherman, S. J., Crawford, M. T., & McConnell, A. R. (2004). Looking ahead as a technique to reduce resistance to persuasive attempts. In B. S. Knowles & J. A. Linn (Eds.), *Resistance and persuasion* (pp. 149–174). Mahwah, NJ: Lawrence Erlbaum Associates.

Thrane, S., & Hald, K. S. (2006). The emergence of boundaries and accounting in supply fields: The dynamics of integration and fragmentation. *Management Accounting Research*, *17*(3), 288–314.

Tomkins, C. (2001). Inter-dependencies, trust and information in relationships, alliances and networks. *Accounting, Organizations and Society*, *26*(2), 161–191.

Trope, Y., & Liberman, N. (2000). Temporal construal and time-dependent changes in preference. *Journal of Personality and Social Psychology*, *79*(6), 876–889.

Trope, Y., & Liberman, N. (2003). Temporal construal. *Psychological Review*, *110*(3), 403–421.

Tucker, L. R., & Lewis, C. (1973). A reliability coefficient for maximum likelihood factor analysis. *Psychometrika*, *38*(1), 1–10.

Van den Abbeele, A. (2006). *Management control of inter-firm relations: The role of information*. Retrieved from https://lirias.kuleuven.be/bitstream/1979/349/1/Encrypted%2BPhd%2Bvoor%2Barchivering.pdf.

Van der Meer-Kooistra, J., & Vosselman, E. G. (2006). Research on management control of interfirm transactional relationships: Whence and whither. *Management Accounting Research*, *17*(3), 227–237.

Venkatesh, V., & Bala, H. (2008). Technology acceptance model 3 and a research agenda on interventions. *Decision Sciences*, *39*(2), 273–315.

Venkatesh, V., & Davis, F. D. (1996). A model of the antecedents of perceived ease of use: Development and test. *Decision Sciences*, *27*(3), 451–481.

Venkatesh, V., & Davis, F. D. (2000). A theoretical extension of the technology acceptance model: Four longitudinal field studies. *Management Science*, *46*(2), 186–204.

Venkatesh, V., Morris, M. G., Davis, G. B., & Davis, F. D. (2003). User acceptance of information technology: Toward a unified view. *MIS Quarterly*, *27*(3), 425–478.

PART IV
RISK MANAGEMENT, MOTIVATION AND PERFORMANCE

ARE PERFORMANCE-DEPENDENT REWARDS A VIABLE TOOL TO ASSURE MANAGERS' COMMITMENT TOWARD FIRMS' GOALS ABOUT RISK MANAGEMENT?

Selena Aureli and Federica Salvatori

ABSTRACT

Purpose – *Since risk management is crucial for achieving strategic objectives in a complex and uncertain environment and its effectiveness relies deeply on efforts to create a risk-conscious culture, this study aims at understanding whether risk management can be promoted and reinforced by the use of performance-based monetary incentives given to Board members and top managers.*

Methodology/approach – *This study is explorative in nature and investigates four case studies based on document analysis and semi-structured interviews with risk managers.*

Findings – *Results show that some companies have already adopted risk measures in incentive schemes. At the same time all interviewees agree*

Performance Measurement and Management Control: Global Issues
Studies in Managerial and Financial Accounting, Volume 25, 249–274

ISSN: 1479-3512/doi:10.1108/S1479-3512(2012)0000025013

with the usefulness of linking traditional performance-based monetary incentives to risk management objectives in order to improve the effectiveness of the latter and to create a risk-aware culture. However, the difficulty in identifying proper measures has been underlined.

Practical implications – *The study confirms the feasibility of linking risk dimensions to reward systems and suggests that firms should move in this direction. The study also outlines and proposes some possible measures to reward managers.*

Limitations – *This study views risk as measurable and managerially actionable and focuses only on incentives while acknowledging the use of other mechanisms that can contribute to the creation of an informed risk culture. Furthermore, the integration of risk management with other management control systems and accounting instruments has not been analyzed.*

Value of the paper – *This study addresses firms and their stakeholders' need to make top managers more accountable for risk in their decision-making.*

INTRODUCTION

Events such as Enron's bankruptcy and British Petroleum's oil disaster have shown that firms' profitability and survival are increasingly affected by the consequences of poor internal management and unexpected events.

Both shareholders and stakeholders now require firms to identify risks and report how risks are being handled (see, e.g., new U.S. Securities and Exchange Commission (SEC) disclosure requirements about risk supervision). At the same time, CEOs and top managers are becoming more aware that effective risk management is fundamental to achieving company objectives and can represent the source of significant competitive advantage (IFAC, 1999).

A good risk management system should be a company-wide system primarily structured to help managers face challenges and improve and guide their decision-making toward value creation in the long term (Frigo, 2009). Furthermore, it should be strictly connected to strategic planning (Puschaver & Eccles, 1996).

In turn, a risk management system can be conceived as a management system devoted to providing information for managers to assist their decision-making and to control strategy implementation (Horngren, Bhimani,

Foster, & Datar, 1999; Ryan, Scapens, & Theobald, 2002). However, this approach requires the creation of an instinctive consideration of risk among all managers involved in planning, execution, and monitoring of a company's strategies as well as strong integration with other management control systems (Cokins, 2009) and accounting instruments that have historically excluded risk from their content (Collier & Berry, 2002).

One possible mechanism that could contribute to creating a risk-aware behavior within organizations consists in linking risk management with performance measurement systems and reinforcing its use by introducing specific rewards related to risks. The reward system can represent a powerful instrument for influencing an organization's culture, although it alone is not sufficient to change cultural values and norms (Kerr & Slocum, 2005); it can also mobilize organizational commitment to successful strategy execution (Thompson, Strickland, & Gamble, 2005).

The aim of this study is to understand whether risk management systems, such as the Enterprise Risk Management (ERM) (COSO, 2004), could be more effective in contributing to a company's achievement of risk-centered goals when reinforced by rewards, and, specifically, by monetary incentives given to Board members and top managers.

The focus on monetary incentives derives from several sources. Monetary incentives represent a common instrument employed for motivating and aligning managers' behavior toward company objectives (Daft, 2006; Latham, 2007) and for rewarding executive directors who sit on company Boards as well as top managers (Hewitt, 2007, 2009). Moreover, their past use has exposed firms to severe risks of bankruptcy instead of advancing shareholders' interests in long-term value creation (Acharya & Richardson, 2009; Roubini & Mihm, 2010). Lastly, very few contributions in the literature underline the need for linking a company's reward system with risk management (Beasley & Frigo, 2010; Crouhy, Galai, & Mark, 2006) and, furthermore, they do not offer proposals.

Nevertheless, monetary incentives cannot be always applied (Simons, 2000) and have some drawbacks (Kohn, 1993; Manzoni, 2010). As a consequence, we question whether traditional performance-based monetary incentives can be related to risk management and whether the link among them reinforces the risk management function and contributes to the development of a risk-aware behavior.

To answer to these questions, the paper is structured as follows. In the next section, we present the research methodology, followed by a literature review on risk management and rewards. Then, we link risk management to performance measurement and reward systems by interpreting risk as measurable and managerially actionable (Williamson, 2007). In the empirical analysis, hypotheses about monetary incentives linked to risk measures are discussed with risk managers of some Italian companies through interviews. In Section "conclusions," the main findings are summarized.

RESEARCH METHODOLOGY

Qualitative methodology appears to be the most appropriate for analyzing the topic under consideration and for understanding managers' reactions to "risk-adjusted" performance incentives, since it allows an in-depth study to collect soft data and managers' perceptions (Corbetta, 2003; Flick, 2002; Silverman, 1997).

The use of quantitative techniques would be hindered by the lack of databases and other accessible sources of information necessary to create a statistical sample and two additional obstacles: a scarce diffusion of risk management systems even among large listed companies – especially in Italy (KPMG, 2010; Protiviti, 2011) – and the usual implementation of risk management systems to limited company areas or their stoppage at the risk assessment phase of the ERM framework (KPMG, 2010). This implies that for the moment it is difficult to undergo a quantitative analysis to explain which factors determine variances in the use and acceptability of incentives based on risk measures.

Since the available information on the topic is limited and the problem has not been clearly defined by the literature, our study can be considered as explorative (Ryan et al., 2002). It does not aim to demonstrate the ability to generalize specific affirmations or to provide definitive conclusions, but rather it explores whether and how performance-dependent rewards can be applicable to motivate executive directors and top managers to effectively use risk management systems.

Accordingly, interviews with some risk managers of four large Italian companies were undertaken in Spring 2011 to get a handle on the situation. Risk managers represent the repository of company knowledge about risks and risk management and provide an independent judgment about possible changes in top managers' reward systems.

Companies were selected from both the industrial sector and the service sector (with the exclusion of financial services) and we considered both listed and unlisted companies, which are treated differently by the Italian legislation in terms of risk management disclosure. With these criteria in mind, the selection of the four specific companies was opportune in the sense that the researchers conducted interviews with those managers that agreed to participate.

The four situations encountered can be seen as four different case studies (Yin, 1994), which are exploratory in nature (Ryan et al., 2002; Yin, 1994) as they are intended to provide insights about how a change in reward systems could be structured.

The analysis of the cases is based on the examination of public documents (e.g., annual reports, reports on corporate governance, and sustainability reports when available) and face-to-face semi-structured interviews, which help capture subjects' point of view (Flick, 2002) and can better ensure that comparable information is obtained compared to unstructured interviews (Corbetta, 2003). In particular, we asked for information concerning how risk is conceived and managed by the company, the features of the implemented reward schemes, existing or potential links that could be put in place between risk management and top managers' remuneration, and personal opinions about the general feasibility of performance risk-dependent rewards in the Italian context.

Attention is actually limited to the remuneration of executive members of the Board of Directors and top managers with strategic responsibilities because their decisions have a large impact on a company's goal achievement. The Board also plays a controlling role in operational and strategic control systems (Zahra & Pearce, 1989) and is ultimately responsible for monitoring the ERM program (Branson, 2010). The Board and the top management share the responsibility of creating a risk-aware culture within the company (Branson, 2010). However, they have been blamed for being one of the causes of the recent worldwide financial crisis as they made choices to increase short-term value of stocks and obtain bonuses, without considering overall company risk exposure (Acharya & Richardson, 2009). Independent and nonexecutive directors of the Board are not investigated since they do not have executive power and their variable remuneration (when it is not completely fixed) is usually based on attendance to the Board and other Committees meetings.

LITERATURE REVIEW

Risk and Its Management

Nowadays, companies operate in such a highly uncertain environment (Chorafas, 2001) that they have to deal with risks when executing their strategy to achieve business objectives and create shareholder value (Andersen & Terp, 2006).

In the managerial perspective, risk has been mainly defined as either the possibility of a negative economic consequence of an event (Crouhy et al., 2006) or a situation with a two-sided outcome, that is, potential loss or gain (Liebenberg & Hoyt, 2003; Rahardjo & Dowling, 1998). Although the concept of risk has been clearly distinguished from the concept of uncertainty (Knight, 1921), the two are closely related. Uncertainty generates the possibility that future events, both positive and negative, might produce a reality with unwanted or unexpected characteristics (McNamee & Selim, 1998; Renn, 1998), which causes the risk that a company might not achieve its objectives (Cokins, 2009).

After many corporate scandals, stakeholders have begun to question the role of Board members and put more pressure on Boards and Committees to effectively carry out their corporate governance and risk management responsibilities. Thus, several guidelines about risk management and risk reporting have been developed in order to improve corporate governance models and promote risk disclosure (Epstein & Buhovac, 2006; Grant & Visconti, 2006).

Most frameworks (i.e., COSO-ERM framework, AS/NZS 4360:2004, and FERMA) issued to help companies manage risks in an integrated manner, recognize the interdependencies within the enterprise and enable the identification of the company's aggregated risk exposure. In fact, all of these frameworks derive from the "enterprise-wide risk management" paradigm (Beasley, Cune, & Hermanson, 2005) that followed the traditional "silos" approach where risks were managed through narrowly focused activities. Moreover, they all suggest specific activities (i.e., risk identification, assessment, responses, and control) and share a common focus on the link between risk management and the strategic process.

Since uncertainty creates the need for strategy, risks are inherent in all strategic decisions and activities. In turn, risk management necessarily copes with most of the long-range decisions and, nonetheless, with the strategy (Baird & Thomas, 1985; Frigo, 2009) that should be adopted. Furthermore, it helps decision-makers make informed choices, supporting the prioritization of actions and the distinction among alternative courses of action as well as decisions about whether a risk is unacceptable and whether a risk treatment will be adequate and effective.

In this perspective, risk management is not a defensive tool, but an offensive and strategic one. It is not about how to avoid or reduce risks, but about the adequateness of risk-taking: how to actively select the type and the level of risk that is appropriate for the firm to assume (Crouhy et al., 2006). Moreover, it has many similarities to a management control system since they both are linked to strategy, aim to provide reasonable assurance regarding a company's goal achievements, involve Board of Directors, management, and other personnel, and tend to operate globally throughout the enterprise (Mikes, 2009).

The Board's commitment is a crucial aspect for the effectiveness of holistic risk management system. It ensures the consistency of the risk management processes designed and implemented by senior executives and risk management professionals while controlling that risk management processes are functioning as designed (Branson, 2010). The Board may assign primary risk oversight responsibility to a Risk Management Committee established within the Board; however, the Internal Audit function, which should monitor the effectiveness of risk management processes designed and implemented by senior management, also plays an oversight role. The involvement of CFOs is important as well since their position allows them to have a holistic understanding of the key activities that drive performance (Branson, 2010).

To manage risks properly, the company's strategy has to be chosen such that its associated risks rest within the stakeholders' risk appetite and the established risk tolerance. Then, the Board approves risk policies to communicate throughout the organization that formalize expectations about how to manage each major source or category of risk, and the senior management should identify events and scenarios that may impact strategy implementation and the achievement of the organizational objectives. In this phase, leading event indicators (or leading risk indicators) help in event identification since they are qualitative and quantitative measures (i.e., consumer confidence) that provide insights into potential events, while interrelation among events and risk drivers can be uncovered through an influence diagram. Later, the closest people to the source of disturbance should carry risk assessment out considering the likelihood of occurrence of each identified event and their impact on the achievement of corporate objectives so as to prioritize events and set risk responses. Overall entity risk or business unit risk should be assessed as well, recurring to aggregated risk measures or translating different risk measures into a common unit of measure (i.e., earning per share).

It is worthwhile to emphasize that risk has been measured in different ways, that is, using quantitative measures, such as the standard deviation of return on assets, variance market-based risk measures (market betas), accounting-based risk indicators (i.e., variability of returns, leverage ratio, accounting beta), and qualitative measures (Baird & Thomas, 1990). However, the mean/variance approach to studying risk in corporate performance measures has been criticized since it does not allow the discrimination of differences in performance (Bettis & Mahajan, 1990). Collins and Ruefli (1992) proposed an ordinal approach to strategic risk measurement based on benchmarks with competitors.

Among the available models, ERM is the most widespread framework for enterprise-wide risk management worldwide (AON, 2010; Hexter & Gates, 2005).

Although several authors argue in favor of its benefits (Meulbroek, 2002; Nocco & Stulz, 2006), others have underlined its obstacles (i.e., the difficulty in establishing a common risk management language and understanding how risks are interconnected as well as the low recognition of benefits) (Hagigi & Sivakumar, 2009; Schrøder, 2006). The recent framework proposed by ISO (31000:2009) tries to overcome some limitations of ERM, such as the objective and negative definition of risk, the lack of integration of ERM in a company's processes, and the consideration of only deliberated strategy (Williamson, 2007), while introducing the possibility to appreciate a company's risk maturity.

Regardless of the kind of framework chosen by a company, the success of risk management strongly depends on efforts to create a risk-conscious culture (AON, 2010), which is a shared responsibility of the Board members and the senior management team (Branson, 2010). Compensation structures may play an important role in creating a culture that allows making strategic as well as day-to-day business decisions while assuring risk-taking behavior, especially when trade-offs between short-term goals and long-term risk-adjusted value are involved in their decisions (Brooks, 2010).

Executives' Rewards

Rewards are a common instrument used to guide and motivate people (Daft, 2006; Latham, 2007). Many studies on rewards draw from the principal-agent problem of agency theory (Jensen & Meckling, 1976), but the greatest amount of research derives from organizational and psychological studies. Researchers such as Vroom (1964) and Herzberg (1987) have developed several behavioral theories concerning employees' motivation, which have been widely employed in management studies (see Locke (1991) for a review). In this field, rewards are held to be fundamental in reinforcing the management systems with which they are associated, since they inform employees about which areas and results are important and motivate and direct their behavior accordingly (Simons, 2000).

Rewards can consist of compensation, benefits, long- and short-term monetary incentives (which represent an extrinsic motivator) as well as nonmonetary incentives. These rewards can be explicit or implicit, depending on the feasibility to verify a result or to identify a measure that approximates the desired result (Milgrom & Roberts, 1992; Simons, 2000). Moreover, companies can decide to reward behavior or performance. This means that companies have to control either employees' activity or results. In the first case, evaluation is subjective and incentives contribute to promote creativity, problem-solving capabilities, and learning. In last case, the measurement method is considered as objective and incentives motivate the achievement of one or more performance indicators (Simons, 2000).

Today top management's rewards mainly consist of monetary incentives (i.e., bonuses, stock options, and company shares), which count for about 2/3 of the managers' entire compensation package (Hewitt, 2009). Such incentives are usually tied to the achievement of specific company and/or individual goals expressed in terms of key performance indicators (KPI), although the literature is not unanimous about the positive association between managerial incentives and company outcomes (Barkema & Gomez-Mejia, 1998; Bonner & Sprinkle, 2002).

The use of explicit rewards based on a pay-for-performance compensation program has been always accepted and welcome for top managers. Executives and top mangers believe in the feasibility of tying pay to strategic performance (Pennings, 1993) and they are more tolerant of putting their own pay at risk compared to those of their employees (Milgrom & Roberts, 1992). Moreover, monetary incentives are preferred by companies because they are flexible on the amount and pay-out frequency and do not create competition among managers as promotions can do (Milgrom & Roberts, 1992). Lastly, this form of compensation seems to be useful for exploiting top managers' skills and informational advantage about how to perform their job, whose complexity does not help specify an expected behavior (Crystal, 1991).

Nevertheless, some critics have recently suggested that monetary incentives may reduce intrinsic motivation (Cameron & Pierce, 1994; Eisenberger & Cameron, 1996), inhibit learning and cooperation (Manzoni, 2010), induce people to ignore important dimensions of performance that have not been specified in a poorly designed reward system (Kohn, 1993) and push managers to choose risky strategies at the expenses of the corporation just to reach the targets set and get the associated incentives.

This last drawback is at the center of today's most important corporate governance discussions. Traditional incentives structures – based on short-term performance and quasi-automatic payments – and particularly stock options, have been criticized for having encouraged executives to make risky bets in the hope of a big return while being aware that their compensations would not carry the negative consequences of bad choices (Acharya & Richardson, 2009; Crouhy et al., 2006; Wright, Kroll, Krug, & Pettus, 2007). When rewards are not properly adjusted for economic risk, it is more likely that managers will be tempted to minimize the potential for unexpected losses and undertake actions that push the share price up – but not necessarily in a sustainable way (Crouhy et al., 2006).

To solve this problem, the banking sector has proposed more long-term incentives, restricted shares, deferred bonuses, and longer stock holding periods, although these mechanisms are scarcely used in Europe (Hewitt, 2009). Similarly, the Conference Board (2009) has suggested using long-term measures such as total shareholder return.

Others have adopted a bonus-malus approach to compensation. Using a bonus pool that is cashed out in an alternate manner over years, bad performances recorded in the future are subtracted from total remuneration (Acharya & Richardson, 2009; UBS, 2008). Thus, reward schemes include information on future risks generated by today actions.

At the same time regulator authorities and government bodies have issued new rules on corporate governance aiming to improve companies' disclosure and accountability on both risk and compensation (i.e., the Definitive Proxy Statement required by the SEC), in addition to the crucial presence of an independent Compensation Committee within the Board (Elson, 1993; Jensen & Murphy, 1990).

However, neither a longer time horizon nor more severe rules and ethical standards alone can ensure that risk will be managed properly. It is more likely that stronger integration of the risk dimension in reward systems will better contribute to obtaining significant improvements in risk management and the diffusion of an internal risk culture.

Linking the Reward System to Risk Management

The key issue in linking a company's reward system to risk management is the choice of performance measures and consequent design of incentive plans. This is never an easy task (Locke, 2004), but it appears even more challenging with reference to risk management.

Following the traditional incentive alignment model (Locke, 2004), rewards could be associated with risk management only in case of clear measurability of expected outcomes, while risks are not always quantifiable. Accuracy in risk quantification tends to decrease when moving from low to top levels (IMA, 2007). Moreover, risks are usually so strongly interrelated that risk outcomes are not always controllable but depend on the work of several managers belonging to different company functions and departments.

Thus, as ISO 31000: 2009 emphasizes, one of the key tasks is determining risk management performance indicators aligned with organizational performance indicators and casual maps that explain which factors (risk drivers) have to be managed (or monitored if not controllable) to curb the company's risks.

Improvements in risk analysis techniques are very helpful here. Instruments such as scenario analysis and/or simulation analysis facilitate the assessment of risk and gain/loss curves capture risk impact on revenues or other metrics, while stochastic economic scenarios depict the aggregate scenario of all risks and their impact on the company's global risk profile. The use of key risk indicators (KRIs) can also prove useful (Hwang, 2010) in establishing more "adequate" KPIs and emphasizing new relevant dimensions, when it is difficult to define a risk measure. KRIs show the potential presence, level, or trend of a risk and signals what can affect performance (and so whether traditional KPIs may be achieved or not) as well as the need for further actions at all organizational levels. They also show the level of stress under which risk management activities may be operating (Beasley, Branson, & Hancock, 2010).

With reference to executive directors or top managers, incentives can be associated with risk goals expressed in terms of synthetic measures such as earnings at risk, calculated for business units or the entire organization on the basis of expected impact of risk drivers on earnings. Thus, managers are not motivated to maximize earnings undertaking whatever decision they prefer but rather to curb the variability of earnings, working directly, or pushing lower managers to work on risk causes measured by KRIs. If casual maps and KRIs are defined correctly, risk-based incentives can be defined for both top and lower managers, for risk owners of all areas and hierarchical levels.

Interestingly, a risk-related reward system has the benefit of pushing managers to consider relevant dimensions that are otherwise usually disregarded – considering that risk is ignored in traditional reward systems. Moreover, since risk identification and assessment are processes that require involvement and confrontation among risk owners of different areas, risk managers, and even the Board, the definition of rewards tied to risk measures reinforces the benefits of a participative process of target-setting such as promoting responsibility, managers' personal involvement, creativity, and collective learning. Participation in questionnaires or interviews to assess risks is not viewed as an additional duty. On the contrary, managers will be more active in participating and refining techniques used to identify and measure risks, benefiting from the efficacy and maturity level of risk management.

Case Studies

Four Italian companies, whose name disclosure has not been authorized, are described in the following sections on the basis of the corporate documents available and interviews with their risk managers (direct quotes from interviewees are displayed in quotation marks).

They all are large companies (see Table 1), but while company D operates in the industrial sector, the rest operate in the service sector. Among the latter, two provide public utilities (although for one of them it is not the main activity) and their equity is held by public sector entities. In one case, a municipality is the only shareholder.

Case A

Company A is one of the largest Italian groups operating in the water, energy, and environmental businesses. It provides gas, electricity, water, waste collection, and disposal to about 3 million customers.

The Board is made up of 18 members, some of whom appointed by municipalities that actually own the 62% of the company's equity. The Chairman and the CEO are the only executive directors.

Table 1. Main Differences Among the Analyzed Companies.

	Case A	Case B	Case C	Case D
No. of employees (2009)	6.500	84.000	700	21.000
Revenues (€) (2009)	4.2 millions	28.6 millions	236.9 millions	5 billions
Ebitda (€) (2009)	567.3 millions	11.1 millions	17.3 millions	957 millions
Listing	Yes	Yes	Yes	No
Internationalization	No	Yes	No	Yes

In regards to risk management, no specific framework or standards have been implemented thus far. Risk is described as the possibility that future events may obstruct the achievement of companies' objectives.

Risks are treated properly and continuously only when they are considered of strategic relevance by the Board, as in the case of the risk of commodity prices, which is managed by a specific committee (the only existent one), presided by the Chairman of the Board. On the contrary, "the management of nonstrategic risks is partially or totally delegated to the lower levels of the company where risk owners manage risks in an unstructured way and related activities are conceived as cost centers." The risk manager is aware that this approach implies that the Board should understand clearly which risks are strategic and which ones are not and they are dynamic over time, while often it only considers daily risks, ignoring, for example, those related to events that have never occurred before.

Risk management is centralized at the corporate level by assigning a wide mandate to the Internal Audit Department, which directly reports to the Group Vice-Chairman of Board and to the Executive Committee. The Risk Management Department operates only in the holding company, not in the subsidiaries. It is in charge of managing pure risks and refers directly to the Legal and Corporate Affairs Department, which, in turn, refers to the Chairman of the Board. Other departments involved in risk management are Internal Audit (compliance risks), CFO (financial risks), and Security and Safety (i.e., asset risks, IT risks).

"The actual silos approach emphasizes the reactive management of risks and the lack of a common risk language obstructs comparison among risk areas as well as the identification of the company's risk profile." As a consequence, the interviewee feels the need for a transversal structure that would have an overall view of the risk exposure and a deep knowledge of the company itself so as to understand the interconnections among risks and among risk responses put in practice by the risk owners.

Remuneration schemes regarding the Chairman, the CEO, and directors who cover specific roles as well as for senior management and executives are mainly designed and proposed to the Board by a Remuneration Committee, which was appointed in 2008 and is composed of four nonexecutive independent directors. Starting from 2006 the incentive system for executives has been linked to the achievement of performance objectives set in the balanced scorecard, which represents the source of Management By Objectives (MBOs).

The aforementioned system was initially introduced for senior executives and has been later extended to all the managers of the Group. The performance bonus calculation is based on a system of profitability and productivity indicators (Group's gross operating margin, per capita gross operating margin of the companies belonging to the Group, and the Territorial Operating Structures), as well as a set of qualitative indicators focused on sustainability (i.e., decrease in number of claims, respect for standards defined in the water service charter) that vary according to the business segment.

Moreover, managers and executives receive an annual bonus based on the achievement of objectives assigned at the beginning of the year (individual balanced scorecard) regarding specific project objectives derived from group's business plan, economic targets set out in the annual budget, and an assessment of compliance with the leadership model. However, the final allocation of the bonus is also determined by the results achieved within certain Group parameters (company results and residential customer satisfaction). Neither stock option plans nor long-term monetary remuneration plans have been set.

Linking the compensation system to risk objectives and measures is considered as a possible advance in improving the risk management effectiveness of the company, since "it may drive managers' and executives' attention toward the risk and, in turn, improve their sensitivity toward the importance of risk management and to keep it always alive." However, the interviewee believes that operating in regulated markets may discourage the implementation of the link between reward and risk management, since the high entry barriers and subsidies from public entities can reduce volatility of a company's results.

Regarding the general feasibility of linking incentives to risk-based performance measurement, the risk manager highlights some conditions to be satisfied wherever applied. First of all, the risk owners have to be clearly identified, which is not always an easy task. Sometimes there are multiple risk owners. In addition, effective incentives should be based on a holistic view of the risk management process, so that the actions of a single owner can be evaluated in relation to a company overall risk exposure. Lastly, he suggests that great difficulties can arise from the definition of generally accepted parameters involving Board members. Although they are risk owners (i.e., competitive and reputational risk), the measurement of risk performance is not easy.

Case B

Company B is a listed group that provides telecommunication services, ICT, and media solutions. The company interprets risk as the possibility that certain events can affect the achievement of the company's objectives and manages it in a strategic and proactive way. Following a top-down approach, risks are initially identified from the objectives set in the strategic plan.

To ensure a global approach to risk management, in 2009 the company shifted from a compliance approach governed by Internal Audit and carried out according to the Control & Risk Self-Assessment (CRSA) methodology to the adoption of ISO 31000 standards. This change implied the establishment of the Risk Management Department, which reports to the CFO.

The Risk Management Committee administers the group's risk management process by coordinating the preventive action plan designed to ensure the operational continuity of the business and monitoring the effectiveness of the responses adopted. It is chaired and coordinated by the head of the "Administration, Finance and Control" Department and is made up of directors of the following departments: "Domestic Market Operations," "Technology & Operations," "International Business," "Corporate and Legal Affairs," "Human Resources," "Purchasing," and "Security."

The risk management approach, as approved by the Risk Management Committee, is based on an assessment of the risk profile by management in relation to both company processes and strategic objectives. This approach involves mapping the risks and focusing on those considered to be the most important, drawing up a Master Plan of mitigating actions and carrying out risk treatment activities involving the establishment of interdepartmental working groups defined by the Committee itself.

With regard to the reward system in practice, the company has share-based forms of remuneration. The Board of Directors (made up of 14 directors, five of whom are independent) has launched a plan for the granting of ordinary shares, free of charge, where the amount of shares depends on the degree of achievement of performance targets expressed in terms of absolute and relative Total Shareholder Return.

Since 2010, the company also promotes a long-term remuneration plan, which involves a selected number of executives who have not previously received remuneration based on financial instruments.

Managers with strategic responsibilities receive a fixed remuneration as well as a variable component linked to company financial results and/or the achievement of specific objectives.

It is the same with the CEO. Specifically, in 2010, the CEO's bonus was related for its 80% to the level of achievement of company's certifiable indicators (i.e., Net Financial Position, consolidated organic Ebitda, and total organic domestic revenues). In addition, a qualitative overall performance evaluation criterion was applied to the CEO, based on the discretional opinion of the Board of Directors and proposed by the Nomination and Remuneration Committee.

Furthermore, the Chairman and the CEO benefit from a stock option plan, which is reserved to top management and to executive directors of the company. Seventy-five percent of stock options assigned to the Chairman and the CEO are not subject to performance conditions and can be exercised for three years from the expiry of the vesting period.

Only at the managerial level was found a link between remuneration and risk performances, restricted to areas where core risks and compliance risks are handled. In those areas (i.e., IT and finance) the perceived importance of the risk leads to a more mature risk management and, in turn, the MBO contains elements coming from the risk management plan. Risk objectives are set for risk managers as well, who are also evaluated on the basis of a "risk maturity index," which measures the level of maturity of the risk management process.

An extension of the mentioned link is under consideration for the lower levels, although "it would be possible to apply it to all corporate levels as long as an adequate risk awareness and maturity in risk management processes exist." Furthermore, "the remuneration can be linked to risk management and work as an incentive when the company sets challenging, but possible objectives."

The two risk managers interviewed appreciate the idea of a Board's remuneration scheme linked to KRI based on objective information or quali-quantitative data. This scheme would not measure the level of a specific risk (i.e., number of accidents), but the results of the risk management actions (i.e., risk mitigation). However, "as strategic risk is difficult to quantify, the identification of KRIs is not an easy task. Often, the only possible indicator that can be used is the development and improvement of risk-reducing actions over time in terms of impact and likelihood, which can be hard to quantify as well."

Nevertheless, they believe that the majority of Italian companies are not ready for a reward system linked to risk performance, because risk management systems (when adopted) are still primitive and the average small-size and family ownership of companies do not contribute to the diffusion of a risk culture.

Case C

Company C operates in the planning and construction of public transport infrastructure. Its activity as general contractor has recently extended to the management of water provision services to an Italian municipality, which owns 100% of the company's equity.

The Board of Directors is made up of the Chairman and four other directors.

Risk is considered the possibility that an event could have negative economic effects that should be prevented and avoided. A risk management framework has not been adopted, nor is a holistic risk management system in place.

When the company activity was limited to the design of infrastructures, the risks were transferred to third parties, that is, insurance companies that offered the best price at a public auction. Although entering into the water service business changed the company risk profile, "risk management still does not seem to be one of shareholders and Boards' concerns."

The Risk Management Department refers to the Legal Affairs Department and it handles pure risks. The director of the water service and his workers are in charge of asset and quality risks, while financial risks are faced by the Administration Department. Each risk owner manages its risk in isolation and there is no coordination at the corporate level. A lack of knowledge about its overall exposure has recently lead the company to ask for an insurance broker to carry out a global risk assessment, although this will not necessarily mean the implementation of a risk management framework.

Rewards are linked only to financial performance measurement. In addition, the interviewed risk manager's pay is not structured to improve risk management effectiveness, that is, one of his past objectives was the reduction of insurance policy' costs, regardless of how this action affected the company's risk exposure.

According to the interviewee, rewards can be effective in improving risk management effectiveness when risks are managed holistically and, thus, the interconnections among risks and among responses taken by the single risk owners are identified and understood. Furthermore, such a link works when there is an adequate risk culture within the company. In this perspective, the public nature of the shareholder can be an obstacle because their representatives neither think in a profit perspective nor do manage by objectives. So "the only way to create the risk culture is through a top-down command and control approach."

This implies a full and deep commitment of the Board, which, in turn, should involve the HR Department. In fact, whoever manages the remuneration schemes should understand very well the risk sources and actions to measure risk-adjusted performances.

Case D

Company D is one of the world's largest cement producers, which operates in 22 countries.

The Board consists of 18 members: the Chairman, the CEO, an Executive Deputy Chairman and 15 directors (ten of whom are nonexecutive). Within the Board, the Remuneration Committee has been appointed the task of making proposals to the Board regarding remuneration, incentives, and stock option plans for directors holding special assignments, the top management, and the head of Internal Control.

The company conceives risk as the possibility that an event could affect the strategy implementation and the achievement of strategic objectives. In this perspective, the Board, who is particularly sensitive to risk issues, decided few years ago to adopt the COSO-ERM framework of which effectiveness is reviewed together with the strategy. Implementing the ERM, the company identified risk scenarios at the group level and a risk management strategy was developed as a consequence.

The Risk Management Department is a staff structure at the corporate level chaired by the Chief Risk Officer (CRO), which reports directly to the CEO and CFO. It was established in 2010 to ensure a holistic approach to risk management integrated into the strategic planning. However, today it still deals with pure risks, which are transferred to third parties and it has a partial overview on the company risk exposure.

In regards to rewards, Directors and the CEO can receive short-term incentives, long-term benefits, and stock options.

To align remuneration with the creation of value in the long term, a three-year stock option plan is addressed to a limited number of directors of the holding and its subsidiaries, which is based on the achievement of objectives linked to the Group economic results and/or to the gradual implementation of strategic projects approved by the Board.

A stock option plan is also set for managers with strategic responsibilities or specific operative tasks, depending on the achievement of annual objectives set by the Board. However, the stock option plan is going to be replaced by a cash-based plan.

A long-term monetary incentive plan is also in practice. It is consistent with the annual incentive plan in terms of scores, targets, and objectives. The plan operates on 3 three-year cycles. The amount of the bonus is proportional to the results achieved in one cycle and is based on the achievement of corporate financial objectives and individual objectives, which are not linked to the company's share price.

Documents specify that managers' long-term incentive schemes are linked to a three-year "Risk and Compliance Program" launched in 2008, which sets clear directives on accountability and boundaries for managing risk areas across the Group, with a particular focus on the allocation of responsibilities at corporate level versus country level. No more details are available, but the incentive plan is linked to the achievement of risk exposure reduction deriving from a risk mitigation action plan launched in 2010.

On the contrary, the interviewed risk managers provided several insights about risk-based performance incentives for directors of cement plants. The company has defined a "Property Preservation Program" whereby activities are analyzed to identify risk sources in the cement plants. Risk areas are monitored through an indicator called a "key performance indicator", which can vary between 0 (worst case) and 1 (best case). At the beginning of the year, the Risk Management Department sets the initial value of the aggregated KPI, while at the end of the year it aggregates the KPI coming from each cement plant and compares it to the initial value.

The KPI is a coefficient made up of three elements:

- the result of an evaluation of 30 factors (i.e., fire prevention plans and training programs) made by an insurance company, which defines the cement plan's risk profile and suggests actions to reduce it;
- the index of the furnaces' reliability; and
- the number of accidents occurring in the plant.

The KPI helps judge the performance of risk responses: if the indicator is stable, then the risk has been managed and monitored. The Risk Management Department reports the aggregated KPI of each cement plant to the HR Department, which connects them to remunerations.

The risk managers interviewed believe that "the link between risk management and reward system is both useful and possible since objectives, no matter whether they are related to risk management activities or to financial results, can be given to specific persons and can be measurable."

DISCUSSION

The four cases reflect a double classification of risk definitions. In fact, two of them understand risk in its negative connotation, that is, the possibility that an uncertain event might produce a loss or a negative effect to the achievement of company objectives. On the contrary, companies B and D describe risk as the uncertainty associated with the achievement of objectives and to the implementation of the strategy, recognizing the possibility that ongoing or future events could have negative as well as positive impact on the company.

How risk is conceived seems to be strictly related with the way it is managed. Those who considers risk as a negative factor use a reactive and silos approach, resulting in absent knowledge both of the existing interconnections among risks and risk responses and of the company overall risk profile and exposure. Risk management helps protect the company from negative occurrences and its main focus is on complying with legal prescriptions. The Risk Management Department reports to Legal Affairs and deals with pure risks, transferring them to third parties. This department is only one of many in charge of managing single risks, without specific coordination.

Conversely, where the Board perceives risk in its duality and judges risk management as a strategic lever, it established a group Risk Committee or Department (cases B and D) lead by the CFO in charge of monitoring and coordinating the risk management. Here the importance of strategic risk has lead to an attempt at integrating its management within the company's strategic planning. Risk management maturity is also shown by the implementation of formalized frameworks: the COSO-ERM and ISO frameworks.

Looking at their remuneration policies, in all cases the reward system includes monetary incentives linked to the achievement of both corporate and individual objectives based on financial measures (such as group Ebitda or ROE) and, to a lesser extent, on qualitative measures (i.e., customer satisfaction index). Listed companies present a more complex reward system for CEO and managers; however, only B and D have a long-term incentive plan.

Interestingly, companies B and D already link managers' pay to risk indicators (these are the two companies which implemented a holistic and strategic risk management framework). While company D uses an aggregated KRI resulting from a structured mix of measures that explains the outcomes of risk measure and the risk exposure in the cement plants, company B uses a risk management maturity index, which has been declared to be more useful in measuring the performance of upper management, since it does not measure the level of risk, but the results of the actions to manage it. However, no indications have been found with reference to a linkage of CEO pay with risk exposure.

The risk managers of the four companies all agree on the usefulness of linking the reward system to risk management in order to improve the effectiveness of the latter and to create a risk-aware culture. Culture plays a really important role and seems to be both a result and a driver. In fact, establishing the mentioned link could lead to a greater sensitivity toward risk issues within the company, while a risk-aware culture also represents a key driver for changing traditional reward systems.

Among other factors that can facilitate or hinder the introduction of a risk dimension in reward systems, the approach to risk management also appears to be fundamental. It cannot be possible to link risk performance to pay where there is no clear understanding of risk objectives and strategies and a clear identification of risk owners as well as of the interconnections among risks and risk responses. The Board's commitment and the difficulty of the measurability of risk performances are also crucial.

It is important to note that none of the subjects interviewed affirmed that linking risk performances to incentives could lead to possible drawbacks such as excessive focus on risk at the expenses of other goals or manager stress due to feeling unable to control uncertainties. Only one manager (company B) emphasized the perils of setting risk targets that could be too challenging.

According to the interviewees, the company's size, ownership structure, and business sector are important factors that can affect the adoption of risk measures in incentive schemes. Small companies as well as family-owned firms tend to suffer from a lack of risk culture, which does not foster the implementation of an advanced risk management process. Operating in regulated markets can also be an obstacle to the implementation of incentive systems linked to risk management. Where monopoly conditions still exist, the volatility of the financial results is reduced and thus the perception of risk may be low. Moreover, companies operating in such sectors are mostly publicly owned and so often lack managerial business practices.

CONCLUSIONS

These discussions highlight the possibility of link between traditional performance-based monetary incentives and risk management objectives. Some companies, and specifically the ones who implemented a strategic and holistic approach to risk management, have already defined risk measures and targets that align managers' operations with the company's strategic risk priorities. Measurability seems to be an easier task with reference to operational and financial risks. Thus, risk performance has been found mainly in top and middle management remuneration, while for CEOs some additional considerations need to be made.

One solution could be to link CEOs' incentives to ERM effectiveness measures, that is, the number of unforeseen events and in-year adjustments or senior management's hours required in strategic risk workshops (Toneguzzo, 2010). This implies some subjective evaluation both in the *ex ante* phase and in the *ex post* assessment phase, which is actually welcome in time of great environmental changes as managers cannot be adequately motivated by bonuses linked to objective indicators difficult to achieve.

A Risk Management Maturity Index could also be used to link rewards to different levels in risk management process implementation or to determine the distance between current conditions and objectives. This index could be based on different dimensions depending on the specific context in which it is applied in terms of risk management maturity, such as knowledge about risk, process management, reporting, awareness, and attitude.

Currently, some companies limit CEOs' risky behaviors through a bonus-malus approach; however, the inclusion of risk performances in managers' evaluation seems to be more effective especially if companies have managerial systems such as Balanced Scorecard in place, which represent a natural framework for identifying risks associated with objectives at all business levels (Kaplan, 2009).

Another important finding emerged from this study is that Italian risk managers believe that company risk management systems can really benefit from a linkage with monetary incentives. Although some organizations are not ready because of a lack of risk culture, this linkage can be useful in reinforcing the function of risk management and none seems to be worried about possible negative consequences such as the possibility of generating extreme cautious behaviors that cause the discharge of good businesses only because they are a little bit more risky.

In addition, this study confirms that a strategic approach to risk management implies that risk managers report to the CFO (Branson, 2010). Thus, risk cannot be considered as independent from management control processes, nor risk managers should appear as specialists who do not communicate with controllers. On the contrary, there can be room for cross-fertilization and risk management can impact accounting practices as well as learning from them.

Lastly, the case studies here analyzed allow some initial hypotheses about which companies may be more open to risk evaluation in managers' performance (i.e., listed companies with international operations, guided by professional managers) and which factors can facilitate the deployment of a reward system based on both risk and financial performance (i.e., culture, risk management approach), thus suggesting where to focus future research.

REFERENCES

Acharya, V., & Richardson, M. (Eds.). (2009). *Restoring financial stability: How to repair a failed system*. Hoboken, NJ: Wiley.

Andersen, K., & Terp, A. (2006). Risk management. In T. J. Andersen (Ed.), *Perspectives on strategic risk management* (pp. 27–45). Copenaghen: CBS.

AON. (2010). *Global Enterprise Risk Management Survey '10*. Retrieved from http://img.en25.com/Web/AON/2010_Global_ERM_Survey_F.pdf. Accessed on March 30, 2012.

Baird, I. S., & Thomas, H. (1985). Toward a contingency model of strategic risk taking. *Academy of Management Review*, *10*, 230–243.

Baird, I. S., & Thomas, H. (1990). What is risk anyway? Using and measuring risk in strategic management. In R. A. Bettis & H. Thomas (Eds.), *Risk, strategy and management* (pp. 21–52). Greenwich, CT: JAI Press.

Barkema, H. G., & Gomez-Mejia, L. R. (1998). Managerial compensation and firm performance: A general research framework. *Academy of Management Journal*, *41*, 135–145.

Beasley, M. S., Branson, B. C., & Hancock, B. V. (Eds.) (December 2010). *Developing key risk indicators to strengthen enterprise risk management*. Retrieved from http://www.coso.org/guidance.htm

Beasley, M. S., Cune, R., & Hermanson, D. R. (2005). Enterprise risk management: An empirical analysis of factors associated with the extent of implementation. *Journal of Accounting and Public Policy*, *24*(6), 521–531.

Beasley, M. S., & Frigo, M. L. (2010). ERM in its role in strategic planning and strategy execution. In J. Fraser & B. J. Simkins (Eds.), *Enterprise risk management. Today's leading research and best practices for tomorrow's executives* (pp. 31–50). Hoboken, NJ: Wiley.

Bettis, R. A., & Mahajan, V. (1990). Risk analysis in corporate performance measurement. In R. A. Bettis & H. Thomas (Eds.), *Risk, strategy and management* (pp. 83–110). Greenwich, CT: JAI Press.

Bonner, S. E., & Sprinkle, G. B. (2002). The effects of monetary incentives on effort and task performance: Theories, evidence, and a framework for research. *Accounting, Organization and Society*, *27*, 303–345.

Branson, B. C. (2010). The role of the board of directors and senior management in enterprise risk management. In J. Fraser & B. J. Simkins (Eds.), *Enterprise risk management. Today's leading research and best practices for tomorrow's executives* (pp. 51–67). Hoboken, NJ: Wiley.

Brooks, D. W. (2010). Creating a risk-aware culture. In J. Fraser & B. J. Simkins (Eds.), *Enterprise risk management. Today's leading research and best practices for tomorrow's executives* (pp. 87–95). Hoboken, NJ: Wiley.

Cameron, J., & Pierce, W. D. (1994). Reinforcement, reward, and intrinsic motivation. A meta-analysis. *Review of Educational Research*, *64*, 363–423.

Chorafas, D. N. (2001). *Managing risk in the new economy*. New York, NY: New York Institute of Finance.

Cokins, G. (2009). *Performance management. Integrating strategy execution, methodologies, risk and analytics*. Hoboken, NJ: Wiley.

Collier, P. M., & Berry, A. J. (2002). Risk in the process of budgeting. *Management Accounting Research, 13*, 273–297.

Collins, J. M., & Ruefli, T. W. (1992). Strategic risk: An ordinal approach. *Management Science, 38*(12), 1707–1731.

Corbetta, P. (2003). *La ricerca sociale: Metodologia e tecniche* (Vol. 3). Le tecniche qualitative. Bologna: Il Mulino.

COSO. (2004). *Enterprise risk management – integrated framework*. New York, NY: COSO.

Crouhy, M., Galai, D., & Mark, R. (2006). *The essentials of risk management*. New York, NY: McGraw-Hill.

Crystal, C. S. (1991). *In search for excess: The overcompensation of American executives*. New York, NY: Norton & Company.

Daft, R. L. (2006). *The new era of management*. Mason, OH: Thomson South-Western.

Eisenberger, R., & Cameron, J. (1996). Detrimental effects of reward: Reality or myth? *American Psychologist, 51*(11), 1153–1166.

Elson, C. L. (1993). Executive overcompensation – a board-based solution. *Boston College Law Review, 34*(5), 937–996.

Epstein, M., & Buhovac, A. R. (2006). *The reporting of organizational risks for internal and external decision-making*. Retrieved from http://www.zdruzenje-ns.si/db/doc/upl/risk_reporting_epstein_rejc_buhovac.pdf. Accessed on March 30, 2012.

Flick, U. (2002). *An introduction to qualitative research* (2nd ed.). London: Sage Publications.

Frigo, M. L. (2009). Strategic risk management: The new core competency. *Management Synergies* (January–February), 7–10.

Grant, R. M., & Visconti, M. (2006). The strategic background of corporate accounting scandals. *Long Range Planning, 39*, 361–383.

Hagigi, M., & Sivakumar, K. (2009). Managing diverse risks: An integrative framework. *Journal of International Management, 15*, 286–295.

Herzberg, F. I. (1987). One more time: How do you motivate employees? *Harvard Business Review, 65*(5), 109–120.

Hewitt. (2007). *While salary increases in 2008 remain modest, variable pay awards reach high record*.

Hewitt. (2009). *Report on Eurotop 100 directors' remuneration*.

Hexter, E. S., & Gates, S. (2005). *From risk management to risk strategy*. Retrieved from http://www.conferenceboard.ca/documents.aspx?DID=1305. Accessed on February 10, 2012.

Horngren, C. T., Bhimani, A., Foster, G., & Datar, S. M. (1999). *Management and cost accounting*. London: Prentice-Hall.

Hwang, S. (2010). Identifying and communicating key risk indicators. In J. Fraser & B. J. Simkins (Eds.), *Enterprise risk management. Today's leading research and best practices for tomorrow's executives* (pp. 125–140). Hoboken, NJ: John Wiley & Sons.

IFAC. (1999). *Enhancing shareholder wealth by better managing business risk*. Report of International Management Accounting Study no. 9. International Federation of Accountants, New York, NY.

IMA. (2007). *Enterprise risk management: Tools and techniques for effective implementation*. Retrieved from http://poole.ncsu.edu/erm/documents/IMAToolsTechniquesMay07.pdf. Accessed on February 10, 2012.

Jensen, M. C., & Meckling, W. H. (1976). Theory of the firm: Managerial behavior, agency costs and ownership structure. *Journal of Financial Economics*, *3*, 305–360.

Jensen, M. C., & Murphy, K. J. (1990). Performance pay and top-management incentives. *The Journal of Political Economy*, *98*(2), 225–264.

Kaplan, R. S. (2009). Risk management and the strategy execution system. *Harvard Business Review*, *11*(6), 1–6.

Kerr, J., & Slocum, J. W. (2005). Managing corporate culture through reward systems. *Academy of Management Executive*, *19*(4), 130–138.

Knight, F. H. (1921). *Risk, uncertainty, and profit*. Boston, MA: Houghton Mifflin.

Kohn, A. (1993). Why incentive plans cannot work. *Harvard Business Review*, *71*(5), 54–63.

KPMG. (2010). *L'enterprise risk management in Italia*. http://www.kpmg.com/IT/it/IssuesAndInsights/ArticlesPublications/Documents/L-Enterprise-Risk-Management.PDF. Accessed on March 30, 2012

Latham, G. P. (2007). *Work motivation: History, theory, research, and practice*. Thousand Oaks, CA: Sage Publications.

Liebenberg, A. P., & Hoyt, R. E. (2003). The determinants of enterprise risk management: Evidence from the appointment of Chief Risk Officers. *Risk Management & Insurance Review*, *6*(1), 37–52.

Locke, E. A. (1991). The motivation sequence, the motivation hub, and the motivation core. *Organizational Behaviour and Human Decision Processes*, *50*, 288–299.

Locke, E. A. (2004). Linking goals to monetary incentives. *Academy of Management Executive*, *18*(4), 130–133.

Manzoni, J. F. (2010). Motivation through incentives: A cross-disciplinary review of the evidence. In M. J. Epstein, J. F. Manzoni & A. Davila (Eds.), *Performance measurement and management control: Innovative concepts and practices* (pp. 19–63). Howard House: Emerald Group Publishing.

McNamee, D., & Selim, G. (1998). *Risk management: Changing the internal auditor's paradigm*. Altamonte Springs, FL: IIA Research Foundation.

Meulbroek, L. K. (2002). Integrated risk management for the firm: A senior manager's guide. *Journal of Applied Corporate Finance*, *14*, 56–70.

Mikes, A. (2009). Risk management and calculative cultures. *Management Accounting Research*, *20*, 18–40.

Milgrom, P., & Roberts, J. (1992). *Economics, organization and management*. Englewood Cliffs, NJ: Prentice Hall.

Nocco, B. W., & Stulz, R. M. (2006). Enterprise risk management: Theory and practice. *Journal of Applied Corporate Finance*, *18*(4), 8–20.

Pennings, J. M. (1993). Executive rewards: A cross-national comparison. *Journal of Management Studies*, *30*(2), 261–280.

Protiviti. (2011). *Strategie e valore a rischio: Quanto il board ne è consapevole*? Retrieved from [illegible] Accessed on March 30, 2012.

Puschaver, L., & Eccles, R. G. (1996). In pursuit of the upside: The new opportunity in risk management. *PwC Review* (December), 1–10.

Rahardjo, K., & Dowling, M. A. (1998). A broader vision: Strategic risk management. *Risk Management*, *45*(9), 44–49.

Renn, O. (1998). Three decades of risk research: Accomplishments and new challenges. *Journal of Risk Research*, *1*(1), 49–71.

Roubini, N., & Mihm, S. (2010). *Crisis economics. A crash course in the future of finance*. New York, NY: Penguin Press.

Ryan, B., Scapens, R. W., & Theobald, M. (2002). *Research method and methodology in finance and accounting* (2nd ed.). London: Thomson.

Schrøder, P. W. (2006). Impediments to effective risk management. In T. J. Andersen (Ed.), *Perspectives on strategic risk management* (pp. 65–87). Copenhagen: CBS.

Silverman, D. (1997). *Qualitative research: Theory, method and practice*. London: Sage Publications.

Simons, R. (2000). *Performance measurement and control system for implementing strategy*. Boston, MA: Harvard Business School Press.

The Conference Board. (2009). *The conference board task force on executive compensation*. Retrieved from http://www.conference-board.org/publications/publicationdetail.cfm?publicationid = 1692. Accessed on February 10, 2012.

Thompson, A. A., Strickland, A. J., & Gamble, J. E. (2005). *Crafting and executing strategy: The quest for competitive advantage*. New York, NY: MacGraw-Hill Irwin.

Toneguzzo, J. P. (2010). How to allocate resources based on risk. In J. Fraser & B. J. Simkins (Eds.), *Enterprise risk management. Today's leading research and best practices for tomorrow's executives* (pp. 189–216). Hoboken, NJ: Wiley.

UBS. (2008). *Compensation report*. Retrieved from http://www.ubs.com/1/e/investors/compensationreport.html

Vroom, V. H. (1964). *Work and motivation*. New York, NY: Wiley.

Williamson, D. (2007). The COSO ERM framework: A critique from systems theory of management control. *International Journal of Risk Assessment and Management*, *7*(8), 1098–1119.

Wright, P., Kroll, M., Krug, J. A., & Pettus, M. (2007). Influences of top management team incentives on firm risk taking. *Strategic Management Journal*, *28*, 81–89.

Yin, R. K. (1994). *Case study research, design and methods* (2nd ed.). Thousand Oaks, CA: Sage Publications.

Zahra, S. A., & Pearce, J. A. (1989). Boards of directors and corporate financial performance. A review and integrative model. *Journal of Management*, *15*(2), 291–334.

ENTERPRISE RISK MANAGEMENT AND PERFORMANCE IMPROVEMENT: A STUDY WITH BRAZILIAN NONFINANCIAL FIRMS

Rodrigo Silva De Souza, Sônia Maria Da Silva Gomes, Adriano Leal Bruni, Gilca Garcia De Oliveira, Márcio Santos Sampaio and Juliano Almeida De Faria

ABSTRACT

Purpose – *This research aimed to study the relationship between Enterprise Risk Management (ERM) and performance improvement.*

Design/methodology/approach – *A questionnaire was used as an instrument of data collection that was passed to managers of nonfinancial companies listed among the 500 largest and best firms in Brazil. The data from this study were analyzed with descriptive statistics and multivariate analysis of correlation and association.*

Performance Measurement and Management Control: Global Issues
Studies in Managerial and Financial Accounting, Volume 25, 275–298

ISSN: 1479-3512/doi:10.1108/S1479-3512(2012)0000025014

Findings – *The results showed that the main drivers of risk management were regulation, stakeholder demands, and business competitiveness. Among the practices that have been used, managers spotlight the utility of basic methods, more subjective, while technical methods, more quantitative, were of secondary importance. It was evidenced that the risks were weighted in the main activities of the organization. As a result, it was demonstrated that improved performance is associated with the maturity level of risk management and the level of stakeholders' involvement in risk management.*

Research limitations/implications – *Other researches could examine how this process was developed in other countries and expand the number of organizations studied.*

Practical implications – *This study provides empirical evidence about theorist assumptions about the relationship between ERM and performance improvement.*

Social implications – *This study demonstrates the importance of human aspects for the processes of risk management and how external factors can influence this process.*

Originality/value – *It gives a broader and deeper comprehension of the process of risk management at nonfinancial firms in Brazil.*

INTRODUCTION

Nothing is static. The more complex an organization becomes, than greater are the exposure to the risks and, therefore, more potential for failure. In other words, Enterprise Risk Management (ERM) must keep pace with the increasing complexity of organization and operating environment.

ERM has attracted the attention of risk management professionals and academics worldwide. Unlike the traditional approach to risk management, organized in "silos," ERM enables companies to benefit from an integrated approach that shifts the focus from a primarily defensive management to one progressively more offensive and strategically focused.

This new paradigm, as described by Barton, Shenkir, and Walker (2002), breaks with the fragmented and standardized idea of risk management, as well as proposes a global concept, shared by all players, about the importance of risk management in all organizational activities. This management should be permanent and continuous. The risks must be understood as a business opportunity and not just a threat. As such, this conception of ERM has become almost a synonymous for good management and is currently seen as an external guarantee of company's ability to sustain a viable pattern of behavior (Miller, Kurunmaki, & O'leary, 2008).

Empirical researches show that risk management practices are influenced by external and internal variables (Collier, Berry, & Burke, 2007). The regulation on financial organizations, for example, requires the disclosure of risk management. Moreover, the positions of stakeholders advocate that the adoption of such practices has shown a better performance in these company's activities.

In most standards devoted to the definition of ERM, it appears that this concept is linked to organizational performance improvement (CAS, 2003; CIMA, 2008; COSO, 2004). However, empirical researches, which demonstrated this relationship, are scarce, especially, those focused on the operations of nonfinancial firms.

Therefore, this research seeks to bring a better comprehension of the benefits generated by the adoption of enterprise risk management practices. As such, we seek to answer the following question: What is the relationship between the practices of enterprise risk management and performance improvement?

Thus, the objective of this work is to show the relationship between the adoption of practices of enterprise risk management and organizational performance improvement in Brazilian nonfinancial firms.

Collier et al. (2007) argue that ERM is conceived as a tool for improving the organizational performance, making organizations more aware of the risks and, possibly, reducing the cost of capital. Clarke and Varma (1999) confirm that an integrated approach to risk management enables companies to deliver a consistently superior performance while proactively managing the risks.

In this perspective, Collier and Berry (2002) described the importance of managers' risk perception and the role of risk management in budget process. Woods (2007) highlights the importance of risk consideration in formulations of strategies of control systems. Furthermore, Chartered Institute of Management Accountants, CIMA (2008) describes the importance of risk management in decision-making.

However, discussions about risk management in nonfinancial companies are still incipient (Guimaraes, Parisi, & Pereira, 2006) and scarce, particularly, in accounting companies (Zhuang, Hou, & Hou, 2008). Collier and Berry (2002) reported that there is a need for more studies on risk management, allowing a better understanding of the implications that broader conceptions about the risks have in the management of organizations. As ratified by Kaplan (2011), one of the most important questions in accounting area today refers to the different ways of managing risk and how it varies in business management, focusing on the importance of management to improve organizational processes.

This work is divided into five sections. The first corresponds to the introduction already carried out, followed by a brief explanation of the enterprise risk management. Then, the third part of this study describes the methodology adopted. From this, we present the results and conclusions of this investigation.

ENTERPRISE RISK MANAGEMENT AND PERFORMANCE

In the financial field, the concept of risk generally refers to aspects related to the negative consequences of an action and to quantify probability of their manifestation (Knight, 1921). Thus, the difference between the risks and uncertainties is the possibility to quantify the first.

Risk management is fundamental today in the dynamic and global environment. In recent years, however, a paradigm is shifting progressively the way risk is managed. This new concept of enterprise risk management was systematized by Barton et al. (2002) as a new paradigm. Instead of looking for risk management in a silo-based perspective, the trend is to have a holistic and integrated view of this management. Thus, this paradigm was called Enterprise Risk Management (ERM).

One of the most popular definitions used in ERM literature (Beasley, Clune, & Hermanson, 2005; Lin & Wu, 2006; Moeller, 2007) was provided by the Committee of Sponsoring Organizations of the Treadway Commission, COSO (2004). According to this Committee, ERM is:

> a process, effected by an entity's board of directors, management and other personnel, applied in strategy setting and across the enterprise, designed to identify potential events that may affect the entity, and manage risk to be within its risk appetite, *to provide reasonable assurance regarding the achievement of entity objectives*.

According to this ERM approach, risk management should be a task of everyone, because risks are present in all activities of an organization. Furthermore, as each company presents a different "risk appetite," management will also change in each entity. This conjecture seeks to ensure the achievement of company objectives.

Risk management has been defined as the process used to understand and manage risks that an organization is inevitably subject to, when it attempts to achieve its corporate objectives (CIMA, 2008). Moreover, the Institute of Risk Management (IRM) in 2002 affirmed that risk management is a central part of any organization's strategic management and defined it in a more detailed manner as:

> the process whereby organizations methodically address the risks attaching to their activities *with the goal of achieving sustained benefit within each activity and across the portfolio of all activities.*

From this definition, the presence of a relationship between risk management and value creation becomes evident for organizations. Similarly, the presumed link between a holistic approach to risk management and performance or value of an organization is clearly observed in the following definition of ERM provided by the Casualty Actuarial Society – Enterprise Risk Management Committee, CAS (2003, p. 8):

> ERM is the discipline by which an organization in any industry assesses, controls, exploits, finances and monitors risks from all sources *for the purpose of increasing the organization's short- and long-term value for its stakeholders.*

Consequently, ERM must provide a "reasonable assurance regarding the achievement of entity objectives" and seeks to achieve a "sustained benefit within each activity and across the portfolio of all activities" "for the purpose of increasing the organization's short- and long-term value for its stakeholders."

In the same perspective, the International Federation of Accountants (IFAC, 1999) published a study entitled Enhancing Shareholder Wealth by Better Managing Business Risk, which defines risk as future uncertain events that may influence the achievement of organizational strategies and operational and financial objectives. As a result, it shifts the focus of risk from a negative concept to a positive way of managing risks as an integral part of sustainable value creation for shareholders. This report argues that the management of corporate risks provides calibration and realigns the relationship between risk, growth, and return.

Likewise, the Turnbull Report of the Institute of Chartered Accountants in England & Wales (1999), which is now part of the Combined Code on Corporate Governance of England and Wales, defines risk as any event that may affect the performance of a company, including environmental risk, ethical issues, and social issues. Therefore, this report extends the categories traditionally used to manage risk and, theoretically, brings more explicitly the relationship between risk management and performance improvement.

To understand better the relationship between risk management and performance improvement, some authors attempted to study the effect of this management in the financial market. Simkins and Smithson (2005), for example, analyzing scholarly literature on the ability of financial risk management to add value to organizations, confirm this idea, but said that its evidence is limited. These authors found a number of recent studies that show a clear positive correlation between stock prices and greater use of derivatives to manage currency risk and interest rate risk. This study provides evidence of the use of derivatives to ensure the price of commodities increases stock prices. However, it appears that the aggregate value hedging by those mechanisms depends on the types of risks that a company is exposed to.

However, the activities of nonfinancial companies differentiate themselves from the activities of financial institutions, especially in regards to the liquidity of assets. Nonfinancial firms have long-term assets such as factories, machinery, and equipment, which have differentiated liquidity of the portfolio of a financial institution. Therefore, the risks in nonfinancial companies are not restricted only to their operational flow.

When the risk is treated outside the financial system by organizations operating in other sectors, its quantification becomes more complex and inaccessible. The risk of a project, for example, arises from several sources, including the project itself, competition, changes in industry, and international and macroeconomic considerations. Thus, the industrial risks can be characterized by technology, law, and market, which may impact the profitability of a project (Damoradan, 2001).

Guimaraes et al. (2006) analyzed through questionnaires sent to risk managers or controllers the practices of risk management in nonfinancial companies located at São Paulo in Brazil. The results showed that risk management is part of these companies' activities as a law adequacy or an important managerial tool that contributes to the decision-making process.

However, the implementation of ERM hasn't been made so simple and linear, as one might imagine. Towers Perrin (2008) made a survey with risk managers and CFOs of insurance companies worldwide, and showed that (a) the incorporation of ERM is a great challenge for these companies, especially about the use of an economic capital in the process of decision-making and performance management, (b) the ERM is influencing the decisions regarding the alteration of key business, (c) a variety of standards for measuring economic capital have emerged, (d) operational risk management is still a deficit point, (e) size is important, larger insurers have significantly more advanced processes in most aspects of the implementation of ERM and increasingly seek to gain competitive advantage through it, and (f) European insurers are better positioned than American insurers.

The factors that impacted the adoption of risk management practices are not fully explained yet. Gordon, Loeb, and Tseng (2009) made an empirical examination in the argument that the ERM is related to business performance. They analyzed 112 U.S. companies that disclose ERM activities in the U.S. Security and Exchange Commission (SEC). The results confirm the positive relationship between ERM and business performance. However, this relationship depends on the compatibility between the ERM and variables such as environmental uncertainty, industry competition, company size, firm complexity, and board support.

Hoyt and Liebenberg (2009) assessed insurers in the United States about the extension that their companies had implemented to specific programs of ERM and the value of implications of this adoption for them. They founded that ERM was positively related to factors such as firm size and institutional ownership, and negatively associated with the use of refinancing, leverage, and lack of transparency in assets. They also showed a positive relationship between firm value and the use of ERM.

One factor that causes difficulties to study the cost-effectiveness of risk management deployment in organizations refers to the difference in maturity of these practices in every sector and firm. Trapp and Corrar (2005) investigated the assessment and management of operational risk in a large Brazilian financial institution, detecting the measurement and analysis instruments and the stage of development of this process. The results suggest that this bank was at an intermediate stage in the management of operational risks; however, it is developing techniques and processes both to suit the requirements of oversight bodies and to improve its results.

The decoupling of organizational reality and the connotation of ERM as a tool of legitimacy led ERM to be applied in different approaches (Mikes, 2005, 2009; Power, 2007), raising the questions of what would be ERM and what it had become in practice. The issue of ERM as a tool of legitimacy remains unexplored (Gephart, Van Maanen, & Oberlechner, 2009; Power, 2009), especially in nonfinancial companies. Therefore, there are possibilities that companies have been deploying ERM only as a compliance tool or as an independent internal control, but without the assimilation of its role into business processes.

Bruno-Britz (2009) argues that, to have an effective ERM, companies should look beyond technology and establish a risk management culture throughout the organization. Thus, ERM should permeate the practices and individual behavior of managers in daily decisions (Standard & Poor's, 2007).

The results of a research conducted by Collier et al. (2007) suggest that risk management was boosted by an institutional response to requests for corporate governance improvement that may reflect both protection and economic opportunities. The main external influences to the practices of risk management, rather than the competitiveness intensity, were the stakeholders and the demands of regulators and legislation, enacted by the boards, which were likely to influence policies and methods adopted for risk management.

Consequentially, ERM can vary from calculating practices, cultural significance, and level of integration (Mikes, 2005, 2009; Power, 2007, 2009). Power (2009) points out the danger of deploying an ERM process as a mechanism for compliance and not as an integrated management for decision-making processes. In the research conducted with CEOs from all continents, PricewaterhouseCoopers (2004) confirms that the ERM is seen as a device for external accountability, which has no impact on decision-making and operations.

ERM intended to benefit management, permeating the approach how each manager makes decisions every day. Inserted in the field of internal control, it emphasizes the values of regulatory compliance and accountability. However, it has been reduced simply to an additional item for internal control and compliance with external regulations, forgetting its managerial perspective (Bowling & Rieger, 2005; Martin & Power, 2007).

METHODOLOGY

The objective of this research is to assess the relationship between enterprise risk management and organizational performance improvement at non-financial companies in Brazil. Aiming to achieve this goal, we carried out a survey research. The population chosen for investigation was the companies registered at the base of "500 Biggest and Best Firms in Brazil" according to Revista Exame (2010).

The sample obtained in this research can be characterized by nonprobability, since the selection of this study sample depends, at least in part, on the researcher's trial (Mattar, 1996). The questionnaires were sent by e-mail and 58 returned responded, representing 65 companies, and could be used in this research. It is important to note that calls were made for a total of 200 companies, and in many cases, the respondents answered that they would not fill the questionnaire, because of the lack of time or the existence of internal policies related to nonparticipation in academic research. This could evidence the difficulties created by this kind of research and that ERM is a strategic issue for those companies.

Data Collection Instrument

This study used a questionnaire developed by Collier et al. (2007). The great contribution of these authors' research was the obtained data of internal activities in UK organizations and accountants' perception about risk, while previous research focused only on relationship disclosure in reports characterized by the analysis of financial data or case studies.

The questionnaire was divided into external drivers such as (a) perceived uncertainty in the environment and (b) external regulation; and internal drivers such as (a) the risk appetite and (b) the risks factored into planning of supporting processes. The risk appetite item was divided on individual and organizational profiles, was diagnosed according to the propensity to take risks, and considered the understanding about the perception of the aims of risk management.

The factor, "perceived uncertainty in the environment," was diagnosed according to the degree and extent of such factors as the competing intensity, perceived uncertainty in the industry sector, the level of risks faced by the organization, the level of risks in the sector.

We attempted to verify when risks were considered and to what extent these risks were identified and evaluated in the formulation of strategic plans, budgets, operational plans, project management, singular events (e.g., mergers), and equity investments. These were the elements considered in the item "supporting processes," and the level of risk considered into this planning was used as a factor to determine the degree to which risks were weighted in these activities.

The external regulation was weighted based on legislation (including regulations of the Central Bank, CVM, BIS, and SOX), regulatory committees, expectations of investors and analysts, the competitive business environment, and the demand of clients and board of directors.

According to Collier et al. (2007), the risk management practices are also a function of the risk appetite of the organization. This aspect was inferred from the degree organization's risk management was designed to take advantage and opportunity of the risks (economic rationality and corporate governance) and the degree it was designed to provide protection against risks (regulated logic).

Techniques for Data Analysis

Aiming to characterize the composition of this research sample, the variables of this study, initially, were treated with descriptive statistics. From the preliminary characterization of the sample, data were statistically tested by the Komogorov–Smirnov test that verify if they obey the normal distribution function (Bruni, 2009; Hair, Tatham, Anderson, & Black, 2005). The set value for the significance of the tests analyzed was 5%, so the results found in the statistical tests performed in this study can be considered correct at the level of 95%.

A total of 152 variables were analyzed, however, to facilitate data interpretation; we chose to use an exploratory factor analysis. According to Malhotra (2001), the exploratory factor analysis is a set of processes used to reduce and summarize data. Thus, the factors can be expressed as linear combinations of observed variables. The first factor accounts for the largest variance in the data, the second factor accounts for the second largest variance, and so on.

We chose to use the rotated matrix to select the factors of this research. According to Malhotra (2001), the rotation matrix could be used to transform an array of variables into simple and easy-to-interpret factors. For this, we opted for the most commonly used rotation method, "varimax" procedure.

In the research, we used the technique of stepwise forward multiple regression to explicit the relationship between enterprise risk management and performance improvement. From this selection method, one variable is entered at a time in the model and consistency of these variables is tested using a partial F test. As a result, explanatory variables that have a low rate were removed from the model. The procedure is repeated until no other variable can enter the model.

The assumption listed by Malhotra (2001), Corrar and Theóphilo (2004) and Fávero, Belfiore, Silva, and Chan (2009) for the application of multiple regressions was met. Multiple regressions serve as a way to describe the relationships among the variables when the researcher has a little knowledge about it. The regression techniques aim to generate a line that best fits to a set of scattered points that are a result of the data collected.

According to Gujarati (2006), regression analysis studies the dependence of one variable, the dependent variable, in relation to one or more variables, the explanatory variables, in order to estimate or predict the average value (of the population) of the first based on known or fixed values (in repeated sampling) of the second. These relationships can be described by the following equation:

$$Y = \beta_0 + \beta_{1\times 1} + \beta_2 X_2 + \cdots + \beta_k X_k + e$$

where Y, dependent variable; β_0, constant; β_1, coefficient of the first independent variable X_1; β_2, coefficient of the second independent variable X_2; β_k, coefficient of the nth independent variable X_k; and e, the difference of predicted and observed value of Y (residual term).

The coefficient of determination Adjusted R^2 is an important measure of quality in multiple regressions. The calculation of Adjusted R^2 allows us to discover how "good" a sample regression line fits the data. Thus, the Adjusted R^2 shows the proportion or percentage of the total variation in the dependent variable Y, which is explained by the explanatory variables (Gujarati, 2006).

Pre-Test

To test the consistency and the effectiveness of the questionnaire, a pre-test research was conducted with consultants, Ph.D.s, masters, and undergraduates. The questionnaire was sent to the e-mail addresses of respondents and was answered by 19 students, who had analyzed and made suggestions to aspects of the structure and content of the questionnaire. In addition, Ph.D.s and consultants were interviewed with the aim of clarifying questions and collecting suggestions and recommendations for improving the questionnaire to Brazilian reality. Thus, some questions have been relocated or upgraded and others entered in the questionnaire. After this consideration, the questionnaire was adjusted to the sample application in this research.

RESULTS

Sample Characterization

Table 1 shows the eight economic activities of the enterprises participating in this research. The group most representative in the sample was Energy, Water and Sanitation sector, with 24.14%, followed by Manufacturing and Construction sector, with 17.24%.

Table 1. Business Sector.

Sector	Frequency	Percentage	Cumulative Percentage
Energy, Water and Sanitation	14	24.14	24.14
Manufacturing and Construction	10	17.24	41.38
Retail and Distribution	5	8.62	50.00
Transport and Consumer	5	8.62	58.62
Oil and Gas	4	6.90	65.52
Services	3	5.17	70.69
Telecommunications and Media	3	5.17	75.86
Other	14	24.14	100.00
Total	58	100.00	

Source: Research Data, 2011.

However, from the data tabulation, the existence of some missing data or missing values was verified. If they were not properly treated, the missing values ruling out the use of 26 responses, for example, and if they were simply excluded from this database, it would represent a loss of 44.83% in the sample obtained in this study. Therefore, we decided to follow recommendations of Saunders et al. (2006) by using the statistical treatment of data imputation based on averages for these cases.

According to Tabachnick and Fidell (1983) and Little and Rubin (2002), missing data can be treated by this imputation technique when limited from 5–20% of each variable. As our missing values are within the range, we use this method. Saunders et al. (2006) demonstrate in their research that the loss of statistical power using data suppression is clearly demonstrated by the decrease in sample size. All other methods of imputation analyzed in their research were able to retain all cases and maximize statistical power, something important especially with small datasets. However, statistical treatment of data imputation based on averages was simpler to use and less extra time was needed to conduct the analysis.

Reduction of Factors

The assumptions listed by Malhotra (2001), Corrar and Theóphilo (2004) and Fávero et al. (2009), for the application of factor analysis, were met. In Table 2, 11 factors have been identified that made the analysis data of this research.

Compared with the findings of Collier et al. (2007) this study showed a factor not related to Effectiveness of Technical Methods. Possibly, this fact is due to the lack of normality evidenced in the variables related to the questions about the effectiveness of "Stochastic Analysis and Statistical Model" and "Software risk management." Thus, these variables could not be reduced and the corresponding factor would be "Effectiveness of Technical Methods."

The results show that the main drivers of risk management are regulation, stakeholder demands, and business competitiveness. Among the practices being used, managers spotlight the use of basic methods, more subjective, while technical methods, more quantitative, were of secondary importance.

Correlation Analyses

We conducted a Pearson correlation test to determine the correlation level between the factors listed in Table 2. The correlation between the variables is a number ranging from −1 to 1. When the correlation is positive, it indicates that the variable varies positively with the other; when it is negative, then the variable varies negatively with the other; and when it is zero, there is no correlation.

So according to the test presented in Table 3, we conclude that there is a significant relationship between Performance Improvement with the

Table 2. Results of Factor Analysis.

No.	Group Construct Description	No. of Items	Cronbach's Alpha	Mean	Std. Dev.
1	External Influences	5	0.768	4.012	0.654
2	Stakeholder Involvement	4	0.801	3.586	0.724
3	Risk Management's Maturity	13	0.955	3.705	0.834
4	Degree of uncertainty and risk faced	3	0.826	3.404	0.824
5	Change in uncertainty and risk faced	4	0.863	3.845	0.677
6	Risks factored into organizational planning	6	0.896	3.233	0.873
7	Usage rate of basic methods	4	0.780	3.205	0.845
8	Usage rate of technical methods	2	0.786	2.505	1.035
9	Effectiveness of basic methods	4	0.777	3.300	0.839
10	Improved performance	9	0.924	3.573	0.710
11	Improved external relationships	3	0.847	3.501	0.862

Source: Research Data, 2011.

Table 3. Pearson Correlation of Performance and Other Factors.

		1	2	3	4	5	6	7	8	9
Performance Improvement	Pearson Correlation	0.053	0.414**	0.719**	0.070	−0.083	0.344**	0.579**	0.461**	0.561**
	Sig. (two-tailed)	0.695	0.001	0.000	0.600	0.534	0.008	0.000	0.000	0.000
	N	58	58	58	58	58	58	58	58	58
Improved external relationships	Pearson Correlation	−0.022	0.464**	0.551**	0.174	−0.022	0.353**	0.386**	0.289*	0.384**
	Sig. (two-tailed)	0.868	0.000	0	0.191	0.868	0.007	0.003	0.028	0.003
	N	58	58	58	58	58	58	58	58	58

**Correlation is significant at the 0.01 level (two-tailed).
*Correlation is significant at the 0.05 level (two-tailed).
Source: Research Data, 2011.

Stakeholders Involvement, Risk Management's Maturity, Risk Factored into Organizational Planning, Usage rate of Basic Methods, and Effectiveness of Technical and Basic Methods.

The performance improvement could be explained by a variety of factors. However, according to these results, the involvement of stakeholders with risk management is an important factor in this enhancement. The Performance is also influenced by the maturity of risk management practices and the way that risk is pondered into organizational planning. Thus, the idea of ERM is important to performance as we saw in the theory. Moreover, both Technical and Basic Methods have a positive influence to the firm's performance improvement.

Table 3 also presents that Improvement on External Relationships has a positively correlation with Stakeholder Involvement, Risk Management's Maturity, Risk Factored into Organizational Planning, and Usage rate of Technical Methods.

According to those results, we could conclude that stakeholders valued the Usage of Technical Methods. Therefore, they wander to be more involved with risk management and incentive for the development of a better risk management. Then, they value the inclusion of risk in organizational planning.

It appears from the results found in the matrix of Pearson correlation that the correlation sought to achieve the third specific objective of this research concerning the examination of the relationship between risk management practices and improvement of organizational performance. This association was evident from the multiple regression analysis conducted in the following section.

Risk Management and Performance Improvement

The association among variables carried out a multiple regression analysis between the factors of performance improvement and other variables. According to the stepwise method, the predictive power of the regression model was evaluated, which measures the performance improvement. It was observed in Table 4 that the model explains 50.8% (Adjusted R^2) of the variance in performance improvement based on the Maturity Level of Risk Management and 55% based on the Maturity Level of Risk Management and Stakeholder Involvement. The other independent variables entered in the model were excluded, because they have significant values lower than 0.05.

Table 4. Improved Performance: Multiple Regression Stepwise Forward.

Model	R^2	F	Coefficient	Not Standardized Coefficients	Std. Dev.	Standardized Coefficients	T	Sig.
	Adjusted R^2	Sig. ANOVA		B		Beta		
1	0.517	59.830	Constant	1.305	0.300		4.344	0.000
	0.508	0.000	Risk Management's Maturity	0.612	0.079	0.719	7.735	**0.000**
2	0.566	35.851	Constant	0.695	0.377		1.843	0.071
	0.550	0.000	Risk Management's Maturity	0.557	0.079	0.654	7.072	**0.000**
			Stakeholder Involvement	0.227	0.091	0.231	2.501	**0.015**

Source: Research Data, 2011.
Note: Bold indicates significance at 0.05.

The Analysis of Variance (ANOVA) showed that chances of this result occurring by sampling error have 0.000 of significance both for the first model and for the second model (Table 4). Therefore, these results confirm the findings from the analysis of Adjusted R^2.

From the results presented in Table 4, it appears that the performance improvement was positively related to the Maturity Level of Risk Management and the Level of Involvement of Stakeholders in Risk Management. Thus, these two factors contribute to nonfinancial Brazilian companies adopt the best practices of risk management.

If a simple regression model is considered, with only one explanatory variable, then the equation that best describes the Performance Improvement in nonfinancial firms is Eq. (1). Then, if the value of the Maturity Level of Risk Management were known, multiplying this by 0.612 and adding 1.305 to it, the value of Performance Improvement perceived by risk managers would be found, with the variations caused by the error of this model.

$$\text{Performance Improvement} = 1.305 + 0.612(\text{ERM Maturity}) + e \quad (1)$$

If the model adopts more than one explanatory variable, then it would be a multiple regression analysis. From Eq. (2), it appears that knowing the value of Risk Management Maturity and the Stakeholders Involvement, and then multiplying these values, respectively, to 0.557 and 0.227 and adding the constant 0.695, the value of Performance Improvement perceived by risk managers would be found, with the variations caused by the error of this model.

$$\begin{aligned}\text{Performance Improvement} = {} & 0.695 + 0.557(\text{ERM Maturity}) \\ & + 0.227(\text{Stakeholders Involvement}) + e\end{aligned} \quad (2)$$

Thus, it appears that the maturity level of risk management is an important factor for the Improvement of Organizational Performance of Brazilian nonfinancial firms. In addition, stakeholders have an important role for this activity, since their level of involvement with risk management also positively influences Performance Improvement and the development of ERM in the organizations studied. Therefore, considering the Performance Improvement factor composed of various variables, these two factors contribute positively to the improvement in Corporate Planning, Resource Allocation and Utilization, Management Reporting, Communication within the Organization, Management of Organizational Change, Reputation,

Recognition and Uptake of Opportunities, and Employee Confidence in Carrying out their Duties.

Risk Management and Improved External Relationships

To analyze the association among variables, a multiple regression analysis was carried out between the factors of Stakeholder Relationship Improvement and other variables. According to the stepwise method the predictive power of the regression model was evaluated, which measures the performance improvement. It was observed in Table 5 that the model explains 30.4% (Adjusted R^2) of the variance in performance improvement based on the Maturity Level of Risk Management and 40.8% based on the Maturity Level of Risk Management and Stakeholder Involvement. The other independent variables entered in the model were excluded, because they have significant values lower than 0.05.

The Analysis of Variance (ANOVA) showed that chances of this results occurring by sampling error have 0.000 of significance both for the first model and for the second model (Tables 4 and 5). Therefore, these results confirm the findings from the analysis of Adjusted R^2.

It appears that the improvement in the relationship with stakeholders is positively related to the Maturity Level of Risk Management and the Level of Involvement of Stakeholders in Risk Management. Thus, it is understood that these two factors contribute to nonfinancial Brazilian companies adopt risk management practices to comply with the demand of stakeholders.

Thus, the model of Improved External Relationships can be explained by Eq. (3), which describes the simple regression model that describes this improvement based on the Maturity Level of Risk Management. Thus, if the value of the Maturity Level of Risk Management were known, multiplying this by 0.570 and adding 1.388 to it, the value of Improved External Relationships would be found, with variations caused by the error of this model.

$$\text{Improved External Relationships} = 1.388 + 0.570(\text{ERM Maturity}) + e \quad (3)$$

If the model adopts more than one independent variable, then it is a multiple regression analysis. From Eq. (4), it appears that knowing the value of Risk Management Maturity and the Level of Stakeholders Involvement and multiplying these values, respectively, to 0.473 and 0.400, adding the constant of 0.313 would clearly show the value of Improved External Relationships, with variations caused by the error in this model.

Table 5. Improved External Relationships: Multiple Regression Stepwise Forward.

Model	R^2 / Adjusted R^2	F / Sig. ANOVA	Coefficient	Not Standardized Coefficients B	Std. Dev.	Standardized Coefficients Beta	T	Sig.
1	0.304	24.469	Constant	1.388	0.438		3.173	0.002
	0.292	0.000	Risk Management's Maturity	0.570	0.115	0.551	4.947	**0.000**
2	0.408	18.966	Constant	0.313	0.534		0.585	0.561
	0.387	0.000	Risk Management's Maturity	0.473	0.112	0.458	4.238	**0.000**
			Stakeholder Involvement	0.400	0.129	0.336	3.110	**0.003**

Source: Research Data, 2011.
Note: Bold indicates significance at 0.05.

$$\text{Improved External Relationships} = 0.313 + 0.473(\text{Maturity GR}) + 0.400(\text{Stakeholders Involvement}) + e \quad (4)$$

Thus, it appears that the maturity level of risk management is an important factor for Improved Relations with Stakeholders; therefore, it is clear that stakeholders appreciate companies that adopt best practices of risk management. Similarly, stakeholders have an important role for this activity, since their level of involvement with risk management also influences positively the Performance Improvement and the Improvement of External Relationships in the organizations studied. Thus, these two factors contribute positively to improving the relationship with shareholders, customers, and suppliers.

According to the models shown both for Performance Improvement and for Improved Relations with Stakeholders, it became clear that these are associated with the Maturity Level of Risk Management and Stakeholder Involvement. This relationship confirms the theoretical assumption that risk management adds value to the organization's activities.

It is noteworthy that the stakeholders play an important role in the development of this approach and, consequently, to improve performance through it. It is understood that external pressures exerted by stakeholders in the organization's activities drive them to the improvement of risk management.

These results confirm the assumptions made by committee bodies such as COSO (2004) about the need for an Integrated Risk Management in organizational culture, therefore, a job of everyone. Moreover, it shows that risk management can be used as an instrument of legitimacy, as recommended by Power (2009). Therefore, when stakeholders appreciate risk management practices, they signal that organizations should adopt it; for that reason this position would increase investment on risk management and its level of maturity, representing a chain reaction. Future researches may seek empirical evidence of these relationships.

CONCLUSION

Recent researches have indicated that the risk management represents a competitive advantage when companies develop a systemic approach, integrated with business strategies (Damoradan, 2009). Therefore, Bernstein

(1997) argues that such management is an important competitive edge and that risks are the drivers for the economy development.

The Enterprise Risk Management could not prove its strategic role until it is implemented and until it has been used for a reasonable period of time. However, it will not be sold to companies, if not proved to be a good investment. Therefore, it is important that ERM be recognized as a tool that represents something far beyond a simple mechanism for compliance and that adds value to an organization's activities.

When the link between the value of risk management and performance improvement is evidenced, ERM will have begun to contribute more significantly to the company's competitive advantage. However, demonstrating the value of prevention is always a difficult task, especially when it comes to future events without the foundation of historical data.

Although largely associated with insurance, compliance, and reduction of losses, the risk management function has been changed recently and is firmly tied to concepts related to process improvement. It was understood that in addition to avoiding negative consequences, risk management provides opportunities for improvement and consequently adds value. However, empirical researches on this subject are still incipient.

Therefore, the objective was to assess the relationship between the practices of Risk Management and Organizational Performance in the 500 Largest and Best Nonfinancial corporations in Brazil.

The sample was formed mainly by companies from the Energy, Water and Sanitation (24.14%) and Manufacturing and Construction sectors (17.24%). From the factor analysis, there was a reduction of 152 variables into 11 factors. The degree of association between organizational performance improvement and other factors was found from the stepwise multiple regressions.

From the multiple regressions conducted, the variance of Performance Improvement was noted as well as Improved External Relationships was explained by Risk Management Maturity and Stakeholders Involvement factors. These results corroborate the theoretical assumptions about the need for a culture of risk in organizations and corroborate the external pressures' influence on the adoption of management practices.

According to our analysis, the maturity of the ERM was identified as an important factor for improving performance in nonfinancial institutions analyzed. This finding demonstrates empirically the importance of ERM to create value and achieve organizational objectives, as described in international guidelines.

Furthermore, we observed the influence that stakeholders represent to the maturity of ERM and consequently in organizational performance. It was shown that stakeholders value the practice of enterprise risk management. Thus, firms would adopt best practices in ERM both to serve the interests of stakeholders in its governance and for the continuous improvement of its activities.

This work also demonstrated its relevance in confirming the theoretical relationship between the ERM and the performance improvement from empirical results. Thus, this research contributes to the development of strategies and business practices to strengthen the Enterprise Risk Management in nonfinancial firms.

As a recommendation for further researches, there was the possibility of studies targeted to specific aspects of this survey questionnaire. Another suggestion is to compare the results of this survey with information from organizations in the financial sector. Additionally, case studies could help to better comprehend about the risk culture at those organizations.

REFERENCES

Barton, T. L., Shenkir, W. G., & Walker, P. L. (2002). *Making enterprise risk management pay off: How leading companies implement risk management*. Englewood Cliffs, NJ: Prentice Hall.

Beasley, M. S., Clune, R., & Hermanson, D. R. (2005). Enterprise risk management: An empirical analysis of factors associated with the extent of implementation. *Journal of Account Public Policy*, *24*, 521–531.

Bernstein, P. L. (1997). Desafio aos Deuses: A fascinante história do risco. Campus, Rio de Janeiro.

Bowling, B. M., & Rieger, L. (2005). Success factors for implementing enterprise risk management. *Bank Accounting and Finance*, *18*(3), 21–26.

Bruni, A. L. (2009). *SPSS aplicado a pesquisa acadêmica*. São Paulo: Atlas.

Bruno-Britz, M. (2009). The age of ERM. *Bank Systems & Technology*, *1*(February), 20.

CAS-Casualty Actuarial Society – Enterprise Risk Management Committee. (2003). *Overview of enterprise risk management*. Retrieved from http://www.casact.org/research/erm/overview.pdf. Accessed on May 6, 2008.

CIMA-Chartered Institute of Management Accountants. (2008). *Fraud risk management: A guide to goof practice*. London: CIMA.

Clarke, C. J., & Varma, S. (1999). Strategic risk management: The new competitive edge. *Long Range Planning*, *32*, 414–424.

Collier, P., Berry, A. J., & Burke, G. T. (2007). *Risk and management accounting: Best practice guidelines for enterprise-wide internal control procedures*. Oxford: Elsevier.

Collier, P. M., & Berry, A. J. (2002). Risk in the process of budgeting. *Management Accounting Research*, *13*, 273–297.

Corrar, L. J., & Theóphilo, C. R. (coordenadores). (2004). *Pesquisa operacional para Decisão em Contabilidade e Administração*. São Paulo: Atlas.

COSO. (2004). *Enterprise risk management – Integrated framework executive summary*. Retrieved from http://www.coso.org/Publications/ERM/COSO_ERM_ExecutiveSummary.pdf. Accessed on December 2010.

Damoradan, A. (2001). *Corporate finance*. New York: John Wiley & Sons.

Damodaran, A. (2009). *Gestão estratégica do risco: uma referência para a tomada de riscos empresariais*. Porto Alegre: Bookman.

Fávero, L., Belfiore, P., Silva, P., & Chan, B. (2009). *Análise de dados: Modelagem multivariada para tomada de decisões*. Rio de Janeiro: Elsevier.

Gephart, R. P., Van Maanen, J., & Oberlechner, T. (2009). Organizations and risk in late modernity. *Organization Studies*, *30*(2–3), 141–155.

Gordon, L. A., Loeb, M. P., & Tseng, C.-Y. (2009). Enterprise risk management and firm performance: A contingency perspective. *Journal of Accounting and Public Policy*, *28*(4), 301–327.

Guimaraes, I. C., Parisi, C., & Pereira, A. C. (2006). Análise das Práticas de Gestão de Riscos nas Empresas Não-Financeiras de Capital aberto da Cidade de São Paulo: Uma Percepç ão dos Gestores de Riscos e Controllers. In XXX Encontro da ANPAD, Salvador/BA. Anais do XXX Encontro da ANPAD.

Gujarati, D. (2006). *Econometria Básica. Tradução Maria José Cyhlar Monteira* (4th ed.). Rio de Janeiro: Elsevier.

Hair, J. F., Tatham, R. L., Anderson, R. E., & Black, W. (2005). *Multivariate data analysis* (5th ed.). Upper Saddle River, NJ: Pearson.

Hoyt, R. E., & Liebenberg, A. P. (2009). *The value of enterprise risk management*. Retrieved from http://papers.ssrn.com/sol3/papers.cfm?abstract_id=1440947. Accessed on August 16, 2011.

IFAC-International Federation of Accountants. (1999). *Enhancing shareholder wealth by better managing business risk*. International Management Accounting Study No. 9. Retrieved from http://www.ifac.org/Committees/FMAC/index.html. Accessed on February 9, 2011.

Institute of Chartered Accountants in England & Wales. (1999). *Internal control: Guidance for directors on the combined code*. Turnbull Report.

IRM-Institute of Risk Management. (2002). *A risk management standard*. London: IRM.

Kaplan, R. S. (2011). Accounting scholarship that advances professional knowledge and practice. *The Accounting Review*, *86*(2), 367–383.

Knight, F. (1921). *Risk, uncertainty and profit*. Boston, MA: Houghton Miffin.

Lin, H. H., & Wu, F. H. (2006). How to manage section 404 of the Sarbanes-Oxley Act: What is wrong with section 404 of the Sarbanes-Oxley Act. *Journal of Accounting and Corporate Governance*, *3*(3), 1–16.

Little, R. J. A., & Rubin, D. B. (2002). *Statistical analysis with missing data* (2nd ed.). New York, NY: John Wiley & Sons.

Malhotra, N. K. (2001). *Pesquisa de marketing: Uma orientação aplicada* (3rd ed.). Porto Alegre: Bookman.

Martin, D., & Power, M. (2007). *The end of enterprise risk management. AEI-Brookings joint center for regulatory studies*. Washington, DC: Brookings Institute.

Mattar, F. N. (1996). *Pesquisa de marketing*. São Paulo: Atlas.

Mikes, A. (2005). *Enterprise risk management in action*. Centre for the Analysis of Risk and Regulation (CARR) Discussion Paper Report Series no. 35.

Mikes, A. (2009). Risk management and calculative cultures. *Management Accounting Research*, *20*(1), 18–40.

Miller, P., Kurunmaki, L., & O'leary, T. (2008). Accounting, hybrids and the management of risk. *Accounting, Organizations and Society*, *33*(7–8), 942–967.

Moeller, R. R. (2007). *COSO enterprise risk management: Understanding the new integrated ERM framework*. Hoboken, NJ: John Wiley & Sons, Inc.

Power, M. (2007). *Organized uncertainty: Designing a world of risk management*. London: Oxford University Press.

Power, M. (2009). The risk management of nothing. *Accounting, Organizations and Society*, *34*(6/7), 849–855.

Pricewaterhousecoopers. (2004). *Managing risk: An assessment of CEO perspectives*. New York, NY: PwC.

Revista Exame (2010, Abril, jul). *Melhores e maiores: as 500 maiores empresas do Brasil*. São Paulo: Abril.

Saunders, J., Morrow-Howell, N., Spitznagel, E., Dork, P., Proctor, E., & Pescarino, R. (2006). Imputing missing data: A comparison of methods for social work researchers. *Social Work Research*, *30*(1), 19–31.

Simkins, B., & Smithson, C. (2005). Does risk management add value? A survey of the evidence. *Journal of Applied Corporate Finance*, *17*(3), 8–17.

Standard & Poor's. (2007). *Request for comment: Enterprise risk management analysis for credit ratings of nonfinancial companies*. Retrieved from http://www.standardandpoors.com/ratingsdirect. Accessed on January 28, 2011.

Tabachnick, B. G., & Fidell, L. S. (1983). Cleaning up your act: Screening data prior to analysis. In B. G. Tabachnick & L. S. Fidell (Eds.), *Using multivariate statistics* (pp. 68–81). New York, NY: Harper & Row.

Towers Perrin. (2008). *Embedding ERM – A tough nut to crack: An ERM update on the global insurance industry*. Global insurance industry ERM survey report. Retrieved from http://www.towersperrin.com/tp/getwebcachedoc?webc=GBR/2009/200901/2008_Global_ERM_Survey_12809.pdf. Accessed on April 21, 2011.

Trapp, A. C. G., & Corrar, L. J. (2005). Avaliação e gerenciamento do risco operacional no Brasil: análise de caso de uma instituição financeira de grande porte. *Revista Contabilidade & Financas, São Paulo*, *1*(37), 24–36.

Woods, M. (2007). Linking risk management to strategic controls: A case study of Tesco plc. *International Journal of Risk Assessment and Management*, *7*(8), 1074–1088.

Zhuang, P., Hou, H., & Hou, H. (2008). Knowledge visualization of most prolific countries and hot-topics of enterprise risk management. *Proceedings of the 4th IEEE international conference on wireless communications, networking and mobile computing*, WiCOM (pp. 1–4).

PART V
PERFORMANCE EVALUATION AND ITS EFFECT ON MOTIVATION

SUBJECTIVE WEIGHTING UNDER ORGANISATIONAL INTERDEPENDENCE CONDITIONS: AN EXPERIMENTAL INVESTIGATION

Daniela Ruggeri

ABSTRACT

Purpose – *Accounting research has long shown the effect of subjectivity in performance evaluation. This study investigates one form of subjectivity in performance evaluation: flexibility in weighting performance measures examining decisions made by supervisors about weighting. Empirical studies show that the performance-measure weights are only partially consistent with the predictions of the agency theory and they are a still outstanding issue.*

Methodology/approach – *We develop an experiment to analyse supervisor decision-making, manipulating two factors: internal organisational interdependence and the level of managerial performance. We derive hypotheses along with both economic and behavioural approaches. The economic approach is based on agency theory predictions and the controllability principle while the behavioural approach is drawn on*

Performance Measurement and Management Control: Global Issues
Studies in Managerial and Financial Accounting, Volume 25, 301–337

ISSN: 1479-3512/doi:10.1108/S1479-3512(2012)0000025015

the organisational justice theory. We argue that in low interdependence contexts the supervisor's decision confirms the agency theory predictions, while in high interdependence conditions weighting decisions could be driven by behavioural considerations of fairness perceptions of the evaluation process and the level of managerial performance.

Findings – *We find that in low interdependence contexts the supervisor's decision confirms the agency theory predictions, while in high interdependence contexts it does not. The results indicate that the supervisor's decision stems from the integration of economic and behavioural perspectives.*

Research and social implications – *The theoretical framework can be useful for interpreting the supervisor decision-making and the weighting process.*

Originality – *The economic and behavioural approaches allow us to understand flexibility in weighting performance measures suggesting that, in addition to economic considerations, a behavioural perspective may also be relevant in explaining subjective weighting.*

INTRODUCTION

Accounting research has generally shown the widespread adoption of multiple performance measurement systems and the effect of subjectivity in performance evaluation (Baiman & Rajan, 1995; Baker, Gibbons, & Murphy, 1994; Gibbs, Merchant, Van der Stede, & Vargus, 2004; Ittner, Larcker, & Rajan, 1997; Ittner, Larcker, Rajan, & Mayer, 2003; Moers, 2005; Murphy & Oyer, 2001). Researchers have suggested that (only using) objective measures can hide important information that subjectivity could reveal. On the one hand, through subjectivity the supervisor can use private and non-contractible information that is not captured by objective measures (Baiman & Rajan, 1995; Baker et al., 1994; Gibbs et al., 2004; Ittner et al., 2003; Prendergast & Topel, 1993). On the other hand, some contributions have highlighted the inefficient use of subjectivity leading to unreliable evaluation and episodes of favouritism inside an organisation (Krishnan, Luft, & Shields, 2005; Lau & Sholihin, 2005; Libby, 1999; Moers, 2005; Otley, 1999). Consequently, the advantage of using subjectivity in managerial performance evaluation has been questioned.

Subjectivity has been explored in all its different forms: (i) use of subjective measures, (ii) flexibility in weighting multiple performance measures and (iii) superior's discretion in allocating rewards (Gibbs et al., 2004; Moers, 2005; Van der Stede, Chow, & Lin, 2006). Highlighting the growing importance that each measure carries in the process of bonus determination, this study investigates one form of subjectivity in performance evaluation: flexibility in weighting performance measures.

Scholars have underlined notable problems relating to how measures should be weighted to get an overall performance evaluation (Hemmer, 1996; Ittner & Larcker, 1998; Murphy & Oyer, 2001). Empirical studies have shown that performance-measure weights are only partially consistent with the predictions of agency theory, highlighting that this form of subjectivity is still an outstanding issue (Krishnan et al., 2005).

Exploring subjective weighting, this study examines supervisor decision-making in weighting performance measures. We derive hypotheses along with agency theory predictions and the controllability principle: the weight of performance measures will depend on how informative and noiseless (controllable) the measures are of a manager's actions (Banker & Datar, 1989; Holmstrom & Paul, 1991). In order to better explain the supervisor's decision, we take two factors into consideration: internal organisational interdependence, where the presence of uncontrollable elements could influence how managers perform, and the level of managerial performance. The first factor reflects the interchange of output and resources between organisational areas and their coordination. Considering internal organisational interdependence is in line with previous studies that proposed it as a potentially important organisational variable for future management accounting studies (Chow, Shields, & Chan, 1991; Keating, 1997; Macintosh & Daft, 1987; Merchant, 1985; Merchant & Van der Stede, 2007) and also with some contributions which recognise its importance in investigating subjectivity in performance evaluation (Gibbs et al., 2004). The second factor, managerial performance level, could play an important role for the supervisor in communicating positive or negative performance evaluation, especially in the case of negative performance, where a supervisor may be fearful of delivering a negative evaluation and thereby damaging the manager's trust in him (Harris, 1994). We argue that in low interdependence contexts the supervisor's decision confirms the agency theory predictions, while in high interdependence context it does not. Drawing on the organisational justice theory, we argue that under high interdependence conditions the weighting decision may be driven by behavioural considerations of perceptions of fairness in the evaluation process and the level of managerial performance.

We develop an experiment to analyse supervisor decision-making, manipulating organisational interdependence and managerial performance levels. The results indicate that supervisor decisions are well explained by the integration of economic and behavioural perspectives. This evidence suggests in addition to economic considerations, a behavioural perspective may also be relevant in explaining subjective weighting. The contribution of this paper is to identify factors which might influence weighting decisions in addition to factors set out by the economic approach.

The remainder of the paper is articulated as follows. The next section discusses the theoretical background on flexibility weighting. The third section argues for subjective weighting under internal organisational interdependence, developing the hypotheses based on the role of controllability principle and the concerns of fairness. The fourth section describes the research method. The last section presents the empirical results, discusses the findings and finally presents our conclusions.

FLEXIBILITY IN WEIGHTING PERFORMANCE MEASURES: LITERATURE REVIEW

Many managerial accounting contributions have proposed analytic models focused on the ex-ante choice or development of performance measures to motivate employees rather than the ex-post use of those measures by evaluators (Feltham & Xie, 1994; Hemmer, 1996). A typical assumption is that performance measures are either mechanistically used in the evaluation process or that evaluators are perfectly rational and optimally use performance measures (Sprinkle, 2003). It is unclear how the evaluators (supervisors) use performance measures information in elaborating the overall assessment of managerial performance and what factors influence the weight placed on performance measures. We focus on analysing the latter aspect regarding the weighting process.

In the accounting literature, flexibility in weighting performance measures has been analysed adopting the informativeness perspective or considering the role played by various factors such as firm strategy, favouritism, distortion and cognitive limitations.

Classifying the academic contributions, we distinguish two theoretical approaches: the economic approach, based on agency theory, and the behavioural approach, based on sociological and psychological theories, used by accounting researchers in order to find new answers to unsolved agency problems. We briefly summarise the prior research of each stream.

Within the economic approach, empirical studies drawn on the informativeness perspective have shown different results in exploring flexibility in weighting (Bushman, Indjejijkian, & Smith, 1996; Murphy & Oyer, 2001). Some researchers demonstrated that the weight assigned to objective measure decreases as its 'noise' rises, allocating a higher weight on the other performance measure (*spillover effect*); while others, using informativeness proxies, show that the weight positively depends on factors which influence the informativeness of performance measures.

Accounting researchers have also analysed some organisational factors which could determine performance-measure weights. In particular, empirical studies have shown the influence of the firm strategy (Ittner et al., 1997): when the firm adopts a cost leader strategy the supervisor assigns a higher weight to quantitative measures (financial measures), while when the firm follows a differentiation strategy the supervisor assigns a higher weight to qualitative (non-financial) measures. Other contributions argue that the weight depends on distortion and favouritism (Prendergast & Topel, 1996): in a multiple performance measurement system the weight associated to discretionary evaluation will be less than the weight assigned to 'hard' measures because of the degree of distortion and favouritism episodes generated by the discretionary evaluation.

Within the behavioural approach some studies have highlighted that performance-measure weights is driven by psychological limitations. Krishnan et al. (2005) argued that determining performance-measure weights is a cognitively difficult task. According to the psychological theory of reasoning, they demonstrate that supervisors are influenced by performance measure precision because they substitute 'the cognitively more accessible but statistically identical attribute of controllability for the precision' (Krishnan et al., 2005, p. 1172). Controllability is more accessible than precision because it 'is based on casual relations between an agent's actions and the related performance measures' (Krishnan et al., 2005, p. 1172). Besides, causality is an accessible attribute that individuals usually adopt in judgments and decisions (Kahneman & Frederick, 2002). Krishnan et al. (2005) also show that individuals do not recognise the spillover effect of changing 'the weight on cost measure, many are likely to stop there rather than consider possible effects of the cost measure's precision on other performance measures' (p. 1174).

Another contribution is given by Ittner et al. (2003), showing that the weight of outcome-oriented measures is higher than other types of measures when the supervisor is affected by an outcome effect distortion. Finally, Frederickson, Pfeffer, and Pratt (1999) argued that sometimes the weight could depend on the supervisor's knowledge of past managerial performance.

Overall, management accounting literature shows that flexibility in weighting does not always depend on the characteristics of measures (precision, informativeness) as the agency theory predicts and even if some researchers have drawn on other social disciplines trying to find a better explanation for the weighting process, it is still an outstanding issue.

SUPERVISOR DECISION-MAKING UNDER ORGANISATIONAL INTERDEPENDENCE CONDITIONS

Agency theory predicts that performance-measure weights will depend on how informative the measure is about agents' actions and how congruent the measure is with the principal's welfare (Banker & Datar, 1989; Datar, Kulp, & Lambert, 2001; Feltham & Xie, 1994).

However, studies testing the association between actual performance-measure weights and proxies for informativeness and congruency have reported mixed results (Bushman et al., 1996; Ittner et al., 1997, 2003). Following the definition by Feltham and Xie (1994), high precision represents low impact of '... uncontrollable events influencing the performance measure'. An increase in precision means an increase in controllability.

The controllability principle, considered a cornerstone in agency setting, postulates that managers should only be evaluated based on what they can control (Atkinson, et al., 1997; Choudhury, 1986; Horngren, Foster, & Datar, 2006; Merchant & Van der Stede, 2007). Applying the controllability principle means neutralising any uncontrollable factors in a manager's performance evaluation which may be generated by unpredictable environmental changes or colleagues' decisions affecting his results. There are two types of neutralisation: ex-ante neutralisation which consists in selecting performance measures which do not include factors that managers cannot control, and ex-post neutralisation, when the supervisor makes year-end adjustments to remove the influence of uncontrollable factors on managers' performances (Giraud, Langevin, & Mendoza, 2008; Merchant & Van der Stede, 2007). We argue that flexibility in weighting is a form of ex-post neutralisation: changing the weight in measures which are affected by uncontrollable factors, balancing the measures' impact in bonus determination. Some studies have demonstrated that the controllability concept is accessible to supervisors who have experience in evaluation process and to them who are not expert on managerial performance evaluation. The concept of controllability readily evokes ideas of causal relations between agent's actions and the related performance measures (Krishnan et al., 2005). Most management accounting textbooks discuss controllability (Atkinson, Kaplan, & Young, 2004; Zimmerman, 2003) and practitioner literature emphasises the desirability of choosing more controllable performance measures (Heneman, 1986).

We decided to analyse flexibility in weighting performance measures in contexts characterised by internal organisational interdependence, considered as a source of internal uncertainty that generates uncontrollable factors. This choice is motivated by the increasing role that this variable has assumed in accounting studies (Chow et al., 1991; Hirst & Yetton, 1999; Keating, 1997; Macintosh & Daft, 1987; Merchant, 1985; Merchant & Van der Stede, 2007). Nowadays many organisational contexts are characterised by internal organisational interdependence and so it is more interesting to investigate this aspect in order to find a better explanation for flexibility in weighting. In particular, we analyse whether under conditions of organisational interdependence the supervisor's weighting decision is driven or not by the controllability principle. The level of internal interdependence increases the number of uncontrollable factors that influence manager's performance – it means that motivation and effort are not the only variables that determine a manager's results. Applying ex-post neutralisation could improve the reliability of performance evaluation and influence managerial behaviour because the controllability principle is related to a reliable evaluation system and to equity concern (Giraud et al., 2008).

In addition to internal interdependence we consider the level of managerial performance. The level of performance expressed by outcome measures has been considered an influenced element in weighting performance measures because it is retained as the origin of the outcome effect (Ittner et al., 2003). Generally in management accounting studies on subjectivity, the communication of negative performance evaluation makes the supervisor fearful of communicating this result to managers trying to protect their trust based relationship (Harris, 1994). Sometimes a supervisor may moderate the evaluation thereby damaging the evaluation process that will be damaged by cognitive bias: leniency and centrality bias (Bol & Smith, 2009). In order to better explain the weighting process, we develop hypotheses considering how the supervisor's decision can change when manager performance level and internal interdependence is high or low.

The Role Played by the Controllability Principle

Considering a context characterised by a low level of organisational interdependence, we expect that when managerial performance level related to performance measure is relatively high or low, the supervisor's decision-making is predicted by the economic approach, based on agency theory propositions. The supervisor places a higher weight on the relatively controllable performance measure than on a less controllable one. In low interdependence context the manager is able to control performance factors captured by performance measures and the supervisor easily recognises causality nexus between manager's action and results. Considering the agency perspective, controllable measures are preferred because agents need to be compensated for the additional risk of uncontrollable factors (Holmstrom, 1979).

The informational value of a performance measure is affected by its noise or the level of precision by which the performance measure provides information about a manager's actions (Banker & Datar, 1989; Feltham & Xie, 1994). The supervisor should attribute higher weight to performance measure when its controllability is higher (Krishnan et al., 2005). It means that the weight placed on performance measures is proportional to the precision of the measures (Lambert & Larcker, 1987).

Following the controllability principle, under low organisational interdependence the manager's performance is determined by his action without the influence of other factors. In this context, the performance measure is informative of the manager's action and the supervisor believes that this measure is explanatory of manager's effort. Under this condition, the weight does not depend on performance level.

H₁. Under low organisational interdependence, the weighting does not depend on managerial performance level.

Supervisor's Decisions Driven by Fairness Concern

Considering a high organisational interdependence context, increased uncertainty reduces the controllability, that is the degree to which an employee can affect an output measure by altering his action inputs (Antle & Demski, 1988).

If supervisor behaves as Feltham and Xie's (1994) model predicts, then he will assign higher (lower) weight to measure when the precision of it is higher (lower) (Krishnan et al., 2005). Agency theory provides that the relative use (weight) of particular performance metrics should be a decreasing function of the noise in the measure (Banker & Datar, 1989; Holmstrom, 1979).

Lambert and Larcker (1987) and Sloan (1992) have provided limited evidence in support of Banker and Datar's (1989) model where weight placed on different performance signals is proportional to their sensitivity and their precision. Some studies associate precision with an abstract statistic attribute of measures that individuals rarely assess (attend to and estimate) unless they are being prompted to do so (Kahneman & Frederick, 2002). We argue that in weighting process other factors can have an influence on supervisor decisions. We suppose that one of these is linked to perceived fairness.

Considering the fairness concern in weighting process could make understandable the supervisor decision-making because 'the evidence suggests that psychology-based explanations may be equally or more relevant than economics-based explanations in understanding measurement practices in some settings' (Ittner et al., 2003). In an organisational context, perceptions of fairness are important since they are a primary antecedent of important work-related outcomes, including employee satisfaction, organisational commitment and job performance, that ultimately affect organisational success (Johnson, Selenta, & Lord, 2006). The veracity and honesty of the principal are expected to have a great impact on the actions of the agent (Prendergast, 1999). The systematic evaluation of the fairness perception in the management accounting context was first undertaken by Hopwood (1972) in analysing performance evaluation methods. Subsequent studies (Lindquist, 1995; Lau & Lim, 2002) have also investigated fairness-related issues in the management accounting context, that is in the use of financial versus non-financial measures (Lau & Sholihin, 2005) and fairness effect on the applicability of the controllability principle (Giraud et al., 2008).

Organisational justice theory provides an interesting framework for analysing the perceived fairness related to performance evaluation and controllability. It suggests that satisfaction with performance evaluation systems is influenced by perceptions of distributive, procedural and interactional justice (Greenberg, 1986): distributive justice refers to the perceived fairness of the ratio outcome/contribution (Adams, 1965), procedural justice concerns individuals' perceptions about the fairness of procedures used to evaluate performance (Thibault & Walker, 1975) and finally, interactional justice theory focuses on the perceived fairness of the way supervisors use the procedure during the evaluation process (Bies & Moag, 1986; Lind & Tyler, 1988). In analysing the implication of fairness concern on performance evaluation process, we focus only on process fairness (procedural justice) and outcome fairness (distributive justice), leaving aside the interactional justice. We do not consider the interactional justice because it is not an independent justice type at all but rather an antecedent of distributive and procedural justice. Interpersonal treatment is a kind of organisational outcome that contributes to the perceptions of distributive justice as well as a part of organisation practices, hence contributing to perceptions of procedural justice (Cohen-Charash & Spector, 2003).

In order to better explain the weighting process and the effect of fairness on supervisor's decision we analyse two different situations: making decisions under high and low level of managerial performance.

Decision-Making Under High Level of Managerial Performance

When the organisational interdependence is high (low controllability) and the level of manager's performance is high, the application of controllability principle could play an important role.

The controllability principle is a condition of fairness, which leads to manager satisfaction and motivation, in turn leading to high performance.

Considering agency perspective, the supervisor should place a higher weight on measure when the level of controllability is higher (Krishnan et al., 2005; Lambert & Larcker, 1987). Increasing organisational interdependence leads to low levels of controllability and consequently to less informative measures. Following the agency perspective the weight placed on measurement should be decreased.

We believe that the supervisor acts in the opposite way. The motivation for this assumption is twofold: the first reason is supported by accounting literature on uncontrollable factors; the second one is supported by behavioural research on organisation justice.

Accounting literature has considered uncontrollable factors in an asymmetric way: what is desired is the prevention of unfairness, the protection of managers against unfavourable factors. Favourable uncontrollable factors, that is 'good luck', have not been really considered. This omission is not formally justified, but can probably be explained by the fact that favourable factors do not raise justice issues, since managers benefit from them. The literature deals with risky situations, not with fortunate circumstances. Larmande and Ponssard (2007) offer a rationale for the fact that the controllability principle is not always applied by firms for external economic factors, as documented by Merchant (1989) and recommended by Jensen and Murphy (2004). They prove that a reward for luck may be the optimal response when top managers have private pre-decision information on these events and can react to it.

According to the behavioural research on organisational justice, evaluation procedures have important behavioural implications for organisational members. Paying attention to procedures is crucial to promote positive employee attitudes and behaviours (Alexander & Ruderman, 1987; Horngren et al., 2006; McFarlin & Sweeney, 1992; Merchant & Van der Stede, 2007; Lindquist, 1995). Looking at the evaluation process, flexibility in weighting consists in changing ex-post procedures used by supervisors to evaluate manager performance.

Earlier studies on procedural fairness were associated with the work of Thibault and Walker (1975) who found that perceptions of procedure fairness in legal settings were positively associated with increased satisfaction. Subsequent studies (Leventhal, 1980) argued that procedural fairness is an important determinant of perceived fairness in almost any allocation decision. This extended procedural justice research from legal settings to organisational settings where procedural fairness judgments were also found to have a major role in affecting organisational members' attitudes and behaviours (Lind & Tyler, 1988). For example, Lindquist (1995) found that the interaction between process control and the fairness of budget (attainable or unattainable) affected task and budget satisfaction.

Regarding the effects of procedural justice, Tang and Sarsfield-Baldwin (1996) argue: 'if managers can apply rules fairly and consistently to all employees and reward them based on performance and merit without personal bias, then employees will have a positive perception of procedural justice, which may lead to a higher satisfaction, commitment and involvement'. Based on the above discussion, it is reasonable to conclude that there is likely to be a positive association between fairness in performance evaluation procedures and job satisfaction (Lau & Sholihin, 2005).

If the supervisor acts coherently to agency perspective, he will change the weight on uncontrollable measure reducing it even if managerial performance level on this measure is high. In this case, the supervisor neutralises the effect of uncontrollable factors adjusting the weight on performance measure.

Paying attention to fairness concern and procedural fairness, the evaluated managers could perceive this adjustment as unfair because the supervisor decreases the weight on performance measure which increases the level of the perceived incentive. Besides, a change to the evaluation procedure could generate some negative consequences such as decreased supervisor credibility, low job satisfaction and low trust in the supervisor.

Under high organisational interdependence context, the high managerial performance level could be attributable to good luck condition or a manager's effort. In the first case, some uncontrollable factors have played in favour of managers increasing the level of performance reached or the manager has used some private information to manage the uncertainty. In the second case, the manager has put more effort into performing the task, reaching a good level of performance. In both cases, the supervisor could reward manager for facing the uncontrollable factors. Giving an incentive to exploit the private information to manage the uncertainty could improve the manager's capacities when facing uncertainty in the future.

We thus expect the supervisor to place a high weight on uncontrollable measure rewarding the manager who was able to manage the uncertainty more and increasing the bonus. This decision increases the incentive cost for the firm in the short term but could guarantee an improvement in future performance if the environment become more uncertain.

> **H_2.** Under high organisational interdependence, the weighting depends on managerial performance level.

> **H_{2a}.** Under high managerial performance level, the weighting does not depend on measure controllability.

Making Decision Under Low Level of Managerial Performance

Some theoretical work indicates that subjectivity can improve incentive contracting because it allows the supervisor to exploit non-contractible information that might be otherwise ignored, and to mitigate the distortion in managerial effort by 'backing out' dysfunctional behaviour induced by incomplete performance measures (Baiman & Rajan, 1995; Baker et al., 1994).

In the considered setting characterised by high interdependence and low performance the supervisor, following the agency predictions, should place low weight on performance measures influenced by uncontrollable factors.

The supervisor, applying the controllability principle, makes an ex-post neutralisation, adjusting the weight on uncontrollable measures, and assigning more relevance to the other adopted performance measures.

We believe that this decision is justified by economic reasons drawn from agency theory and fairness concern linked to distributive fairness.

Principal-agent research has often been criticised for not paying attention to the effect of personal action choices on personal reputation for trust and fairness (Baiman, 1990). Allowing for jointly observable but unverifiable information and incorporating incomplete contracts in a multi-period setting, allows concerns for trust and fairness and, more generally, a reputation to endogenously arise. A supervisor's reputation has to do with employees' beliefs, based upon an individual's history of action, as to whether he will choose to act in a certain way in the future. How employees believe he will act in the future influences how they will act in the future.

The supervisor might be reluctant to adjust for 'good luck' as this might be perceived as increasing the performance standard when performance is better than expected (Gibbons, 1987). The use of subjectivity to adjust weighting is therefore only likely to occur when 'bad luck' makes performance measures unreachable.

In this sense, distributive fairness plays an important role in supervisor's decision-making because if the managers perceive that the allocation of incentives is unfair due to uncontrollable factors, they decrease the effort on task and consequently their future performance could decrease. The measure used for evaluating a manager's performance may not capture the manager's effort or may fail to take into consideration the influence of factors that he cannot control.

The concept of distributive fairness is drawn on the equity theory that was developed by Adams (1965) in a broader context of social exchange. Adams represents social exchange as a two-way transaction in which each side provides something to the other and, therefore receives something in return. Employees for example have an exchange relationship with employers. Employees expend time and effort in working for employers; in the language of equity those are some of the contributions that employees make as their input to the exchange. In return, employers provide wages and other forms of compensation as the output employees receive for their work. Adams (1965) suggests that because equity is such a fundamental norm, individuals will experience inequity stress when the allocation of outcomes among group members is not proportional to contribution. Lind and Tyler (1988) conceptualise fairness of outcomes as judgments on how fair are those norms which deal with the outcomes people receive.

Providing employees with negative feedback might therefore be simply unpleasant (Harris, 1994) and damage the supervisor–employee relationship based on mutual trust.

Given these consideration, we believe that the supervisor should apply subjectivity to protect managers from uncontrollable factors making an ex-post adjustment of weights on performance measure. This decision should be motivated by the informativeness and controllability principle and the perception of outcome fairness.

H_{2b}. Under low managerial performance level, the weighting depends on measure controllability.

EXPERIMENTAL METHOD

Properly designed experiments are useful mechanisms for studying cause–effect relations under pure and uncontaminated conditions (Kerlinger & Lee, 2000). The experimental method allows testing theory predictions, behavioural validity and assumptions, offering evidence regarding why actual behaviour deviates from what is predicted by a theoretical model. Our experiment research provides evidence regarding how organisational interdependence and managerial performance level affect the supervisor's decision-making about the weighting process.

In order to design our experiment we first focus on the characteristics that the experimental context should have and the kind of participants it should involve.

The experimental context is designed considering three factors: organisational setting, principal–agent relationship and task characteristics.

The organisational setting is defined by interdependence between departments, which 'depend upon each other and exchange information and resources to accomplish their respective tasks' (Macintosh & Daft, 1987). The principal–agent relationship is characterised by the moral hazard problem arising when owners cannot observe the actions or their employee's effort level and must therefore evaluate employees' performance, basing compensation contracts on imperfect surrogates of behaviour (Arrow, 1985). We assume that the supervisor, for example the director of the unit, can incorporate non-contractible information about subordinates' actions in performance evaluation (Baiman & Rajan, 1995;

Baker et al., 1994), changing the weight assigned to performance measures. The supervisor is a 'special witness' of subordinates' actions, who uses the information from his observations in the evaluation process when some uncontrollable factors influence subordinates' results. Finally, we consider a task characterised by quantitative and qualitative aspects whereas quantitative measures capture what is measurable, and qualitative measures give some information about how the subordinates act.

Regarding the participants, we decided to design our experiment for professionals. Conducting an experiment with professionals is useful when 'you feel that the kinds of skills needed in the experimental task may be difficult to acquire in few laboratory sessions ... (and) it may become necessary to recruit professional subjects' (Friedman & Sunder, 1994, p. 42). In line with Burns (1985) 'it is unproductive to choose as subjects those whose prior experience is contrary to the current design requirements, for they will have difficulty in adjusting to a new frame of reference with consequent suboptimal behaviour' (p. 152). In order to explain the weighting process, our participants have to be familiar with evaluation procedures. We design a hospital context where the department director is the supervisor of physicians in his unit, evaluating them using two performance measures: quality index and patient's satisfaction index. We focus on physicians as evaluated subjects because they have an important role in determining hospital performance and they are considered a relevant group in healthcare organisation[1] (Gomez-Mejia & Balkin, 1992). Our attention to the healthcare context is justified by the growing interest regarding the evaluation process in this kind of organisation.[2] The nature of healthcare service and the healthcare organisational interdependence makes it interesting to explore the weighting decision in this context.

Participants

The sample was taken from a population composed of 295 subjects working in 33 units of a large teaching hospital in the Southern Italy. The top management sent a letter to physicians in order to encourage them to participate in the experiment on evaluation process. The letters were sent by mail and physicians, interested into the experiment, communicated their availability via e-mail. At the end of the recruiting process, the sample was composed of 75 subjects. Selected physicians were familiar with the evaluation process adopted into the teaching hospital.

Participants assume the role of department director evaluating physician performance. In particular, their task consists in assigning weight to two performance measures: quantitative and qualitative performance measures. We report some top management notes in order to help participants understand their role in the experiment. In particular, we highlight that they have to consider performance measures, stressing their importance and suggesting attributing equal weight to both measures. This information is given with the aim of attributing the same importance to both performance measures.

In order to motivate participants to accomplish the required task, we use an incentive. Considering that in the real context participants receive high bonus, a monetary incentive may not be sufficiently effective in motivating them to perform the task required.

According to Maslow (1970), individual needs can be classified in five categories: physiological, safety, love and belonging, esteem, self-actualisation. Physiological and safety needs are satisfied indirectly by the pay system, clarity of the evaluation procedures, pay, career prospects; love and belonging are satisfied by working environments that facilitate socialisation and cooperation between individuals. The experimental incentive will focus on esteem and self-actualisation. We choose as incentive a social donation to the Italian Red Cross. This incentive should encourage individuals to perform the required task in order to obtain the donation in his/her name. Individuals should be motivated by a desire to improve their social condition, in order to have a good reputation within their working group. In this case, the donation is a social recognition gained through the commitment and effort demonstrated performing the experimental task. At the beginning of the experiment, participants write down the number casually assigned to them by a yellow card on the top of their experimental instructions. This procedure will allow us to associate the experimental result to this number, ensuring anonymity. We tell participants that at the end of the experiment we will communicate the classification of results and the first two classified will be awarded with a donation of 100 Euros to the Italian Red Cross for the mission 'Haiti emergency'. At the end of the experiment, participants were informed by a letter that the classification was not processed because the task could not be classified, given to the subjective nature of it. The idea of 'classification' was used to motivate the participants to carry out the task required. The awards were assigned randomly. The communication of the winners was announced via e-mail and the participants with the corresponding number sent their data in order to make the donation in their name.

Experiment Design and Independent and Dependent Variables

The experiment employed a 2 × 2 between-subjects design in which the two levels of the within-subjects factor were difference in organisational interdependence and managerial performance level on quantitative measure (hereafter called performance level).

Regarding the organisational interdependence, we manipulated the controllability of quantitative measure by varying the degree of the interaction between departments. Under low organisational interdependence condition, the level of performance on quantitative measure is described by information about the autonomy of each department. Under high interdependence condition, the level of performance on quantitative measure is described as affected by uncontrollable interactions between departments. Regarding the quantitative performance measure, we manipulated the physician's score on quality index. This index is composed of three performance dimensions: average stay in bed, the ratio days of 'stay in bed'/ theoretical days of 'stay in bed' and hygiene quality (number of infections contracted by patients in hospital). In particular, we manipulated only one performance dimension, the average stay in bed, in order to make the experimental task easy. The quality index was scored on a 0–15 point scale. We manipulated it by varying the physician's score on it: a physician's score has a value of 4 when the level of quantitative performance measure is low and a value of 12 when the level of it is high. We manipulated the performance level only on the quantitative measure keeping fixed the score on the qualitative measure at 8 (medium level of performance).

According to the aim of this study, the dependent variable is the weight assigned by the supervisor to performance measures. After reading the experimental instruction, participants provided a subjective weighting, placing a weigh on the two performance measure on a scale from 0% to 100%, reflecting the importance that they assigned to the performance measures. The two measures are averaged together to arrive at the overall payoff that the department manager will receive.[3]

RESULTS

Descriptive Statistics

'Manipulation checks' were included at the end of each section in the experiment instructions. In particular, participants were asked to express their degree of agreement or disagreement with the statements about the manipulated variables. Manipulation checks permitted us to assess whether participants had assimilated the information correctly. Regarding the level of performance on the quantitative performance measure, we manipulated the 'quality index' by varying the score obtained on this measure. Participants were asked to express their opinion about the level of performance on a 1–5 point scale. Participants recalled that the physician's score on quantitative performance measure was higher under high performance level condition than under low performance level condition ($p<0.01$). Regarding the level of organisational interdependence, we manipulated the presence (or absence) of uncontrollable factors which affects the physician's results (score) on quantitative performance measure. Participants also judged the controllability of quantitative performance measure higher under low organisational interdependence than under high organisational interdependence ($p<0.01$). Thus, our manipulations of the level of performance on quantitative measure and organisational interdependence were successful.

Table 1 shows the mean and the standard deviation of weight on quantitative measures by experimental cells. Fig. 1 graphically shows the means of the dependent variable under conditions of high and low interdependence, and high and low performance level.

Table 1. Least Square Means (Standard Errors) of Weighting by Performance Measure Level and Organisational Interdependence.

Level of Performance Measure	Low Organisational Interdependence	High Organisational Interdependence	Row Mean
High	63.33	61.67	62.50
	(11,882)	(8,575)	(10,247)
	$N=18$	$N=18$	$N=36$
Low	62.63	31.50	46.67
	(11,471)	(8,751)	(18,685)
	$N=19$	$N=20$	$N=39$
Column mean	62.97	45.79	54.27
	(11,514)	(17,496)	(17,099)
	$N=37$	$N=38$	$N=75$

Level of performance measure is manipulated by varying whether the physician achieved a relatively high or low performance score during the period (where each performance score was associated with a particular range of performance on quantitative measure). Organisational interdependence is manipulated by varying whether an event outside the physician's control significantly affected his performance during the period.

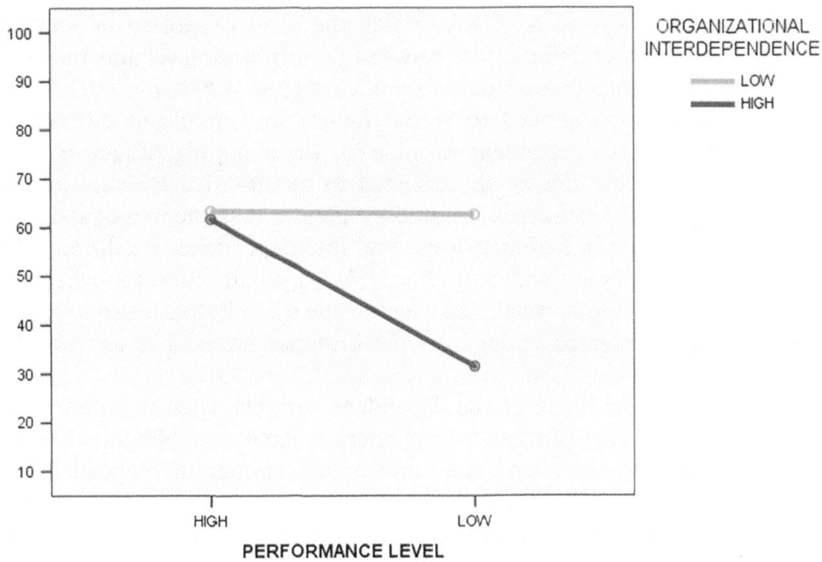

Fig. 1. Mean Weighting in the High and Low Organisational Interdependence Conditions.

Analysis of the 'means-treatment plot' (Fig. 1) shows that under low organisational interdependence, weight assigned to the quantitative measure is similar under high and low performance levels. However, under high organisational interdependence the weight assigned to the quantitative measure decreases under low performance level. Graphically, we note an interaction effect between the two manipulated factors that is a combined effect of both independent variables on the dependent variable.

Hypothesis Tests

We test the interaction effect by applying the two-way ANOVA model. In order to verify the appropriateness of using this model, we test the homogeneity of variance between groups. The Levene's test shows a *p*-value of 0.552, so we accept the null hypothesis that the variances between the groups are equal at a significance level $\alpha = 0.01$. We also verify the normality distribution by experimental cells[4] applying the nonparametric Kolmogorov–Smirnov test, and we accept the null hypothesis that the distribution of the dependent variable is normal.

Applying the two-way ANOVA model, the results reported in Table 2 indicate that the interaction effect between performance level and the level of organisational interdependence is significant ($F = 38.589$, $p < 0.01$). Post-hoc Tukey's test was applied to verify if there are significant differences between means of the dependent variable for the manipulated factors.

Table 3 shows that the weight assigned to quantitative measure in low organisational interdependence condition is greater than the weight assigned to the same measure in high organisational interdependence condition. The test for the contrasts is significant ($F = 47.813$, $p < 0.01$). Besides, when the performance level is low, weight assigned to the quantitative measure is less than the weight assigned under high performance levels. The test for the contrasts indicates significant values ($F = 42.352$, $p < 0.01$).

Table 4 shows the mean of the dependent variable when organisational interdependence and performance level interact. Analysis highlights that the mean values are similar when the level of organisational interdependence is low, demonstrating that there are no differences between the weights assigned to quantitative performance measure under low and high performance conditions. This result confirms H_1: *Under low level of organisational interdependence, the weighting does not depend on managerial performance level.* The supervisor assumes the economic perspective and gives greater weight to the quantitative measure than the qualitative

Table 2. Analysis of Variance for Performance Level and Organisational Interdependence.

Dependent Variable: Weighting

Source	Type III sum of squares	df	Mean square	*F*-stat.	*p*-value
Corrected model	14,161,246(a)	3	4,720,415	44,846	0.000
Intercept	224,656,333	1	224,656,333	2,134,310	0.000
Organisational interdependence	5,032,803	1	5,032,803	47,813	0.000
Performance level	4,457,974	1	4,457,974	42,352	0.000
Organisational interdependence × performance level	4,061,805	1	4,061,805	38,589	0.000
Error	7,473,421	71	105,259		
Total	242,500,000	75			
Corrected total	21,634,667	74			

$R^2 = 0.655$ (adjusted $R^2 = 0.640$).
a, signifcance at 0.01.

Table 3. Pairwise Comparisons.

Dependent Variable: Weighting

Organisational interdependence (*I*)	Organisational interdependence (*J*)	Mean difference (*I*–*J*)	SE	Sig. (a)	99% confidence interval for difference (a)	
					Lower bound	Upper bound
Low	High	16,399*	2,372	0.000	10,122	22,676
High	Low	–16,399*	2,372	0.000	–22,676	–10,122

Based on estimated marginal means.
*The mean difference is significant at the 0.01 level.
Adjustment for multiple comparisons: Least significant difference (equivalent to no adjustments).

Univariate Tests

Dependent Variable: Weighting

	Sum of squares	df	Mean square	*F*	Sig.
Contrast	5,032,803	1	5,032,803	47,813	0.000
Error	7,473,421	71	105,259		

The *F* test shows the effect of organisational interdependence. This test is based on the linearly independent pairwise comparisons among the estimated marginal means.
a, signifcance at 0.01.

Table 3. *(Continued)*

Dependent Variable: Weighting

Performance level (*I*)	Performance level (*J*)	Mean difference (*I*–*J*)	SE	Sig. (a)	99% confidence interval for difference (a)	
					Lower bound	Upper bound
Low	High	−15,434*	2,372	0.000	−21,712	−9,157
High	Low	15,434*	2,372	0.000	9,157	21,712

Based on estimated marginal means.
*The mean difference is significant at the 0.01 level.
Adjustment for multiple comparisons: Least significant difference (equivalent to no adjustments).

Univariate Tests

Dependent Variable: Weighting

	Sum of squares	df	Mean square	*F*	Sig.
Contrast	4,457,974	1	4,457,974	42,352	0.000
Error	7,473,421	71	105,259		

The *F* test shows the effect of performance level. This test is based on the linearly independent pairwise comparisons among the estimated marginal means.

Table 4. Organisational Interdependence × Performance Level.

Dependent Variable: Weighting

Organisational interdependence	Performance level	Mean	SE	99% confidence interval	
				Lower bound	Upper bound
Low	Low	62,632	2,354	56,402	68,862
	High	63,333	2,418	56,933	69,734
High	Low	31,500	2,294	25,428	37,572
	High	61,667	2,418	55,266	68,067

measure because he considers it more controllable and informative of physician's actions, working in a context characterised by low uncertainty. This result also confirms the validity of the controllability principle because the supervisor identifies a causal link between physician's action and the performance measure.

Table 4 also shows difference in means under high level of organisational interdependence. In particular, when the level of performance is high, the weight assigned to quantitative performance measure is higher than under low levels of performance. This result confirms H_2: *Under high organisational interdependence, the weighting depends on the managerial performance level.* According to the economic perspective, the weight on quantitative measure decreases when performance measure is less controllable. Consistent with the controllability principle, the supervisor neutralises the effect of events beyond physician's control, adjusting weights on performance measures. Our findings show that participants decided to assign different weights to performance measures based on the level of performance achieved by the physician. In particular, under high levels of performance, weight assigned to the quantitative performance measure takes similar values in low and high interdependence conditions. This aspect corroborates H_{2a}: *Under high managerial performance level, the weighting does not depend on measure controllability.* In this case the supervisor's decision is not driven by economic theory but by other factors. We argue that the supervisor adopts a behavioural perspective because he is worried about how the evaluated physician could perceive his decision and in particular, if he perceived it to be unfair. For example, this happens when the supervisor reduces the weight on the performance measure whereas the physician has reached high performance (score), in turn reducing the assigned bonus. A change in evaluation procedures may also generate negative consequences such as reducing the supervisor credibility. Additional analysis is required in order to demonstrate the validity of the theoretical arguments that supported procedural justice as possible explanatory variable of change in the perspective adopted by the supervisor in the weighting decision.

Finally, Table 4 confirms H_{2b}: *Under low managerial performance level, the weighting depends on measure controllability*. In fact, the weight assigned to the quantitative measure is significantly different in low and high organisational interdependence. In this case, the supervisor decides to adopt an economic approach based on the agency theory where weight depends on the level of controllability, neutralising the impact of factors outside the physician's control. In this way, the supervisor reduces the risk factors related to the incentive. From the theoretical perspective we argue that supervisor's decision is also influenced by behavioural considerations about the consequences of his weighting decision on the perception of assessed subjects. In particular, the adjustment of the weights may be driven by considerations on the perceived distributive justice.

Overall, H_1 confirms the economic literature while H_2 questions the applicability of the controllability principle. The contribution of this study consists in revealing that the supervisor's decision on weighting does not reflect the expectation of the economic approach under high organisational interdependence and high levels of performance conditions. Other factors might explain the supervisor's decision. In the next section, we will analyse the effects of the two manipulated factors first on the perceived procedural justice and later on the perceived distributive justice.

Supplementary Analysis

To support the hypotheses related to conditions of high organisational interdependence, we adopt a behavioural perspective based on organisational justice theory. In particular, it was assumed that under high organisational interdependence conditions the supervisor's decision about weighting is driven by organisational insights related to procedural and distributive justice.

The results of previous analysis have revealed that under high organisational interdependence weighting decision does not always follow agency predictions. These findings are coherent with our considerations about procedural and distributive justice presented in the hypotheses section. Our supplementary analysis investigates whether weight of quantitative measure was consciously influenced by considerations relating to the two types of justice.

We submit post-experiment questions to participants in order to catch the effect of procedural and distributive justice on their weighting decisions. Participants were asked to express their opinion about the importance of two type of justice on a 1–5 point scale. Regarding to procedural justice, 'response mean' was 3.5 and 58% of the participants responded that procedural justice was an important factor that they took into consideration in weighting decision. Regarding to distributive justice, 'response mean' was 3.69 and 66% of the participants responded that distributive justice was also an important factor in weighting decision.

Statistically, distributive and procedural justice were individually treated as dependent variables and analysed by the two-way ANOVA model.

Procedural Justice in Weighting Decision

We proceed to analyse the importance of procedural justice in weighting decision manipulating two independent variables, organisational interdependence and performance level, and considering their interaction effect.

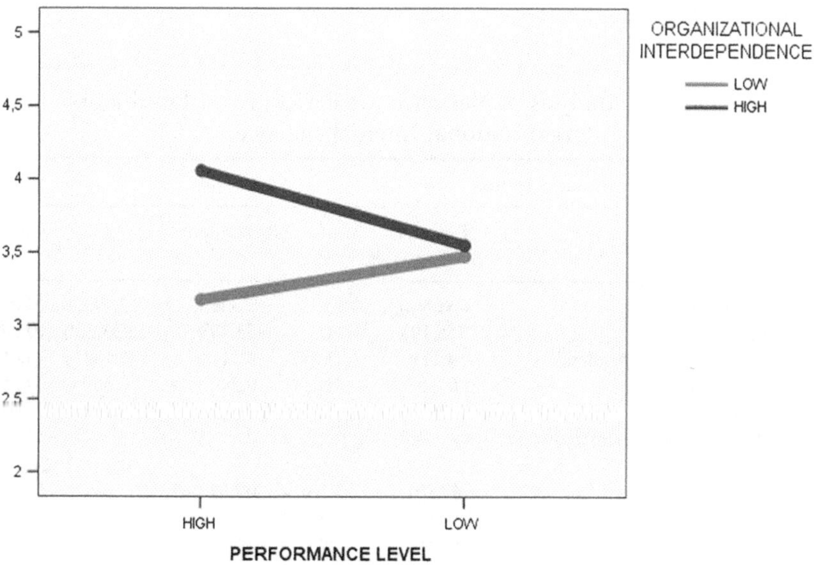

Fig. 2. Means of Procedural Justice in High and Low Organisational Interdependence Conditions.

Analysis of the 'means-treatment plot' (Fig. 2) shows that under low organisational interdependence, participants recognise procedural justice as an important factor in their weighting decision in both high and low performance level conditions. Otherwise, under high organisational interdependence procedural justice plays a greater role under high performance level than low performance level. Graphically, we note an interaction effect between the two manipulated factors.

We test the interaction effect by applying the two-way ANOVA model. In order to verify the appropriateness of using this model we test the homogeneity of variance between groups. The Levene's test shows a p-value of 0.046, so we accept the null hypothesis that the variances between groups are equal at a significance level $\alpha = 0.01$. We also test the normality distribution by experimental cells applying the nonparametric Kolmogorov–Smirnov test, and we accept the null hypothesis that the distribution of the dependent variable is normal.

Applying the two-way ANOVA model, the results reported in Table 5 indicate that the interaction effect between performance level and the level of organisational interdependence is significant ($F = 4.334$, $p < 0.05$). Post-hoc Tukey's test was applied in order to verify if there are significant differences between the means of the dependent variable for the manipulated.

Table 5. Analysis of Variance for Performance Level and Organisational Interdependence.

Dependent Variable: Procedural Justice

Source	Type III sum of squares	df	Mean square	F	Sig.
Corrected Model	6,966(a)	3	2,322	3,493	0.020
Intercept	953,193	1	953,193	1,433,806	0.000
Organisational interdependence	4,111	1	4,111	6,184	0.015
Performance level	0.240	1	0.240	0.360	0.550
Organisational interdependence × performance level	2,882	1	2,882	4,334	0.041
Error	47,201	71	0.665		
Total	1,008,250	75			
Corrected total	54,167	74			

$R^2 = 0.129$ (adjusted $R^2 = 0.092$).
a, signifcance at 0.05.

Table 6. Pairwise Comparisons.

Dependent Variable: Procedural Justice

Organisational interdependence (*I*)	Organisational interdependence (*J*)	Mean difference (*I*–*J*)	SE	Sig. (a)	95% confidence interval for difference(a)	
					Lower bound	Upper bound
Low	High	–0.469*	0.188	0.015	–0.845	–0.093
High	Low	0.469*	0.188	0.015	0.093	0.845

Based on estimated marginal means
*The mean difference is significant at the 0.05 level.
Adjustment for multiple comparisons: Least significant difference (equivalent to no adjustments).

Univariate Tests

Dependent Variable: Procedural Justice

	Sum of squares	df	Mean square	*F*	Sig.
Contrast	4,111	1	4,111	6,184	0.015
Error	47,201	71	0.665		

The *F* test shows the effect of organisational interdependence. This test is based on the linearly independent pairwise comparisons among the estimated marginal means.
a, signifcance at 0.05.

Table 6. *(Continued)*

Dependent Variable: Procedural Justice

Performance level (*I*)	Performance level (*J*)	Mean difference (*I–J*)	SE	Sig. (a)	95% confidence interval for difference (a)	
					Lower bound	Upper bound
Low	High	−0.113	0.188	0.550	−0.489	0.263
High	Low	0.113	0.188	0.550	−0.263	0.489

Based on estimated marginal means
Adjustment for multiple comparisons: Least significant difference (equivalent to no adjustments).

Univariate Tests

Dependent Variable: Procedural Justice

	Sum of squares	df	Mean square	*F*	Sig.
Contrast	0.240	1	0.240	0.360	0.550
Error	47,201	71	0.665		

The *F* test shows the effect of performance level. This test is based on the linearly independent pairwise comparisons among the estimated marginal means.

Table 6 indicates that procedural justice has greater importance under low organisational interdependence rather than under high organisational interdependence. The test for contrasts is significant ($F = 6.184$, $p < 0.05$), showing a main effect of organisational interdependence. Regarding the performance level, there are no significant differences under low and high performance level, the test for contrasts is not significant ($F = 0.360$, $p > 0.05$).

Table 7 shows the mean of the dependent variable in each single experimental group. In particular, participants recognised procedural justice as important under high organisational interdependence and high performance level.

This result confirms the relevance of the behavioural approach to the economic approach. In the previous section, our analysis confirmed H_{2a}. An additional analysis confirms that supervisors are afraid to change the weight on uncontrollable performance measure because their decision could be perceived as unfair by the evaluated physician (Harris, 1994). The adjustment is perceived as unfair when supervisor, following the controllability principle, reduces the weight on performance measure where the assessed subject has reached high results. In this case, supervisor's decision leads to a reduced bonus. Changing evaluating procedures may lead to negative consequences such as reducing supervisor credibility. In order to avoid such negative consequences, the supervisor should take into consideration how his weighting decision could be perceived in terms of procedural justice. When the level of performance is high the supervisor does not reduce the weight on uncontrollable performance measure as is expected by agency predictions but he maintains a higher weight on it.

Table 7. Organisational Interdependence × Performance Level.

Dependent Variable: Procedural Justice

Organisational interdependence	Performance level	Mean	SE	95% confidence interval	
				Lower bound	Upper bound
Low	Low	3,474	0.187	3,101	3,847
	High	3,194	0.192	2,811	3,578
High	Low	3,550	0.182	3,186	3,914
	High	4,056	0.192	3,672	4,439

According to our supplementary analysis procedural justice can be considered a possible explanatory variable of weighting process.

Distributive Justice in Weighting Decision

We follow the same analysis for testing the importance of distributive justice in weighting decision that we adopt for procedural justice. We assume the distributive justice as dependent variable, and organisational interdependence and performance level as the two manipulated independent variables.

To verify the appropriateness of ANOVA model we test the normality distribution by experimental cells applying the nonparametric Kolmogorov–Smirnov test, and we accept the null hypothesis that the distribution of the dependent variable is normal. We also test for the homogeneity of variance between groups. Levene's test shows a p-value of 0.001, so we reject the null hypothesis that the variances between groups are equal at a significance level $\alpha = 0.01$. There is no homogeneity of variances between groups, so we cannot apply the analysis of variance. So, collected data on distributive justice are not significantly different between the groups, and we cannot analyse the main effects of two independent variables on distributive justice and their interaction effect. We do not verify H_{2b}.

Probably, under high organisational interdependence and low performance level there will be other factors which lead the supervisor to adjust weights on performance measures.

CONCLUSION

In recent years, management accounting literature has shown a relevant interest in investigating subjectivity in performance evaluation systems (Gibbs et al., 2004; Ittner et al., 2003; Moers, 2005).

The core concern of this study was to investigate the subjectivity expressed in the form of weighting of performance measures, explaining the supervisor decision-making in assigning weight to performance measures.

The contribution of this paper is to identify factors which might influence weighting decisions in addition to factors set out by the economic approach. In particular, we argue that the supervisor's decision should be considered as the result of the predictions of both economic and behavioural perspectives, whereas the latter bases itself on the organisational justice theory.

Our results show that in a context characterised by a low level of interdependence, the weight assigned to performance measures is independent of the performance level achieved by the evaluated subject. The supervisor assumes an economic perspective and applies the controllability principle because he recognises the causal link between an individual's actions and results reached on quantitative performance measure, so the results are a direct consequence of individual effort. The weight on performance measure assigned by the supervisor will be directly proportional to the degree of informativeness and controllability of the measure (Antle & Demski, 1988; Banker & Datar, 1989; Datar et al., 2001; Feltham & Xie, 1994; Holmstrom, 1979).

Under high interdependence conditions the results show that assigning weights to performance measures depends on the level of performance achieved. According to the economic approach, the weight assigned to the measure decreases when the degree of its controllability is low. In fact, the measures are less sensitive and informative of individual's actions (Banker & Datar, 1989). Consistently with the controllability principle, the supervisor should neutralise the effect of events beyond individual's control, adjusting the weights on performance measures and assigning higher weight to less uncontrollable measures. Our results show that supervisors make decisions not in line with the economic approach. Under high level of performance, weighting is independent of the measure controllability. The supervisor assumes a behavioural perspective. We demonstrate by an additional analysis the influence of 'procedural justice' variable on supervisor decision-making. In particular, when the performance level is high, the importance given to procedural justice in weighting decision is greater under high organisational interdependence than low organisational interdependence conditions. According to the organisational justice theory, managers are satisfied with the evaluation systems when they perceive the assessment procedures as fair (Greenberg, 1986). Changing weight of performance measure could be perceived as unfair by assessed subject because the supervisor reduces the weight of performance measure in which the evaluated subject reaches high levels of performance. If the adjustment is not carried out, the manager would receive a higher bonus. In particular, the importance given to procedural justice shows that the supervisor considers the effects of his evaluation decisions on the future individual's behaviour. In fact, if the process of performance evaluation is perceived as an unfair system, this could lead to conflict, tension and anxiety (Hopwood, 1972). The choice of supervisor is supported by consistent positive association between procedural justice and job satisfaction highlighted by the accounting literature (Lau & Sholihin, 2005).

Some insights come from the hypothesis on weighting decision under low performance conditions, where the supervisor's decision depends on measure controllability. In line with our initial analysis, the supervisor decides to adopt an economic approach assigning weight to performance measures based on the level of its controllability. In this way he reduces the risk factor related to the incentive. Our supplementary analysis will try to find an explanation for that decision in the perceived distributive justice. However, this is not confirmed by our supplementary analysis.

Our study offers two contributions to the managerial performance evaluation literature. The first is to confirm the economic literature on weighting where the weight of performance measures depends on their degree of informativeness and controllability. The second contribution is to question the applicability of the controllability principle under high organisational interdependence. In particular, our findings highlight that the supervisor's decision is the result of the two approaches: economic and behavioural. This integration of the two approaches takes place under high interdependence context.

This study is subject to the usual limitation of experimental research. Our participants were homogeneously grouped and the information given to them was exactly what they needed for making their decision. However, in natural settings, the characteristics of individuals (decision makers) and the information available to them can differ considerably, and the combinations derived from individuals and information used could lead to different results from those obtained by this research. The experimental method has the advantage of isolating variables and studying their effect on the dependent variable, highlighting the consequences of decisions generated by the presence of a specific set of information. Our results need further empirical confirmation in order to increase their generability.

The flexibility in weighting performance measures still offers extensive research perspectives. This work has shown that the supervisor's decision on weight does not always follow the economic approach based on the agency theory but other factors may drive his decision. Considering the causal relationship incentive–effort–performance, the factors which influence the weighting of performance measures may be manifold. Bonner (1999) identifies three categories of variables that determine individual performance: individual variables, task variables and environmental variables. In the present study we investigated the effect of an environment variable, organisational interdependence, which defines the degree of uncertainty within the firm and also represents an element of complexity of required task. Future research could interestingly investigate the effect of individual variables on supervisor's decision (Bonner & Sprinkle, 2002). The motivation, the different personality profiles, skills, risk preferences, cognitive style could represent potential issues which influence the weighting process.

NOTES

1. Gomez-Mejia and Balkin (1992) argue that not all employees are equally important for the organisation and compensation strategy must focus on the needs and expectations of the groups considered most important. The importance of a particular group depends on three conditioning factors: (1) the difficulty of replacing the group, (2) the potential impact of their performance on the achievement of firm mission and (3) its membership to core technical group. According to these factors, the replacement of physicians is related to their specific expertise and knowledge, and also once recruited, physicians enjoy the privileges of public work, including the safety of the workplace. The physicians influence hospital productivity and healthcare demand. Finally, physicians are part of the technical core group. They have decision-making autonomy and adopt technical language with which they quantify and qualify their performance, making difficult for the administrative staff evaluate and monitor their performance.

2. A growing interest in management accounting have been shown towards heath care organisations, in particular regarding performance measurement and evaluation system adopted in these organisations (Abernethy & Chua, 1996; Lehtonen, 2007; Modell, 2001).

3. Materials on experimental instructions are available upon request to the author.

4. The results of the *p*-value obtained in each group were 0.333, 0.181, 0.235, 0.222, so we accept the null hypothesis that the distribution of the dependent variable is normal. We also verified the normal distribution of error terms. We proceed with the test of normality of error terms by calculating the time correlation between the residuals and the values expected in normal conditions. The correlation coefficient obtained were significant (0.964, $p < 0.01$). This result confirms the assumption on the normality of the residuals.

ACKNOWLEDGMENTS

This paper is drawn on the author's PhD dissertation and results from the author being invited as visiting scholar to the University of Illinois at Urbana-Champaign, January–May 2010. Thanks are due to Jasmijn Bol for her encouragement and constructive comments. The author is also grateful for the suggestions provided by Rosa Alba Miraglia and Davide Rizzotti.

This paper results from the authors being involved into the national research project 'Strategic cost and performance management for the challenges of competitiveness and sustainability' – Progetto di RIcerca Nazionale (P.R.I.N.), financially supported by the Italian Government, directed by Lino Cinquini from Scuola Superiore Sant'Anna di Pisa and coordinated by Riccardo Giannetti from the University of Pisa and Rosa Alba Miraglia from the University of Catania.

REFERENCES

Abernethy, M. A., & Chua, W. F. (1996). A field study of control systems "redesign": The impact of institutional processes on strategic choice. *Contemporary Accounting Research, 13*(2), 569–606.

Adams, J. S. (1965). Inequity in social exchange. In L. Berkoswisk (Ed.), *Advances in experimental social psychology* (Vol. 2, pp. 267–299). New York, NY: Academic Press.

Alexander, S., & Ruderman, M. (1987). The role of procedural and distributive justice in organizational behaviour. *Social Justice Research, 1*, 117–198.

Antle, R., & Demski, J. S. (1988). The controllability principle in responsibility accounting. *The Accounting Review, 63*, 700–718.

Arrow, K. (1985). The economics of agency. In J. Pratt & R. Zeckhauser (Eds.), *Principals and agents: The structure of business*. Boston, MA: Harvard Business School Press.

Atkinson, A. A., Balakrishnan, R., Booth, P., Cote, J. M., Groot, T., Malmi, T., ... Wu, A. (1997). New directions in management accounting research. *Journal of Management Accounting Research, 9*(1), 70–108.

Atkinson, T., Kaplan, R., & Young, S. M. (2004). *Management accounting* (4th ed.). Upper Saddle River, NJ: Prentice Hall.

Baiman, S. (1990). Agency research in accounting: a second look. *Accounting Organizations and Society, 15*, 341–371.

Baiman, S., & Rajan, M. V. (1995). The informational advantage of discretionary bonus schemes. *The Accounting Review, 70*, 557–579.

Baker, G. P., Gibbons, R., & Murphy, K. J. (1994). Subjective performance measures in optimal incentive contracts. *Quarterly Journal of Economics, 109*(November), 1125–1156.

Banker, R., & Datar, S. M. (1989). Sensitivity, precision and linear aggregation of signals of performance evaluation. *Journal of Accounting Research, 27*(Spring), 21–39.

Bies, R. J., & Moag, J. S. (1986). Interactional justice: Communication criteria of fairness. *Research on Negotiation in Organizations, 1*, 43–55.

Bol, J. C., & Smith, S. D. (2010, July 1). Spillover effects in subjective performance evaluation: Bias and the asymmetric influence of controllability. *The Accounting Review*. Retrieved from http://ssrn.com/abstract=1427584

Bonner, S. E. (1999). Judgment and decision-making research in accounting. *Accounting Horizons, 13*, 385–398.

Bonner, S. E., & Sprinkle, G. B. (2002). The effects of monetary incentives on effort and task performance: Theories, evidence, and a framework for research. *Accounting, Organizations and Society*, *27*, 303–345.

Burns, P. (1985). Experience and decision-making: a comparison of students and businessman in a simulate progressive auction. In V. L. Smith (Ed.), *Research in experimental economics* (Vol. 3, pp. 139–157). Greenwich, CT: JAI Press.

Bushman, R., Indjejijkian, R., & Smith, A. (1996). CEO compensation: The role of individual performance evaluation. *Journal of Accounting and Economics*, *21*(April), 161–193.

Choudhury, N. (1986). Responsibility accounting and controllability. *Accounting and Business Research*, *16*(63), 189–198.

Chow, C. W., Shields, M. D., & Chan, Y. K. (1991). The effects of management controls and national culture on manufacturing performance: An experimental investigation. *Accounting, Organizations and Society*, *16*(3), 209–226.

Cohen-Charash, Y., & Spector, P. E. (2003). The role of justice in organizations: A meta-analysis. *Organization Behavior and Human Decision Processes*, *86*, 278–321.

Datar, S., Kulp, C. S., & Lambert, R. A. (2001). Balancing performance measures. *Journal of Accounting Research*, *39*, 75–92.

Feltham, G. A., & Xie, J. (1994). Performance measure congruity and diversity in multi-task principal/agent relations. *The Accounting Review*, *69*(3), 429–453.

Frederickson, J. R., Pfeffer, S. A., & Pratt, J. (1999). Performance evaluation judgments: Effects of prior experience under different performance evaluation scheme and feedback frequencies. *Journal of Accounting Research*, *37*(1), 151–165.

Friedman, D., & Sunder, S. (1994). *Experimental methods, a primer for economists*. New York, NY: Cambridge University Press.

Gibbons, R. (1987). Piece-rate incentive systems. *Journal of Labor Economics*, *5*, 413–429.

Gibbs, M., Merchant, K. A., Van der Stede, W., & Vargus, M. E. (2004). Determinants and effects of subjectivity in incentives. *The Accounting Review*, *79*(2), 409–436.

Giraud, F., Langevin, P., & Mendoza, C. (2008). Justice as a rationale for the controllability principle: A study of managers' opinions. *Management Accounting Research*, *19*, 32–44.

Gomez-Mejia, L. R., & Balkin, D. B. (1992). *Compensation, organizational strategy and firm performance*. Cincinnati, OH: South Western Publishing Co.

Greenberg, J. (1986). Determinants of perceived fairness of performance evaluations. *Journal of Applied Psychology*, *71*(2), 340–342.

Harris, M. M. (1994). Rater motivation in the performance appraisal context: A theoretical framework. *Journal of Management*, *20*(4), 737–756.

Hemmer, T. (1996). Allocations of suck capacity costs and joint costs in a linear principal-agent model. *The Accounting Review*, *71*(3), 419–432.

Heneman, R. L. (1986). The relationship between supervisory rating and result-oriented measures of performance: A meta-analysis. *Personnel Psychology*, *39*, 811–826.

Hirst, M. K., & Yetton, P. W. (1999). The effects of budget goals and task interdependence on the level of and variance in performance: A research note. *Accounting, Organizations and Society*, *24*, 205–216.

Holmstrom, B. (1979). Moral hazard and observability. *Bell Journal of Economics*, *10*, 74–91.

Holmstrom, B., & Paul, M. (1991). Multi-task principal-agent analyses: Linear contracts, asset ownership and job design. *Journal of Law, Economics and Organization*, *7*, 24–52.

Hopwood, A. G. (1972). An empirical study of the role of accounting data in performance evaluation, empirical research in accounting: Selected studies. *Journal of Accounting Research*, *10*, 156–182.

Horngren, C. T., Foster, G., & Datar, S. M. (2006). *Cost accounting: A managerial emphasis* (12th ed.). Upper Saddle River, NJ: Prentice Hall.

Ittner, C., Larcker, D., & Rajan, M. (1997). The choice of performance measures in annual bonus contracts. *The Accounting Review*, *72*, 231–255.

Ittner, C., Larcker, D., Rajan, M., & Mayer, M. (2003). Subjectivity and the weighting of performance measures: Evidence from balanced scorecard. *The Accounting Review*, *78*(July), 725–758.

Ittner, C. D., & Larcker, D. F. (1998). Innovations in performance measurement: Trends and research implications. *Journal of Management Accounting Research*, *10*, 205–239.

Jensen, M. C., Murphy, K. J., & Wruck, E. G. (2004). *Remuneration: Where we've been, how we got to here, what are the problems, and how to fix them.* Harvard NOM working 30 Paper No. 04-28; ECGI-Finance Working Paper No. 44/2004. http://ssrn.com/abstract=561305

Johnson, R. E., Selenta, C., & Lord, R. G. (2006). When organizational justice and the self-concept meet: Consequences for the organization and its members. *Organizational Behavior and Human Decision Processes*, *99*, 175–201.

Kahneman, D., & Frederick, S. (2002). Representativeness revisited: Attribute substitution in intuitive judgment. In T. Gilovich, D. Griffin & D. Kahneman (Eds.), *Heuristics and biases: The psychology of intuitive judgment* (pp. 49–81). Cambridge: Cambridge University Press.

Keating, A. S. (1997). Determinants of divisional performance evaluation practices. *Journal of Accounting and Economics*, *24*, 243–273.

Kerlinger, F. N., & Lee, H. B. (2000). *Foundations of behavioral research*. Orlando, FL: Harcourt, Inc.

Krishnan, R., Luft, J., & Shields, M. D. (2005). Effects of accounting-method choices on subjective performance-measure weighting decisions: Experimental evidence on precision and error covariance. *Accounting Review*, *80*(4), 1163–1192.

Lambert, R., & Larcker, D. (1987). An analysis of the use of accounting and market measures of performance in executive compensation contract. *Journal of Accounting Research*, *25*(Suppl.), 85–125.

Larmande, F., & Ponssard, J. P. (2007). *Responsibility accounting with a privately informed agent*. Retrieved from http://ssrn.com/abstract=993761. Accessed on February 10, 2012.

Lau, C. M., & Lim, E. W. (2002). The effect of procedural justice and evaluative styles on the relationship between budgetary participation and performance. *Advances in Accounting*, *19*, 139–160.

Lau, C. M., & Sholihin, M. (2005). Financial and nonfinancial performance measures: How do they affect job satisfaction? *The British Accounting Review*, *37*, 389–423.

Lehtonen, T. (2007). DRG-based prospective pricing and case-mix accounting – Exploring the mechanisms of successful implementation. *Management Accounting Research*, *18*, 367–395.

Leventhal, G. S. (1980). What should be done with equity theory? New approaches to the study of fairness in social relationships. In K. J. Gergen, M. S. Greenberg & R. H. Willis (Eds.), *Social exchange: Advances in theory and research* (pp. 27–53). New York, NY: Plenum press.

Libby, T. (1999). The influence of voice and explanation on performance in participative budgeting setting. *Accounting, Organizations and Society*, *24*, 125–137.

Lind, E. A., & Tyler, T. R. (1988). *The social psychology of procedural justice*. New York, NY: Plenum.

Lindquist, T. M. (1995). Fairness as antecedent to participative budgeting: Examining the effects of distributive justice, procedural justice and referent cognitions on satisfaction and performance. *Journal of Management Accounting Research*, *7*, 122–147.

Macintosh, N. B., & Daft, R. L. (1987). Management control systems and departmental interdependencies: An empirical study. *Accounting, Organizations and Society*, *12*(1), 49–61.

Maslow, A. H. (1970). *Motivation and personality*. New York, NY: Harper & Row Publications.

McFarlin, D. B., & Sweeney, P. D. (1992). Distributive and procedural justice as predictors of satisfaction with personal and organizational outcomes. *Academy of Management Journal*, *35*, 626–637.

Merchant, K. A. (1985). *Control in business organizations*. Cambridge, MA: Ballinger Publishing Company.

Merchant, K. A. (1989). *Rewarding results: Motivating profit center managers*. Boston, MA: Harvard Business School Press.

Merchant, K. A., & Van der Stede, W. A. (2007). *Management control systems: Performance measurement, evaluation and incentives* (2nd ed.). Harlow, UK: Prentice Hall.

Modell, S. (2001). Performance measurement and institutional process: A study of managerial responses to public sector reform. *Management Accounting Research*, *12*, 437–464.

Moers, F. (2005). Discretion and bias in performance evaluation: The impact of diversity and subjectivity. *Accounting, Organizations and Society*, *30*, 67–80.

Murphy, K. J., & Oyer, P. (2001). *Discretion in executive incentive contracts*. Working Paper No. 48. Marshall School of Business, University of Southern California, California.

Otley, D. T. (1999). Performance management: A framework for management control systems research. *Management Accounting Research*, *19*, 363–382.

Prendergast, C., & Topel, R. H. (1993). Discretion and bias in performance evaluation. *European Economic Review*, *37*, 355–365.

Prendergast, C. (1999). The provision of incentives in firms. *Journal of Economic Literature*, *37*, 7–63.

Prendergast, C., & Topel, R. H. (1996). Favoritism in organizations. *Journal of Political Economy*, *10*(5), 958–978.

Sloan, R. G. (1992). Information in prices about future earnings: Implications for earnings response coefficients. *Journal of Accounting and Economics*, *15*(2/3), 143–171.

Sprinkle, G. B. (2003). Perspectives on experimental research in managerial accounting. *Accounting, Organizations and Society*, *28*(February/April), 287–318.

Tang, T. L. P., & Sarsfield-Baldwin, L. J. (1996). Distributive and procedural justice as related to satisfaction and commitment. *SAM Advanced Management Journal*, *61*(3), 25–31.

Thibault, J., & Walker, L. (1975). *Procedural justice*. Hillsdale, NJ: Lawrence Erlbaum.

Van der Stede, W. A., Chow, C. W., & Lin, T. W. (2006). Strategy, choice of performance measures, and performance. *Behavioral Research in Accounting*, *18*, 185–205.

Zimmerman, J. L. (2003). *Accounting for decision-making and control*. New York, NY: McGraw-Hill.

BUY ONE, GET ONE FREE: BENEFITS OF FOLLOWING THE CONTROLLABILITY PRINCIPLE FOR INTRINSIC MOTIVATION?

Jennifer Kunz and Stefan Linder

ABSTRACT

Purpose – *The controllability principle has attracted considerable attention in performance management research. Whereas this discussion hinges centrally on a conception of human motivation, findings from psychological motivation research on intrinsic motivation have not received attention yet. Addressing this research gap, this chapter analyzes the effect of perceived controllability on extrinsic* and *intrinsic motivation and thereby on work effort.*

Methodology/approach – *We conducted a vignette study with undergraduate students of an accounting course in a German University in early 2009. Each participant was confronted with a written description of a hypothetical business situation (a vignette) on which the participants should base their further answers. These vignettes also contained a description of a remuneration system. Among other things, the degree of this remuneration system's controllability was varied from vignette to vignette. After reading the vignette each participant was instructed to*

Performance Measurement and Management Control: Global Issues
Studies in Managerial and Financial Accounting, Volume 25, 339–362

ISSN: 1479-3512/doi:10.1108/S1479-3512(2012)0000025016

answer some questions regarding his/her willingness to exert additional effort in the described situational context and the motivational reasons behind this willingness.

Findings – *The study exhibits that a lack of perceived controllability has a negative impact on extrinsic, but also on intrinsic, motivation and thereby on work effort.*

Originality – *The findings put the current discussion into a broader perspective, as they suggests that adherence to the principle of controllability is not only warranted given the impact on extrinsic motivation as discussed in large parts of the literature but also for intrinsic motivation.*

INTRODUCTION

Accounting research – especially the principal-agent-based research stream – has dealt extensively with the notion of controllability and suggests that it is one of the major aspects to consider when designing an effective performance-based incentive system (e.g., Giraud, Langevin, & Mendoza, 2008; Horngren, Foster, & Datar, 2006; Solomons, 1965). The literature commonly defines controllability as "the ability of the manager to anticipate behavior-outcome-reward contingencies" (Choudhury, 1986, p. 189n) and calls for including only factors in performance evaluations that can be influenced by the evaluated persons (e.g., Ferrara, 1964; Gordon, 1963; Horngren et al., 2006). The rationale underlying this advice is the idea that holding individuals responsible for factors beyond their control will cause them to perceive the evaluation as unfair and to diminish their motivation (Giraud et al., 2008).

Although this suggestion sounds reasonable, recent, more behaviorally oriented research calls it into question (Giraud et al., 2008; Simons, 2007), as empirical results show that in practice this principle is frequently violated (e.g., Atkinson, 1987; Bushman, Indjejikian, & Smith, 1995; Drury & El-Shishini, 2005; Fremgen & Liao, 1981; Merchant, 1987, 1989; Moon & Fitzgerald, 1996; Otley, 1990; Ugras, 1994). This violation can be ascribed to two potential reasons (Giraud et al., 2008): On the one hand, firms operate in such complex and dynamical environments that most activities cannot be measured in a way that follows the controllability principle. On the other hand, firms also want their employees to keep an eye on developments that they cannot control completely but which they can at least influence partly. Following these observations and putting more emphasis on the individual manager, several empirical studies have been undertaken to examine the importance of the controllability principle: for example, the study conducted by Giraud et al. (2008) uncovered different effects depending on whether the uncontrollable factors influencing the performance evaluation are external or internal to the company.

The present chapter adds to this research a more differentiated perspective on motivation. As mentioned, the adherence to the controllability principle is especially called for by the more modeling-based accounting research that centers on extrinsic motivation. However, the current behaviorally oriented discussion highlights the importance of incorporating behavioral aspects. One of these aspects that are key to the effectiveness of incentive systems is their effect on human motivation. Findings from psychological research stress the fact that there are *different kinds of motivation*, namely intrinsic and extrinsic motivation (e.g., Bonner & Sprinkle, 2002; de Charms, 1968; Deci, 1971, 1975; Deci & Ryan, 2000; Frey, 1997; Kunz & Pfaff, 2002). This separation of the motivation construct proposes that humans perform activities either because of an internal impetus (intrinsic motivation) or external pressure (extrinsic motivation) (e.g., Deci & Ryan, 2000; Ryan & Connell, 1989). Moreover, some studies show an interaction between the two forms of motivation. This observation has sparked a vivid and continued debate about the existence and magnitude of potential dangers of pay-for-performance systems for an individual's intrinsic motivation, i.e., the presence of a so called "crowding-out" of intrinsic motivation by extrinsic incentives (see, e.g., Frey & Jegen, 2001; Kohn, 1993; Kunz & Pfaff, 2002; Weibel, Rost, & Osterloh, 2010). This raises the question of *whether the adherence to the controllability principle influences the intrinsic part of human motivation.*

As the controllability principle is applied to performance measures that are part of extrinsic rewards, one would expect that it actually has a negative effect on the intrinsic part of motivation, counteracting the findings from principal-agent-based accounting research. However, a look into the literature regarding this topic reveals that one of the dominant theories explaining the mentioned interactional effects between intrinsic and extrinsic motivation – self-determination theory – suggests at least a *less negative* effect of an incentive scheme on intrinsic motivation, if this scheme follows the controllability principle. Adhering to the principle can be expected to enhance the perception of self-determination and competence, which, in turn, fosters intrinsic motivation (Deci & Ryan, 2000).

With the present chapter, we aim to advance knowledge on this proposition. We focus on combining the concept of intrinsic versus extrinsic motivation and the notion of controllability. That is, we strive to mold the explanations of human behavior from the two strings of literature, thereby showing the role of *perceived* controllability on intrinsic and extrinsic motivation and work effort. Using a vignette experiment, we provide first evidence that a lack of perceived controllability has a negative impact on both intrinsic and extrinsic motivation and thereby on work effort. Our study thus supports the long-standing call for the consideration of this principle. Moreover, we find the impact of controllability to be positive on both extrinsic and intrinsic motivation. Our results thereby suggest that including controllability might also help in understanding better the crowding-out phenomenon.

THEORETICAL BACKGROUND AND HYPOTHESES

Antecedents to Perceived Controllability

The notion of controllability has received a lot of attention in the accounting literature, particularly in publications dealing with responsibility accounting: It has been discussed with respect to its tension with other aspects of responsibility accounting (Choudhury, 1986), regarding the notion of information content (Antle & Demski, 1988), and concerning its interplay with the outcome effect (Gosh, 2005). Moreover, empirical surveys were conducted to judge its applicability in complex modern organizations (e.g., Merchant, 1989; Moon & Fitzgerald, 1996; Otley, 1990; Simons, 2007) and regarding managers' and employees' points of view on this principle (Giraud et al., 2008; Merchant, 1987, 1989).

Although some authors have expressed doubt that the controllability principle can be introduced without problems into today's complex organizations and that it might even be favorable for firms to violate it under specific circumstances (Giraud et al., 2008), mainstream accounting literature assumes a positive impact of adhering to the principle on employees' effort. However, the rich body of research on this topic exhibits that there is not a direct link between actual controllability and employee effort (e.g., Choudhury, 1986; Giraud et al., 2008): There are several intervening variables in the causal chain between the degree of objective controllability, i.e., the degree to which performance outcomes, which are the basis of a remuneration system, are objectively controllable by an employee, and that employee's exhibited effort.

Although the starting point of this causal chain is the actual controllability of an incentive scheme, Choudhury (1986) already points out that, in order to induce any behavioral reaction, the degree of controllability also needs to be perceived as such by the employee. Hence, the first intervening variable between objective controllability and effort is the perceived controllability.

As individuals are characterized by bounded rationality (Simon, 1977) and thus limited perceptional and cognitive capacities, perceived controllability might differ from the actual one. Individuals may either overrate or underestimate their controllability due to cognitive biases. Moreover, besides cognitive biases, the relationship between actual and perceived controllability should also be moderated by an individual's learnt self-efficacy beliefs: Following Bandura (1997), individuals form a general belief about the effectiveness of their activities. An individual who has learnt a low self-efficacy belief over the course of its life should thus experience a low perceived controllability, even though the actual controllability in the situation at hand may be higher. Thus, both cognitive biases as well as self-efficacy beliefs of an individual should lead to a significant impairment between objective and perceived controllability. However, it seems still reasonable to assume that some (potentially weak) positive linkage between actual and perceived controllability should nevertheless exist, i.e., a higher degree of objective controllability should in general result in a higher degree of perceived controllability. This gives rise to our first hypothesis:

H1. Actual controllability has a positive impact on perceived controllability.

Individual cognitive biases and self-efficacy beliefs are, however, only half of the story for arriving from actual controllability at perceived controllability. An employee's perception of controllability of his/her remuneration will also strongly depend on his/her past experience with the organization's trustworthiness in providing the compensation for good performance. If the employee learnt in the past that promised rewards are not or not truly reliably provided (e.g., through ex post resetting target levels by the firm or through multiple kinds of economic or noneconomic "excuses"), he/she will not fully orient his/her effort according to the incentive scheme as he/she will doubt that a high effort level will actually result in a high reward. In case there are any such doubts, the perceived controllability should be lower than the actual one (as seen from the perspective of the firm). Hence, a lack of trust most probably results in the same behavioral effect as a lack of objective controllability, leading to the second hypothesis:

H2. Trustworthiness regarding the provision of incentives has a positive effect on the degree of perceived controllability.

Controllability and Human Motivation

Even though human motivation has remained a thorny construct until today, psychological research has identified the existence of different kinds of motives and motivations (e.g., Alderfer, 1972; Deci & Ryan, 2000). One of the most widely discussed differentiations of the motivation construct is the one into intrinsic and extrinsic motivation. It can be traced back to de Charms (1968), who proposed that humans perform activities either because of an internal incentive or external pressure. Internal incentives place the *locus of causality* on the acting individual, whereas with external pressure the locus of causality is external to the individual (*ibid.*). Behavior triggered by such an external locus of causality is commonly referred to as extrinsically motivated (e.g., Deci & Ryan, 2000; Ryan & Connell, 1989). In contrast, behavior that is due to an internal locus of causality, can be either *enjoyment based* or *internalized* (i.e., *based on integrated regulation* or *obligation based*) (e.g., Deci & Ryan, 2000; Lindenberg, 2001; Ryan & Connell, 1989). While the latter one is not always put under the *rubrum* of intrinsic motivation, the prior one in general always is. In the present chapter, we follow the arguably more prominent stream of literature referring only to the prior one as intrinsic motivation (e.g., Deci & Ryan, 2000).

Ever since the first rough structuring of motivation into its intrinsic and extrinsic forms has been presented in literature (de Charms, 1968), experimental research has investigated the factors that influence these two kinds of motivation, with a particular focus on the impact of extrinsic monetary rewards on intrinsic motivation. In a frequently cited experiment on this question, Deci (1971) found a result that sparked some uproar and discussion: a crowding-out of intrinsic motivation by extrinsic incentives. Most of the experiments triggered by this intriguing result were conducted in social psychology and education (e.g., Deci, 1975; Deci, Koestner, & Ryan, 1999; Lepper, Greene, & Nisbett, 1973). Yet even though researchers have investigated the role of different experimental designs, measures, and definitions of the underlying constructs in producing the relation between extrinsic rewards and intrinsic motivation, no commonly accepted opinion on the subject has emerged so far (see for overviews, Cameron & Pierce, 2002; Deci et al., 1999; Reiss, 2005). Similarly, a significant number of partially competing, partially complementary explanations for these effects has been presented. At present, the arguably most prominent and widely known explanation for this interaction is based on Deci and Ryan's self-determination theory of human motivation (see Kunz & Pfaff, 2002, for an overview of explanations).

Self-determination theory builds on de Charms' (1968) concept of *locus of causality* for analyzing human motivation and has received considerable experimental support (see Deci et al., 1999). According to Deci (1975) and Deci and Ryan (2000), behavior that is truly intrinsically motivated is self-determined; i.e., "activities that people do naturally and spontaneously when they feel free to follow their inner interests" (Deci & Ryan, 2000, p. 234). These authors have established that this behavior is aimed at satisfying three central (intrinsic) human needs that are key to psychological well-being: the need for competence, the need for autonomy, and the need for relatedness (*ibid.*). The *need for competence* comprises "a propensity to have an effect on the environment as well as to attain valued outcomes within it" (Deci & Ryan, 2000, p. 231). This need also is related to the wish to be able to reliably predict outcomes of one's actions. Otherwise, the individual cannot derive the feeling of competence from the received performance feedback, since that feedback is also influenced by aspects beyond the individual's reach and – hence – competent action. The *need for relatedness* refers to "the desire to feel connected to others" (Deci & Ryan, 2000, p. 231). Finally, the *need for autonomy* refers to the need to experience one's behavior as self-determined. A feeling of autonomy requires "acting with a sense of volition and having the experience of choice" (Gagné & Deci, 2005, p. 333).

Self-determination theory neatly allows fitting the idea of individuals being driven both by intrinsic as well as extrinsic motivation to the concept of controllability:

First, it explicitly allows for extrinsic rewards – like the ones from a firm's performance measurement and reward system – to influence an individual's extrinsic part of motivation in line with what standard economics and accounting theories would also suggest. Providing an individual with extrinsic rewards is considered to lead to an increase in extrinsic motivation (Deci & Ryan, 2000; Gagné & Deci, 2005). It thus seems to assume the same underlying mechanisms between extrinsic rewards and extrinsic motivation to be present as do economics and accounting literatures. Consequently, it allows drawing on the findings from existing economics and accounting literature on controllability already discussed in the previous part of this chapter to assess controllability's influence on extrinsic motivation. This literature suggests a positive influence of controllability – thereby implicitly assuming a kind of motivation that would fall into the extrinsic category following the psychological motivation research by Deci and Ryan's (2000) self-determination theory. We pick up this aspect and only introduce the perceived controllability as one additional intervening variable, resulting in the third hypothesis:

H3. Perceived controllability has a positive effect on extrinsic motivation.

Second, self-determination theory provides a basis for analyzing the influence of controllability on intrinsic motivation. As outlined above, self-determination theory proposes humans to possess three major innate and internal needs. A relationship of perceived controllability on intrinsic motivation (and thus finally on behavior) can be established based on the one hand on the need for competence, since as outlined above, this need is linked to the wish to be able to (precisely) predict the outcomes of one's actions. On the other hand, also the satisfaction of the need for autonomy is positively influenced by the feeling of being in control of one's destiny, i.e., the incentive outcomes of one's actions, and hence being autonomous from external influences. These aspects encourage us to propose a positive relationship between perceived controllability and intrinsic motivation leading to the following hypothesis:

H4. Perceived controllability has a positive effect on intrinsic motivation.

Economic theory typically assumes that employees incur disutility from exerting effort. Hence, to entice managers or employees to exert more effort at work, economic models call for providing extrinsic incentives that more than compensate for this disutility (e.g., Lindenberg, 2001; Milgrom & Roberts, 1992). Consequently, motivation stemming from extrinsic sources should lead to an increase in employees' effort. Hence, we propose the following hypothesis:

H5. Extrinsic motivation has a positive effect on employee's effort.

Moreover, several experiments show a positive effect of intrinsic motivation on human effort in different situations (e.g., Frey, Eichenberger, & Oberholzer-Gee, 1996; Frey & Götte, 1999; Frey & Jegen, 2001). Additionally, the recent discussion of intrinsic motivation, especially in the business context, underscores the importance of this motivational type for employees' activity levels also in an economic context (e.g., Frey, 1997; Osterloh & Frey, 2000; Weibel et al., 2010). This leads to our last hypothesis:

H6. Intrinsic motivation has a positive effect on employee's effort.

Fig. 1 summarizes these hypothesized relationships.

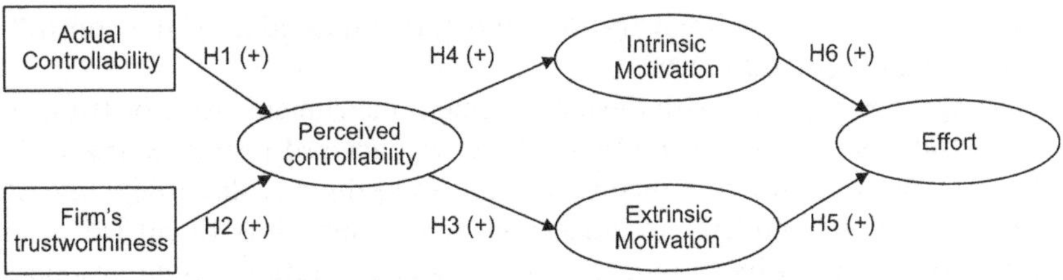

Fig. 1. Hypothesized Relationships.

METHOD AND DATA

Method

To test the hypotheses, we used a vignette experiment that employs a factorial survey design similar to conjoint analysis (Kuhfeld, 2005). Vignette experiments are especially suitable for analyzing the influence of social context on individual decisions and behavior (Taylor, 2006) and allow for the examination of not directly observable personal aspects, like the degree of motivation. Vignettes are "short descriptions of a person or social situation that contain precise references to what are thought to be the most important factors in the decision-making or judgement-making processes of respondents" (Alexander & Becker, 1978, p. 94). They are constructed by deliberately changing the parameters of the independent variables. The various situations resulting therefrom are then combined with test questions to examine the participant's reaction to the situation. Hence, this method combines characteristics of experimental and survey research (Rost, Weibel, & Osterloh, 2010).

Data

We collected data from a broad sample of business students without prior exposure to the theoretical concept of controllability at a large metropolitan university in Germany. Using these students was thought to be the best guard against socially desirable response behavior, since they had not yet been introduced in their classes to the ideas of the controllability principle and would thus respond as natural as possible and representative of the majority of "ordinary" employees, i.e., employees without a formal education in business administration (and thus knowledge of the controllability principle), within firms.

The vignettes together with the measurement instruments were distributed to 261 students in early 2009. Out of these, we received responses from 253 subjects. Two were severely incomplete and were directly discarded, and 24 were partially incomplete or ambiguous and were thus discarded after careful inspection, which indicated no important differences in response behavior with respect to the fully complete ones. Thus, a total number of 227 usable responses resulted. In order to test for presence of common-method bias, we applied the Harman one-factor test (Podsakoff & Organ, 1986), which indicated multiple dimensions, thereby suggesting that common-method bias should not be a serious issue in our data sample.

Measurement Instruments

Situational Variables

The situational variables are intended to provide participants with an understanding of the real-world situation they should be thinking about. They were told that they should think of themselves as working in a project-oriented setting, such as an advertising or marketing campaign or a capital expenditure project. We chose the six arguably most prominent factors from prior research on working conditions as situational variables: employee participation (Osterloh & Frey, 2000), felt proficiency and competence (Deci & Ryan, 2000), work autonomy (Deci & Ryan, 2000), work climate (Deci & Ryan, 2000; Frey, 1997), monetary incentives (Cummings, 1984; Lawler, 1969), and predictability of firm behavior, i.e., locus of control over rewards (Rotter, 1954).

In order to prevent bias resulting from an arbitrarily selected constellation of the situational factors, we created multiple situations by varying the values we assigned to these variables (see Appendix A1).

Controllability of Performance-Based Remuneration

Whereas actual controllability was modeled by using different sentences in the vignettes (see Appendix A1) – each describing a different level of controllability (fixed pay, i.e., no controllability = 0 to individual pay for performance = 3) – perceived controllability was captured using a multi-item instrument inspired by Greenberg (2003) and McDaniels, Axelrod, Cavanagh, and Slovic (1997). Cronbach's alpha for the four-item instrument was .96 (see Appendix A2 for details). All items loaded on a single factor with eigenvalue greater than 1 (3.58).

Additional Work Effort

For measuring participants' willingness to exert work effort, we chose a four-item instrument used by Kunz and Linder (2009) before (see Appendix A2). Respondents were asked to state on Likert scales how much they agree to four statements relating to the exertion of additional work effort. Cronbach's alpha of the four-item instrument in the present sample was .85 and factor analysis suggested that the four items represented one underlying factor (eigenvalue: 2.77).

Intrinsic and Extrinsic Motivation

The measures of extrinsic and intrinsic motivation followed the question on additional effort. The items were taken from Ryan and Connell (1989) and covered intrinsic joy-based motivation (3 items, Cronbach's alpha = .82) and extrinsic motivation (3 items, Cronbach's alpha = .81). Factor analysis yielded, as expected, two factors with eigenvalue greater than 1 (2.29 for factor one, i.e., intrinsic motivation, and 2.11 for factor two with the extrinsic items loading onto), and no item had cross-loadings above the common threshold of .30. Confirmatory factor analysis showed factor loadings of .58 to .80 (extrinsic motivation) and .72 to .75 (intrinsic motivation) on the expected theoretical constructs.

Control Variables

As a vignette experiment relies heavily on the quality of the vignettes, we included a control variable that deals with the ecological validity. It was measured by a single item inspired by Weibel et al. (2010). Participants were asked to rate ecological validity of each vignette in their questionnaire on a seven-point Likert scale. The mean of the normally distributed responses on the ecological validity item is 4.29, which is similar to the results obtained in other vignette experiments in literature (e.g., Weibel et al., 2010).

Another set of items was added to control for the potential influence of work experience on the answers to other items. They included an open-ended question on months of work experience in paying jobs and a binary choice question regarding exposure to project-oriented work in their job (0 = none; 1 = past exposure). We also asked participants to report on a seven-point Likert scale their own perceived exposure to performance-related rewards. Additionally, we asked them to provide information about their age and gender. Table 1 reports the results of the questions.

Table 1. Descriptive Statistics.

Variable	Mean	SD	Min.	Max.	1	2	3	4	5	6	7	8	9	10	11	12	13
1 Additional work effort	.00	1.00	−2.21	2.09	*.85*												
2 Intrinsic motivation	.00	1.00	−2.08	2.53	**.47*****	*.82*											
3 Extrinsic motivation	.00	1.00	−2.30	2.12	**.46*****	.05	*.81*										
4 Perceived controllability	.00	1.00	−1.74	1.56	**.64*****	**.29*****	**.45*****	*.96*									
5 Actual controllability	1.44	1.12	.00	3.00	**.32*****	**.14*****	**.22****	**.43*****									
6 Past trust experience	.50	.50	.00	1.00	**.29*****	**.20*****	.12	**.28*****	**.14***								
7 Closeness to reality	4.29	1.32	1.00	7.00	**.21*****	**.24*****	.05	**.21*****	.06	**.15****							
8 Gender	1.55	.50	1.00	2.00	−.10	**−.13****	.01	**−.11***	−.10	−.07	−.03						
9 Age	21.90	2.01	19.00	32.00	−.04	−.05	.01	.06	**.14****	−.08	−.05	**−.13****					
10 Work experience (months)	17.65	21.16	.00	120.00	.00	.08	−.03	.06	.08	.00	.01	**−.22*****	**.43*****				
11 Project experience	1.68	.47	1.00	2.00	−.04	.00	.02	.00	.08	−.05	−.09	−.01	.00	−.03			
12 Pay-for-performance exposure	2.88	2.09	1.00	7.00	.02	.07	.06	.08	.06	.06	.00	−.02	.03	.06	**−.25*****		
13 Socially desirable responses	.00	.82	−2.43	2.00	−.04	**−.16****	.07	.04	.01	.01	−.09	**.17****	−.05	−.03	−.02	−.02	*.61*

$n = 227$. Significance of correlations: $^{***}p<.01$; $^{**}p<.05$; $^{*}p<.10$ (two-tailed test). Cronbach alphas of constructs along the diagonal.

Finally, we also employed a translated version of the Strahan and Gerbasi (1972) "short scale B" for assessing social desirability bias in response. While this scale has been used before in many Anglo-American studies in social-psychological research, Cronbach's alpha for our data set is only .61 – and thus below the typically mentioned threshold of .70. However, exploratory factor analysis suggested a single factor. A potential explanation for the low alpha might be cultural differences: respondents in the German sample might have reacted differently to the scales originally developed for a US context than an American sample population would. Since the scale, however, only serves us to check for a bias by inspecting its correlation with the constructs of interest to the present study, the quality of the measurement overall thus seems acceptable.

RESULTS

Table 1 provides the descriptive statistics. As expected, both types of motivation correlate significantly with the intention to engage in additional work effort. Similarly, there are significant correlations of actual and perceived controllability, motivation, and effort. Past trust experience, however, seems not to matter for extrinsic motivation as it correlates only with intrinsic motivation, perceived controllability, and work effort.

Work experience, project experience, and experience with pay-for-performance schemes do not exhibit statistically significant relationships with any of the dependent variables in our study, suggesting that results should be free from biases from these factors. Also social desirable response behavior seems not to be a major issue as there is only a single significant relationship with the constructs of interest to the present study – namely with intrinsic motivation – thus leading to a potential underestimation of the level of intrinsic motivation. Similarly, based on the correlations, age seems not to play a major role. The only exception being that, surprisingly, a small correlation between age and actual controllability seems to exist – suggesting that the allocation of our vignettes to the participating students is not fully random. A potential explanation might be different preferences in seating in the class room used between differently aged students. Again, however, since correlation analysis suggests no relationship of age with the dependent variable of interest in the present study, this nonrandomness can be neglected.

Overall, however, the descriptive statistics suggest that the dataset is reasonably suited to a further analysis testing our structural model.

Out of the growing number of approaches for doing structural equation modeling, we chose the Partial Least Squares (PLS) algorithm and used the SmartPLS software (Ringle, Wende, & Will, 2005) to calculate our proposed structural model for the following reasons:

First, since we do not truly test a theory or hypotheses directly derived from it, but rather try to investigate the potential for extending a theory beyond its current realm based on plausible reasoning, PLS as a "soft" structural equation modeling approach seems very much in line with our research project.

Second, in contrast to other approaches of structural equation modeling based on covariance (e.g., LISREL), PLS make less restrictive assumptions about the underlying distributions of the variables (Henseler, Ringle, & Sinkovics, 2009). Whereas skewness and kurtosis of our data are small, the Shapiro–Wilk and Shapiro–Francia tests nevertheless rejected the assumption of normal distribution for about half of our items – thus suggesting not to rely on an approach assuming multinormality like LISREL. Whereas choosing other, more robust estimators as implemented, e.g., in Mplus can help to address the nonnormality of items, this comes at the cost of either a requirement for very large sample sizes (i.e., sample sizes beyond our size) or the inapplicability of chi-square comparisons and the chi-square difference test to assess the relative fit of different models. As PLS offers an (admittedly imperfect) overall goodness-of-fit index to judge alternative models, allows to treat a broader range of sample sizes, and does not rely on (the in our sample problematic) covariances (Henseler et al., 2009), it seems best suited to early stage investigations into relationships about which not much prior knowledge is available – as is the case with the link of controllability to intrinsic motivation. Moreover, selecting PLS allows for easy comparison of our findings with findings from laboratory studies that given the massive effort needed for laboratory studies typically have an even more limited sample size.

The model exhibits a highly significantly positive influence of both, objective controllability and trustworthiness on perceived controllability, supporting hypotheses H1 and H2. Moreover, the model shows a significant impact of perceived controllability on both types of motivation and also a positive impact of these types of motivation on employee's effort, backing the other hypotheses suggested. The graph in Fig. 2 summarizes our results for the proposed model.

To corroborate our findings, we also tested alternative models (see Table 2 for details). A model leaving out perceived controllability resulted in lower goodness of fit – as did a model following "classical" agency-theoretical

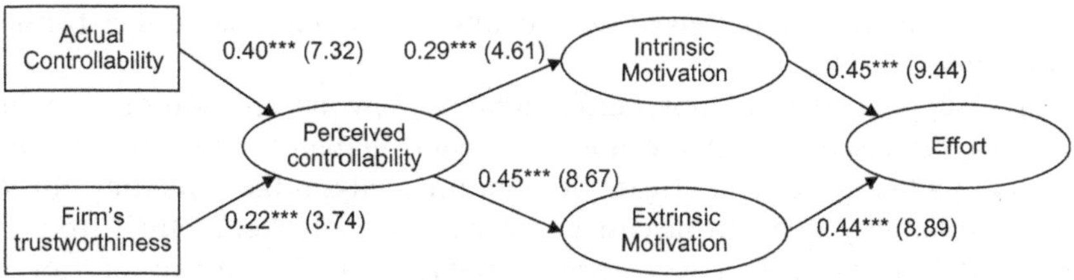

Fig. 2. PLS Path Model.

Table 2. Model Fit.

Alternative Models	Average AVE	Average R^2	GOF
Model 1 (Fig. 2)	.76	.23	.42
Model 2 (w/o trust experience)	.76	.22	.41
Model 3 (model 1 w/o intrinsic motivation)	.77	.22	.41
Model 4 (only actual controllability on work effort)	.69	.10	.26
Model 5 (w/o actual controllability)	.76	.20	.38
Model 6 (only actual controllability on motivation and work effort)	.71	.16	.34

Note: The well-known goodness-of-fit (GOF) metrics of structural equation modeling using a covariance-based approach (e.g., LISREL) are not applicable to variance-based approaches (e.g., PLS) (Hulland, 1999). The closest assessment of an overall GOF under PLS can be achieved by a comparison of alternative models' GOF calculated as follows: GOF = sqrt(average(AVE) × average(R^2)). The model with the largest GOF overall fits the data best (up to now there are no recommendations for a minimum GOF available).

arguments by leaving out intrinsic motivation. Since the same holds true for the other alternative models, the hypothesized model seems to best fit the present data.

DISCUSSION AND CONCLUSION

Our results in general support the hypothesized relationships. Controllability indeed seems to increase the willingness to exert effort via the mediating factors of perceived controllability and intrinsic and extrinsic motivation.

This suggests that a more differentiated view on motivation should enhance the discussion of the notion of controllability within behavioral accounting literature. This stream of literature analyzes the mediating effect of controllability between incentive systems and employee's effort from a behavioral perspective. It pays special attention to its influence on "soft" factors like the perception of fairness of these systems. The results of the present chapter add to this literature by incorporating and analyzing particularly the influence of controllability on intrinsic motivation, i.e., another presumably "soft" factor intervening between incentive systems and an employee's work effort. Thereby, it enlarges the set of mediating factors and enriches the perspective on the mentioned cause–effect chain.

Moreover, our findings also put the discussion regarding controllability in the literature relying on agency-theory into a broader perspective: This stream of literature to date focuses mainly on extrinsic motivation, and, hence, assumes that controllability influences effort via an extrinsic motivational path. In contrast, our results suggest that controllability can work also via intrinsic motivation – and that adherence to the principle of controllability may thus even be beneficial when dealing with individuals who are intrinsically motivated.

However, a number of caveats apply to our findings that call for further investigation in future research.

While we included an employee's past trust experience with a firm's promises as a separate variable into our model to predict perceived controllability, our study does not separately account for the role of cognitive biases versus self-efficacy beliefs in determining perceived controllability. As we did neither collect data on the self-efficacy beliefs of our participants nor on their perceptional and judgmental capabilities, our study thus confounds the two factors (the competence experience included in the situational variables only covers short-term, situational impressions). Future studies should collect respondents' generalized self-efficacy beliefs and try to assess their cognitive perceptional and judgmental abilities.

Like other experimental studies in the area of management accounting (e.g., Dilla & Steinbart, 2005; Lipe & Salterio, 2000), we relied on students for conducting our study. Even though most of them had some prior job experience, the complexity of the subject analyzed in our study may raise issues of how appropriate students truly are as a group of subjects. A replication of our study with subjects having more work experience thus seems of prime importance. We expect that the effects measured should even be more pronounced than in our present study as more experienced subjects should find it easier to put themselves into the hypothetical work situations of performance-based pay. At the same time, however, we feared that using a sample of more experienced employees within firms might easily lead to an issue of social desirability bias, as some employees of such a sample might have been exposed to the concept of controllability that could easily lead them to answer in a socially desirable manner.

Another potential limitation of the present study concerns its laboratory character, which may cast doubt on the ecological validity of the results (Colquitt, 2008). First, it should be noted that in a meta-analytic study a high correlation of .73 was found between effects in experimental studies and field studies (Anderson, Lindsay, & Bushman, 1999). Second, vignettes expose participants to a written description of a situation that models a field situation more closely than is possible in a normal laboratory experiment. Unfortunately, to the authors' knowledge, there is currently no meta-analysis on the correlation between effects observed in factorial vignettes experiments and field studies that address this suggestion. Thus, we are currently unable to fully assess the ecological validity of our findings. However, we submit that the potential limitations of our methodology are compensated for by its advantages, such as a richer environmental stimulus than laboratory experiments are likely to offer, and the greater possibility to infer the direction of causality as compared to the common cross-sectional field surveys. Given that "inferring causality is one of the most difficult aspects of scientific research" (Colquitt, 2008, p. 616), we propose that research benefits from the use of different methods. Vignette experiments can be seen as one way to generate new empirical results on controllability, which, of course, leads in turn to the need for further testing and methodological refinements.

These limitations notwithstanding, we nevertheless feel confident that the effects found in our study should hold and that further investigations of the issue of controllability therefore should take both extrinsic and intrinsic motivation into account. This even more so as following the principle of controllability seems to foster employee motivation and effort even for mainly intrinsically motivated individuals.

ACKNOWLEDGMENTS

We are indebted to Ralf Ewert, Nicolai Foss, as well as workshop and conference participants at Tilburg University, WHU-Otto Beisheim School of Management, and the 2nd European Reward Management Conference for helpful comments on earlier versions of this manuscript. Mirko Tinz provided help with coding the data. The authors gratefully acknowledge the participation of the students involved in this study. This work originated during the second author's time with Copenhagen Business School's Department of Strategic Management & Globalization, Porcelænshaven 24B, 2000 Frederiksberg/Denmark.

REFERENCES

Alderfer, C. P. (1972). *Existence, relatedness and growth: Human needs in organizational settings*. New York, NY: The Free Press.

Alexander, C. S., & Becker, H. J. (1978). The use of vignettes in survey research. *Public Opinion Quarterly*, *42*, 93–104.

Anderson, C. A., Lindsay, A. J., & Bushman, B. J. (1999). Research in the psychological laboratory: Truth or triviality? *Current Directions in Psychological Science*, *8*, 3–9.

Antle, R., & Demski, J. S. (1988). The controllability principle in responsibility accounting. *The Accounting Review*, *63*, 700–718.

Atkinson, A. A. (1987). *Intra-firm cost and resource allocations: Theory and practice*. Toronto, Canada: The Canadian Academic Accounting Association.

Bandura, A. (1997). *Self-efficacy: The exercise of control*. New York, NY: W. H. Freeman & Company.

Bonner, S. E., & Sprinkle, G. B. (2002). The effects of monetary incentives on effort and task performance: Theories, evidence, and a framework for research. *Accounting, Organizations and Society*, *27*, 303–345.

Bushman, R. M., Indjejikian, R. J., & Smith, A. (1995). Aggregate performance measures in business unit manager compensation: The role of intrafirm interdependencies. *Journal of Accounting Research*, *33*, 101–128.

Cameron, J., & Pierce, W. D. (2002). *Rewards and intrinsic motivation: Resolving the controversy*. London: Bergin and Garvey.

Choudhury, N. (1986). Responsibility accounting and controllability. *Accounting and Business Research* (Summer), 189–198.

Colquitt, J. A. (2008). From the editors. *Academy of Management Journal*, *51*, 616–620.

Cummings, L. L. (1984). Compensation, culture, and motivation: A systems perspective. *Organizational Dynamics* (Winter), 33–43.

de Charms, R. (1968). *Personal causation: The internal affective determinants of behavior*. New York, NY: Academic Press.

Deci, E. L. (1971). Effects of externally mediated rewards on intrinsic motivation. *Journal of Personality and Social Psychology*, *18*, 105–115.

Deci, E. L. (1975). *Intrinsic motivation*. New York, NY: Plenum Press.

Deci, E. L., Koestner, R., & Ryan, R. M. (1999). A meta-analytic review of experiments examining the effects of extrinsic rewards on intrinsic motivation. *Psychological Bulletin*, *125*, 627–668.

Deci, E. L., & Ryan, R. M. (2000). The 'what' and 'why' of goal pursuits: human needs and the self-determination of behavior. *Psychological Inquiry*, *11*, 227–268.

Dilla, W. N., & Steinbart, P. J. (2005). Relative weighting of common and unique balanced scorecard measures by knowledgeable decision makers. *Behavioral Research in Accounting*, *17*, 43–53.

Drury, C., & El-Shishini, H. (2005). *Divisional performance measurement: An examination of the potential explanatory factors*. London, UK: The Chartered Institute of Management Accountants.

Ferrara, W. L. (1964). Responsibility accounting – a basic control concept. *N.A.A. Bulletin* (September), 11–19.

Fremgen, J. M., & Liao, S. S. (1981). *The allocation of corporate indirect cost*. New York, NY: National Accounting Association.

Frey, B. S. (1997). *Not just for the money: An economic theory of personal motivation*. Cheltenham, UK: Edward Elgar Publishing.

Frey, B. S., Eichenberger, R., & Oberholzer-Gee, F. (1996). The old lady visits your backyard: A tale of morals and markets. *Journal of Political Economy*, *104*, 193–209.

Frey, B. S., & Götte, L. (1999). *Does pay motivate volunteers?* Switzerland: Institute for Empirical Research in Economics, University of Zürich. Retrieved from http://www.research-projects.uzh.ch/p1871.htm

Frey, B. S., & Jegen, R. (2001). Motivation crowding theory: A survey of empirical evidence. *Journal of Economic Surveys*, *15*, 589–611.

Gagné, M., & Deci, L. E. (2005). Self-determination theory and work motivation. *Journal of Organizational Behavior*, *26*, 331–362.

Giraud, F., Langevin, P., & Mendoza, C. (2008). Justice as a rationale for the controllability principle: A study of managers' opinions. *Management Accounting Research*, *19*, 32–44.

Gordon, M. J. (1963). Toward a theory of responsibility accounting systems. *N.A.A. Bulletin* (December), 3–10.

Gosh, D. (2005). Alternative measures of managers. Performance, controllability, and the outcome effect. *Behavioural Research in Accounting*, *17*, 55–70.

Greenberg, J. (2003). Creating unfairness by mandating fair procedures: The hidden hazards of a pay-for-performance plan. *Human Resource Management Review*, *13*, 41–57.

Henseler, J., Ringle, C. M., & Sinkovics, R. R. (2009). The use of partial least squares in path modeling in international marketing. *Advances in International Marketing*, *20*, 277–319.

Horngren, C. T., Foster, S. M., & Datar, G. (2006). *Cost accounting* (12th ed). Englewood Cliffs, NJ: Prentice Hall International.

Hulland, J. (1999). Use of partial least squares (PLS) in strategic management research: A review of four recent studies. *Strategic Management Journal*, *20*, 195–204.

Kohn, A. (1993). *Punished by rewards*. New York, NY: Houghton Mifflin.

Kuhfeld, W. F. (2005). *Marketing research methods in SAS: Experimental design, choice, conjoint, and graphical techniques*. Cary, NC: SAS Institute.

Kunz, J., & Linder, S. (2009). Crowding-out of intrinsic motivation by extrinsic rewards: A vignette study. Paper presented at the Academy of Management Meeting 2009, Chicago, IL.

Kunz, A. H., & Pfaff, D. (2002). Agency theory, performance evaluation, and the hypothetical construct of intrinsic motivation. *Accounting, Organizations and Society*, *27*, 275–295.

Lawler III., E. E. (1969). Job design and employee motivation. *Personnel Psychology*, *22*, 426–435.

Lepper, M. R., Greene, D., & Nisbett, R. E. (1973). Undermining children's intrinsic interest with extrinsic rewards: A test of the overjustification hypothesis. *Journal of Personality and Social Psychology*, *28*, 129–137.

Lindenberg, S. (2001). Intrinsic motivation in a new light. *Kyklos*, *54*, 317–342.

Lipe, M. G., & Salterio, S. E. (2000). The balanced scorecard: Judgmental effects of common and unique performance measures. *The Accounting Review*, *75*, 283–298.

McDaniels, T. L., Axelrod, L. J., Cavanagh, N. S., & Slovic, P. (1997). Perception of ecological risk to water environments. *Risk Analysis*, *17*, 341–352.

Merchant, K. A. (1987). How and why firms disregard the controllability principle. In W. J. Bruns & R. S. Kaplan (Eds.), *Field study in management accounting* (pp. 316–338). Cambridge, MA: Harvard Business School Press.

Merchant, K. A. (1989). *Rewarding results: Motivating profit center managers*. Cambridge, MA: Harvard University Press.

Milgrom, P., & Roberts, J. (1992). *Economics, organization & management*. Upper Saddle River, NJ: Prentice-Hall.

Moon, P., & Fitzgerald, L. (1996). Delivering the goods at TNT: The role of the performance measurement system. *Management Accounting Research*, *7*, 431–457.

Osterloh, M., & Frey, B. S. (2000). Motivation, knowledge transfer, and organizational forms. *Organization Science*, *11*, 538–550.

Otley, D. T. (1990). Issues in accountability and control: Some observations from a study of colliery accountability in the British coal corporation. *Management Accounting Research*, *1*, 101–123.

Podsakoff, P. M., & Organ, D. W. (1986). Self-reports in organizational research: Problems and prospects. *Journal of Management*, *12*, 531–544.

Reiss, S. (2005). Extrinsic and intrinsic motivation at 30: Unresolved scientific issues. *Behavior Analyst*, *28*, 1–14.

Ringle, C. M., Wende, S., & Will, A. (2005). *SmartPLS 2.0 M3*. Retrieved from http://www.smartpls.de. Accessed on February 10, 2012.

Rost, K., Weibel, A., & Osterloh, M. (2010). *Good organizational design for bad motivational disposition*. Working paper.University of Zürich, Zurich Open Repository and Archive, Zurich. Retrieved from http://www.zora.uzh.ch/44815/4/GoodV.pdf. Accessed on February 10, 2012.

Rotter, J. B. (1954). *Social learning and clinical psychology*. Englewood Cliffs, NJ: Prentice-Hall.

Ryan, R. M., & Connell, J. P. (1989). Perceived locus of causality and internalization – examining reasons for acting in 2 domains. *Journal of Personality and Social Psychology*, *57*, 749–761.

Simon, H. A. (1977). *The new science of management decision (Revised Version)*. New York, NY: Prentice-Hall.

Simons, R. (2007). *Revisiting the controllability principle in the twenty-first century*. Working paper, Cambridge, MA, Harvard University. Retrieved from http://www.hbs.edu/units/am/pdf/Rethinking%20the%20Controllability%20Principle.pdf. Accessed on February 10, 2012.

Solomons, D. (1965). *Divisional performance: Measurement and control*. Homewood, IL: Irwin.

Strahan, R., & Gerbasi, K. C. (1972). Short, homogeneous versions of the Marlow-Crowne social desirability scale. *Journal of Clinical Psychology*, *28*, 191–193.

Taylor, B. J. (2006). Factorial surveys: Using vignettes to study professional judgement. *British Journal of Social Work*, *36*, 1187–1207.

Ugras, J. Y. (1994). Factors affecting allocation of noncontrollable costs for performance evaluation use: A survey. In M. J. Epstein (Ed.), *Advances in management accounting* (Vol. 3, pp. 255–278). Greenwich: JAI Press.

Weibel, A., Rost, K., & Osterloh, M. (2010). Pay for performance in the public sector – benefits and (Hidden) costs. *Journal of Public Administration Research and Theory*, *20*, 387–412.

APPENDIX A1: SITUATIONAL VARIABLES AND CONTROLLABILITY*

Gamma Company is an international firm that organizes its processes in a project-oriented way. Currently, you are a member of a large internal project. In this project you are **in charge of three cost centers**. The other two sub-projects are headed by two other colleagues. The whole project is supervised by a manager who is situated in hierarchy above you. Your working environment is characterized by the following aspects.

Participation

- Your supervisor **emphasizes the participation** of employees in decision processes and often asks for your opinion when setting work targets.
- Your supervisor commonly decides on work-related aspects **on their own** and rarely asks for your opinion or tries to understand your perspective.

Competence Acceptance

- However/Moreover, **your work often gives you the feeling of being proficient**, as you frequently experience success in attaining your project's objectives.
- However/Moreover, **your work seldom gives you the feeling of being proficient**, as you hardly experience success in attaining your project's objectives.

Autonomy

- Additionally, in many situations you **can decide on your own** on the time and the way of performing your work and do not have to consult your supervisor.
- Additionally, you often **have to follow preset steps and procedures** or to consult your supervisor if you want to deviate from them.

Climate of Work

- You feel as an accepted part of the company, as it is characterized by a **close and cooperative** relationship between employees.

*Translated into English by the authors. Original vignettes in German.

- In your company, you feel just as one out of many employees, as it is characterized by a rather **anonymous** relationship between employees.

Predictability/Reliability

- Your company has a **notable tradition of reliability and trustworthiness** toward its employees and you can trust the company on getting a promised reward (e.g., salary increases).
- Your company has **no tradition of reliability and trustworthiness** toward its employees: It rarely provides promised rewards (e.g., salary increases) and you do not truly trust the company on getting a promised reward.

Monetary Incentives

- Your wage has no variable, performance-based component. It is a fixed salary.
- Your wage is based to a high degree on your performance. Only 50% are fixed, the other 50% are variable.

Controllability of Variable Pay (only used in situations where variable incentives exist)

- In your firm the variable part of the incentives depends mainly **on individual performance**. In your case, it is influenced by the degree to which you keep your costs in your cost center under control.
- In your firm the variable part of the incentives depends mainly on the **total performance of those projects in which a particular employee participates** and also on the degree to which the whole costs of projects are kept under control.
- In your firm the variable part of the incentives depends mainly on the **total success of the firm**.

APPENDIX A2: PLS CONSTRUCT QUALITY INDICATORS

Construct	Items	Factor Loadings	Indicator Reliability	Cronbach's Alpha	Average Variance Extracted	Composite Reliability	R^2
Effort	*I will spend an additional 5 hours per week on the project.*	.85	.72	.85	.69	.90	.41
	I will continue working until I have finished a time-critical part of the project, although I have already passed my normal daily working hours.	.82	.67				
	I will work also during weekends to finish a time-critical part of the project.	.88	.77				
	I will skip parts of my holidays to finish a time-critical part of the project.	.77	.60				
Extrinsic motivation	*A different behavior would involve many disadvantages for me.*	.89	.80	.81	.71	.88	.21
	I would get into trouble, if I did not do it.	.87	.76				
	This is expected from me.	.76	.58				
Intrinsic motivation	*I like doing this.*	.86	.73	.82	.73	.89	.08
	It satisfies me.	.87	.75				
	It makes me personally happy.	.85	.72				
Perceived controllability	*The amount of my remuneration heavily considers my individual share to company performance.*	.94	.88	.96	.89	.97	.23
	My remuneration significantly mirrors how well I have worked.	.95	.91				
	The amount of my remuneration strongly depends on my effort level.	.95	.91				
	I have a strong influence on the degree of my remuneration.	.94	.88				

$n = 227$. All items measured on seven-point Likert scales (1 = Don't agree at all, 7 = Fully agree). *Note*: Common quality thresholds for constructs in PLS are (Henseler et al., 2009): Factor loading $\geq .7$; indicator reliability $\geq .5$; alpha $\geq .7$; AVE $\geq .5$; composite reliability $\geq .7$; R^2 of outer endogenous variable (here: "Effort") $\geq .19$ (weak) $\geq .33$ (mediocre).

DOES THE RELIANCE ON NONFINANCIAL MEASURES FOR PERFORMANCE EVALUATION ENHANCE MANAGERS' PERCEPTIONS OF PROCEDURAL FAIRNESS?

Jane Q. He and Chong M. Lau

ABSTRACT

Purpose – *The adoption of performance measurement and evaluation systems comprising nonfinancial measures has rendered the investigation of behavioral consequences of such measures an increasingly important research issue. The purpose of this study is to investigate the process by which the use of nonfinancial measures affects employees' perceptions of procedural fairness. It proposes that the effects of nonfinancial measures on procedural fairness are indirect through the mediating variables of (1) job relevant information and (2) role clarity.*

Methodology – *Data collected from a survey of 276 managers in different functional areas are used to test the models. The data are analyzed using structural equation modeling (Partial Least Square).*

Performance Measurement and Management Control: Global Issues
Studies in Managerial and Financial Accounting, Volume 25, 363–388

ISSN: 1479-3512/doi:10.1108/S1479-3512(2012)0000025017

Findings – *Results from structural models indicate that the use of nonfinancial measures has a positive impact on job relevant information, role clarity, and procedural fairness. In addition, the findings suggest that the use of nonfinancial measures is indirectly related to procedural fairness through job relevant information and role clarity. Specifically, the results indicate that the use of nonfinancial measures affects job relevant information. Job relevant information then influences role clarity. Role clarity, in turn, is positively related to procedural fairness.*

Value of paper – *This study provides systematic empirical evidence on how the use of nonfinancial measures for performance measurement and evaluation can affect employee perceptions of procedural fairness. It helps organizations to understand how this process occurs and provides them with some assurance that the adoption of nonfinancial measures may be beneficial particularly through the higher information content of such measures and the consequential enhancement of employee role clarity and perception of fairness. By studying the effects of nonfinancial measures, in isolation, this study also helps to demonstrate to organizations that some of the beneficial effects on employee outcomes found by prior management accounting studies involving a combination of financial and nonfinancial measures may be achievable from the use of nonfinancial measures alone without the need of financial measures. This may assist organizations in designing simpler performance measurement systems.*

INTRODUCTION

This study investigates how the use of nonfinancial measures affects employee behaviors. Specifically, it examines how the use of nonfinancial measures for performance measurement and evaluation affects job relevant information and role clarity and how such effects translate into favorable employees' perceptions of procedural fairness.

The study contributes to the management accounting literature in several ways. First, performance evaluations are not only ubiquitous in organizational setting but are also important to employees. Not only are employees' compensation packages and careers strongly affected by their performance evaluations, but performance evaluations are reflections of employees' successes or failures. This may ultimately also influence the intrinsic values of employees including their self-esteem and self-confidence. It should therefore not be surprising that employees are concerned with their performance evaluations and generally pay a great attention to the procedures employed to measure and evaluate their performance (Burney, Henle, & Widener, 2009; Hartmann & Slapnicar, 2009; Lipe & Salterio, 2000)

In this regard, it is likely that employees' perceptions of the fairness of performance evaluation procedures will have imperative behavioral implications for employees. As far back as the 1970s, researchers had begun to investigate systematically the importance of fairness of performance evaluation procedures on employee behaviors. Hopwood (1972) and Otley (1978) both hypothesized that performance evaluative methods affect employees' perceptions of fairness. Hopwood (1972, p. 174) argued that performance evaluations based on an inflexible use of financial measures might be a "biased indicator of managerial performance ... potentially inequitable" and concluded that the performance evaluation methods might affect justness of evaluation, which in turn might affect employee satisfaction.

This study also contributes to the management accounting literature by adopting an alternative viewpoint in the investigation of the link between performance evaluation and procedural fairness. In the management accounting domain, prior studies on procedural fairness have focused largely on the *consequences* of procedural fairness (e.g., Lau & Tan, 2006a; Lau, Wong, & Eggleton, 2008; Sholihin & Pike, 2009). For those few studies that have considered the antecedents of procedural fairness, the variables of interest are generally related to budgeting primarily budgetary participation (e.g., Lau & Lim, 2002; Libby, 1999; Lindquist, 1995; Wentzel, 2002). Some recent management accounting studies on fairness have begun to pay some attention to the effects of performance evaluation systems on procedural fairness (e.g., Burney et al., 2009; Hartmann & Slapnicar, 2009; Lau & Moser, 2008). However, these studies have generally treated the relationship between performance evaluation system and procedural fairness as a black box. Hence, while these studies hypothesized that performance evaluation systems may be related to procedural fairness, they have however not addressed the important issue of how performance evaluation systems influence employees' perceptions of procedural fairness. Precisely, how performance evaluation systems affect procedural fairness? Are the effects direct or are they indirect through some mediating variables? An investigation of these issues may provide valuable insights into these research issues.

Some recent management accounting research has provided empirical evidence on the effects of performance measurement systems on a number of important outcomes including job relevant information and role clarity.

A recent study by Burney and Widener (2007) has found that the use of Strategic Performance Measurement Systems comprising both financial and nonfinancial measures affected managerial performance through *job relevant information*. A study by Hall (2008) found that comprehensive performance measurement systems comprising both financial and nonfinancial measures affected managerial performance through *role clarity*. These two studies therefore provide empirical evidence to suggest that the use of performance measurement systems comprising both financial and nonfinancial measures for performance evaluation may affect the extent of employee job relevant information as well as their role clarity. However, both of these studies have not investigated if and how the two relationships found in the two studies (performance measurement systems and *job relevant information* for Burney and Widener (2007), and performance measurement systems and *role clarity* for Hall (2008)) are related. More importantly, these two studies have not investigated if and how the relationships found in their studies are related to employees' perceptions of procedural fairness.

In summary, there is empirical evidence to suggest that performance evaluation system is related to procedural fairness (Burney et al., 2009; Hartmann & Slapnicar, 2009). There is also empirical evidence to support the relationship between performance measurement systems and the extent of employee job relevant information (Burney & Widener, 2007) as well as the relationship between performance measurement systems and employees' role clarity (Hall, 2008). However, precisely how performance measurement system, job relevant information, role clarity, and procedural fairness are related has not been addressed by any prior studies. By developing a comprehensive model that integrates all the aforementioned four variables in a single model, our study may provide valuable insights into how these four variables are interrelated and better understood.

Finally, our study also contributes to this research area by investigating the effects of nonfinancial measures *in isolation* on employees' fairness perceptions. With the exception of Lau and Moser (2008), all other prior studies (Burney et al., 2009; Burney & Widener, 2007; Hall, 2008; Hartmann & Slapnicar, 2009) have conceptualized performance measurement systems as systems comprising both financial and nonfinancial measures. It is therefore unclear if the effects of performance evaluation systems found in these studies were due to financial measures, nonfinancial measures, or a combination of financial and nonfinancial measures. By focusing only on nonfinancial measures, our study is able to isolate the effects of nonfinancial measures on role clarity, job relevant information, and procedural fairness. While nonfinancial measures are not new, the use of nonfinancial measures

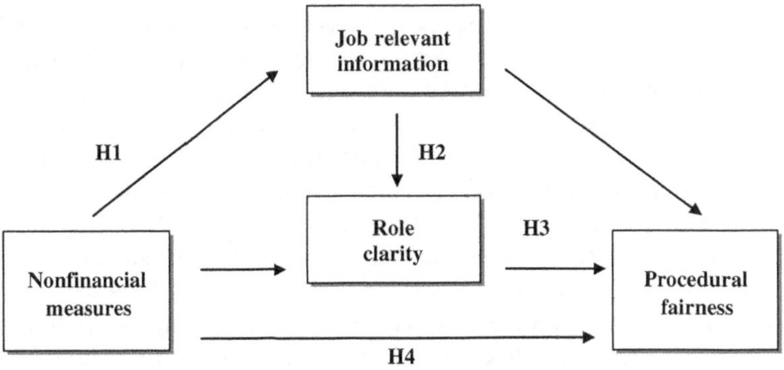

Fig. 1. Conceptual Model.

as criteria for employee performance evaluation may be becoming increasingly pervasive in contemporary organizations. Hence, it may be beneficial to both practitioners and researchers if we are able to ascertain if the use of such measures affect employee outcomes and indeed the extent of their effects on employee outcomes without the confounding effects of financial measures.

Fig. 1 presents the model used in this study. The model first ascertains if the use of nonfinancial measures affects procedural fairness. Second, it ascertains if (1) the use of nonfinancial measures increases the extent of employees' job relevant information and (2) if job relevant information in turn enhances employees' role clarity. Lastly, the effect of role clarity on employees' perceptions of procedural fairness is investigated.

The remainder of the paper is organized as follows. In the next section, we discuss the relevant literature and develop the hypotheses. This is followed by a discussion on the sample and the research method. In the subsequent sections, we present the statistical analyses and discuss the results. Finally, in the last section, we present the conclusions and limitations of the study.

HYPOTHESIS DEVELOPMENT

Nonfinancial Measures

In this study, the concept of the nonfinancial measures is derived from the three nonfinancial perspectives of the Balanced Scorecard (Kaplan & Atkinson, 1998; Kaplan & Norton, 1996). These are learning and growth perspective, internal business process perspective, and customer perspective. The learning and growth perspective focuses on long-term growth and improvement of organizations. As a result, the measures include training provided by the organizations to employees, employee morale, turnover rate, and employees' satisfaction rate. Furthermore, the adoption of new technology and the number of new innovations used are also included in this perspective. The internal business process perspective (the internal perspective, hereafter) identifies the critical internal processes in which the firm must excel (Kaplan & Norton, 1996). This perspective covers comprehensive organizational internal processes including setup times for manufacturing processes, defect rates of output, quality of manufacturing output, inventory level, and manufacturing cycle time. The customer perspective generally identifies the target customer and market segments. It is associated with a number of core or generic measures including the number of new customers acquired (market share), customer satisfaction, response time to customer response time, the number of customer complaints, and the number of overdue deliveries to customers (Kaplan & Norton, 1996).

According to Kaplan and Norton (1996), while the learning and growth perspective, the internal perspective, and the customer perspective are related, they are nevertheless distinct perspectives. As such, they will be investigated as separate models in our study. In addition, if these three perspectives were combined into a single variable, it will be based on the assumption that the three perspectives are equally weighted, which may not be realistic.

Nonfinancial Measures and Job Relevant Information

Job relevant information refers to the extent to which the information necessary for managers to complete their jobs or tasks is available (Kren, 1992). In the management accounting literature, job relevant information is referred to as decision-facilitation information that is intended to improve employees' knowledge and abilities to make decisions that are in line with organizational interests, thereby enabling employees to achieve the organization's goals and objectives (Kren, 1992; Sprinkle, 2003). Providing job relevant information to employees to make decisions helps reduce "pre-decision uncertainty" (Sprinkle, 2003, p. 302). This means that job relevant information can be viewed as the use of information to reduce ex ante uncertainty (Tiessen & Waterhouse, 1983), and is analogous to the problem-solving use (Simon, Guetskow, Kozmetsky, & Tyndall, 1954).

Performance measurement systems comprising nonfinancial measures may be characterized by measures that are expressed in nonmonetary terms. The ability to express employee performance in nonmonetary terms may result in several benefits. First, it may enable a much broader spectrum of employee performance to be measured. Most organizational activities are nonfinancial. The benefits of employee efforts expended to reduce setup time, reduce defect rate, or improve customer satisfaction are usually very difficult to measure in monetary terms. However, difficulties associated with the translation of nonfinancial activities into monetary terms are easily resolved by the use of nonfinancial measures and nonmonetary terms. This is likely to enable organizations to use a much wider variety of measures than would be possible if financial measures are used. It will also result in greater accuracy in performance measurement. The measurement of nonfinancial activities (e.g., effort to reduce defects) with nonfinancial terms (e.g., defect rate), will likely be more accurate as there will be no need to convert the improvement in defect rate into dollar terms.

The use of nonmonetary terms and hence the possibility of a wider range of possible measures may also make it easier to customize performance measures to fit the individualized situations of each employee. Performance measures that are customized to suit the individualized situations of individual employees are likely to be more meaningful to the employees. Finally, because nonfinancial measures are based on a variety of nonmonetary terms (e.g., defect rate, setup time, customer satisfaction rate), they cannot be aggregated. A common criticism of financial measures is that they are too aggregated. Financial measures are expressed in a common dollarized term that can be easily aggregated. When measures can be aggregated, accountants do aggregate them. This may result in aggregated financial measures (e.g., variable overhead efficiency variance) that may comprise many items under the responsibilities and control of many individuals. Difficulties in ascertaining which items are causing the variance and who are responsible may lead to information loss and confusion. In contrast, because nonfinancial measures are difficult to aggregate, they are less aggregated. Consequently, they are likely to be more meaningful and hence may be more actionable and easier to trace to the individuals who are responsible for the measures. This may provide employees with information on the causal relationship between the measurement of their performance and their actual performance as well as valuable feedback on the effort they expended. This suggests that the use of nonmonetary terms for nonfinancial measures is likely to be associated with a greater extent of job relevant information for employees.

With respect to time period considerations, for many organizations, financial reporting is constrained by the annual reporting cycle, generally annually, following the taxation year. Hence, while there is no requirement for all aspects of the accounting systems to comply with this annual financial and fiscal reporting cycle, many aspects of the accounting systems of many organizations do follow this annual cycle as a matter of convenience and efficiency (Johnson & Kaplan, 1986; Kaplan, 1984). In contrast, the use of nonfinancial measures is less constrained by the annual reporting cycle. The measurement and tracking of employee performance in terms of reduction in defect rates, reduction in setup time, and increases in customer satisfaction rates are not directly related to the annual reporting and fiscal cycle. Hence, a time horizon, which may be shorter or longer than the annual reporting period, would generally not pose any undue difficulties to the systems. Organizations may therefore select the periods of measurement and reporting of nonfinancial measures that are most appropriate and accurate for the measures concerned. This may lead to a reduction in the extent of inaccuracies in allocations and measurement caused by arbitrary cutoff dates to meet annual fiscal requirements. The use of nonfinancial measures may therefore lead to more accurate job relevant information for employees.

The ability of nonfinancial measures to target longer-term organizational goals and perspectives may also lead to a need for additional information. In order to perform better and achieve long-term goals, employees evaluated by nonfinancial measures are more likely to seek information as they believe that the cause-and-effect relationships among the nonfinancial measures are helpful for them to achieve better performance in the long term. Hence, the use of nonfinancial measures may trigger the search for additional job relevant information (Chong & Chong, 2002; Renn, 2003).

In summary, this study argues that the contents of nonfinancial measures provide employees with job relevant information. Nonfinancial measurement systems communicate the performance evaluation criteria and objectives of organizations from a more complete, more accurate, less aggregated, more customized, and long-term perspective. They provide more meaningful information and feedback to employees. They also trigger the search for more information. Consequently, a positive relationship between the use of nonfinancial measures (customer, internal process, learning, and growth) as evaluation criteria and job relevant information is expected from the discussion above. Accordingly, we propose:

Hypothesis 1. The use of nonfinancial measures is positively associated with job relevant information.

Job Relevant Information and Role Clarity

Role clarity refers to individuals' beliefs about the expectations and behaviors associated with their work role (Kahn, Wolfe, Quinn, Snoek, & Rosenthal, 1964). Kren (1992, p. 513) states that while job relevant information and role clarity are similar constructs, they are different since job relevant information reflects the extent of "information available to employees" and role clarity reflects the "extent to which employees understand their duties and responsibilities." A lack of role clarity means there is a gap between the information needed for employees to properly perform a job and the information available. Role clarity has played an important role in the literature of performance systems (Burney & Widener, 2007; Figler, 1980; Hall, 2008) and is an important indicator of behavioral consequences as a result of adopting certain performance measurement systems. Prior research shows that dysfunctional individual and organizational consequences result from the existence of role ambiguity, namely a lack of role clarity (Rizzo, House, & Lirtzman, 1970). Therefore, from organizational theory, roles must be precisely defined for a superior to know which subordinate to hold accountable, and in turn, to better assess their performance. With this information, the superior can delegate tasks, provide instruction and guidance, and monitor and measure performance against expectations (Rizzo et al., 1970).

Employees require sufficient information to perform tasks effectively. A lack of information regarding the descriptions and the objectives of the job can result in a reduction of role clarity. Kaplan and Norton (1996) argue that a clear and focused strategy will increase job relevant information, which in turn will reduce the potential for receiving incompatible requests and hence will clarify role expectations. Parasuraman and Alutto (1981) similarly purport that inadequate information can lead to increased role stress. If roles are not precisely defined, subordinates face decreased role clarity. From the discussion above, we propose:

Hypothesis 2. Job relevant information is positively related to role clarity.

Role Clarity and Procedural Fairness

Role clarity is likely to be positively associated with employees' perceptions of procedural fairness. Lind and Tyler (1988, p. 216) conceptualize procedural fairness as the judgments on how fair are those social norms that deal with "how decisions are made" and "how people are treated by authorities and other parties." As a result, procedural fairness is concerned with the fairness of the means or the rules and processes that superiors use to make decisions (Konovsky & Folger, 1989). McFarlin and Sweeney (1992) consider procedural fairness as the subordinates' perceptions of how fair are the procedures employed by their superiors to evaluate their performance, to communicate performance feedback, and to determine their rewards and compensation.

According to Lind and Tyler (1988), there are several fairness criteria. They include (1) completeness and accuracy of information, (2) retention of control over decisions, (3) correctability of incorrect decisions, and (4) consideration of the interests of all parties. In line with these fairness criteria, role clarity is likely to increase employees' perceptions of procedural fairness. First, role clarity refers to the extent to which employees understand their job-related duties and responsibilities (Kren 1992). In situations where employees have a clear understanding of their roles, various aspects associated with the employees' task assignments, task completions, task performance, and performance evaluations are likely to be far more accurately done than in situations where employees have a poor and unclear understanding of their roles and responsibilities. When roles are clear, superiors are also able to design clearer plans, clearer targets, and clearer criteria of performance evaluation. Employees assessed against criteria that are clear to them are likely to perceive the performance evaluation procedures as fair.

High role clarity is also likely to be associated with better control on the part of the employees. The larger the extent to which employees understand their roles, the better the control they will have over the tasks or jobs. Employees with a clear understanding of their roles are likely to feel that they are in control because they are in a position to focus and direct their attention to where it is needed for them to carry out their responsibilities. According to Kahn et al. (1964), role clarity is associated with control since employees will have confidence over their tasks and their authority to make decisions.

Employees with a clear understanding of their roles are likely to be in a better position to seek explanations for their performance evaluation because they understand clearly what their roles are at work and what constitute their roles. In the event of any incorrect evaluations of their performance, employees who have clear understanding of their roles are able to identify these mistakes because of their clear understanding of their roles. They would be in a strong position to explain to their superiors why they perceive the performance evaluations to be incorrect and hence have any mistakes in performance evaluations to be rectified. The ability on the part of those employees with high role clarity to be aware of mistakes in performance appraisals as well as the ability to appeal for such mistakes to be rectified constitutes an important aspect of procedural fairness. According to Leventhal (1980), correctability of incorrect decisions is an essential criterion of procedural fairness judgment. Hence, role clarity may assist in improving the correctability of incorrect decisions and therefore the enhancement of procedural fairness. Role clarity is therefore likely to be positively related to procedural fairness. We therefore propose:

Hypothesis 3. Role clarity is positively associated with employees' perceptions of procedural fairness.

The Indirect Effects Through Job Relevant Information and Role Clarity

The discussion in the preceding sections suggests the following. First, nonfinancial measures may be related to the increasing amount of job relevant information (Hypothesis 1). Second, job relevant information is positively associated with role clarity (Hypothesis 2). Third, role clarity is likely to be positively associated with procedural fairness (Hypothesis 3). These relationships indicate that the effects of nonfinancial measures as evaluation criteria on procedural fairness may be indirect via both job relevant information and role clarity. Thus, we propose:

Hypothesis 4. The relationship between the use of nonfinancial measures and procedural fairness is indirect through job relevant information and role clarity.

METHOD

The data were collected by a mailed survey questionnaire in Australia and the United Kingdom. The data collection in the United Kingdom was conducted from a sample of organizations from Kompass UK, the United Kingdom online database. The Australian sample was selected from the Who's Who of Australia business directory published in the electronic version of *Jobson's Year Book of Public Companies*. The manufacturing sector was selected for two reasons. First, it allows for some degree of control for industry. Second, manufacturing industry is a large sector of the economy of both the countries. This allows sufficient number of organizations to be contacted to participate in the study. Only manufacturing organizations with more than 100 employees and an annual turnover of more than 75 million pounds and Aus 250 million Australian dollars were chosen for the United Kingdom and Australian samples, respectively. These two criteria were chosen to provide some degree of control over the size of the organizations. The other reason for choosing relatively large size organizations was because accounting and control procedures tend to become more sophisticated and specialized with increasing firm size. Developing and implementing nonfinancial measures by themselves or in conjunction with financial measures are relatively complex and time-consuming processes. Accordingly, it is reasonable to assume that only large organizations with strong managerial expertise and resources will be capable of implementing such systems.

These selection criteria resulted in 142 organizations in the United Kingdom and 162 organizations in Australia. A random sampling technique was employed to select only 100 manufacturing organizations in Australia. Among the UK organizations, six companies could not be contacted and 33 were not willing to participate. This left the study with 103 UK organizations. For the purpose of avoiding a company bias, no more than four managers were selected from each organization. Telephone calls were made to the organizations. The names of 276 Australian managers and 296 UK managers were obtained for the survey. A total of 158 responses were returned in the United Kingdom. This constitutes a response rate of 53.5%. A total of 134 responses were returned in Australia, generating a response rate of 48.5%. The overall response rate is 51%.

A comparison of the means for the early respondents and the late respondents was carried out (Oppenheim, 1992). As the data entry was conducted based on the response dates, we compared early respondents (early 33%) to late respondents (late 33%) for all constructs of interest (demographic and model variables). The results indicate that there are no significant differences with respect to the proportion of early and late respondents in each variable from the data collected from both the United Kingdom and Australia. Therefore, it is unlikely that the results in this study will be affected by nonresponse bias.

The demographic features of the respondents indicate that the average functional head in the sample is about 51 years of age, has spent 9 years in the current job, and has 141 employees in his/her unit. They possess on average a tertiary education degree. These data suggest that the respondents are generally highly qualified and experienced managers in their organizations. This provides some degree of confidence for the quality and reliability of their responses.

Measurement of Variables

Nonfinancial Measures

The nonfinancial measures variables were measured with an instrument developed by Lau and Moser (2008). The instrument asks respondents: "When your superior is evaluating your performance, how much importance do you think he or she attaches to the following items?" This instrument contained 15 items for nonfinancial measures variables, which were based on the generic measures suggested by Kaplan and Atkinson (1998). According to Kaplan and Atkinson (1998), there are three perspectives of nonfinancial measurement – the learning and growth perspective, the internal business perspective, and the customer perspective. The five learning and growth measures in the instrument are employee satisfaction rate, number of employees trained, employee turnover rate, number of innovations developed, and the adoption of new technology. Quality of output, defect rates, setup time, manufacturing cycle, and inventory level are the measures for the internal perspective. The five customer perspective measures are number of new customers acquired, response time to customers, number of customer complaints, number of overdue deliveries, and customer satisfaction rate. A seven-point scale instrument anchored from "1 – never important" to "7 – always important" was used. A factor analysis of the 15 items indicates that the items load on three factors, each with an eigenvalue greater than 1. The three factors correspond with the three different perspectives of the Balanced Scorecard. All the items load onto their respective perspectives with loadings between 0.636 and 0.897, which are above the 0.50 cutoff point (Nunally, 1978).

Job Relevant Information

Job relevant information (JRI) was measured using a three-item instrument developed by Kren (1992). This construct assesses the extent to which managers perceive that adequate information is available for effective job-related decision making. The instrument was selected because it has been used extensively in other relevant accounting studies (Chong & Chong, 2002; Kren, 1992; Lau & Tan, 2006b; Magner, Welker, & Campbell, 1996). The instrument has a seven-point scale ranging from "1 – strongly disagree" to "7 – strongly agree." A factor analysis indicates that the factor loadings for the three items loaded satisfactorily (above 0.50) on a single factor. The Cronbach alpha is 0.773.

Role Clarity

Role clarity was measured through the use of Rizzo et al. (1970) instrument that comprises six items in a seven-point Likert-type scale ranging from "1 – strongly disagree" to "7 – strongly agree." One of these six items asks respondents to indicate the extent to which they perceive that they know how much authority they have at work. The main reasons for using Rizzo et al. (1970)'s instrument is its wide usage in the literature of role stress (Bernardin, 1979; House & Rizzo, 1972) as well as its successful application in the accounting literature (Chenhall & Brownell, 1998; Gregson, Wendell, & Aono 1994). The factor loadings from a factor analysis of the six items range between 0.735 and 9.917, which are above the benchmark of 0.50. The Cronbach alpha coefficient of 0.918 is also well above the benchmark of 0.7 (Nunally, 1978).

Procedural Fairness in Performance Evaluation
The instrument by McFarlin & Sweeney (1992) was used to measure the respondents' perceptions of procedural fairness. There are four items for which respondents indicate on a five-point Likert scale ranging from "1 – very unfair" to "5 – very fair." The items include fairness of the procedures used to evaluate performance, communicate feedback, and determine pay raises and promotions. The factor analysis results support the unidimensional nature of the four items. The rotated factor loadings for these four items range between 0.833 and 0.859. The instrument also has a Cronbach alpha of 0.873, which supports the reliability of the instrument (Nunally, 1978).

Table 1 presents the descriptive statistics for the indicators in this study. It includes the minimum and maximum values of each indicator reported by respondents, the mean and standard deviation values.

RESULTS

Since the variables in our study are all latent variables, structural equation modeling (SEM) is chosen for data analysis. There are two methods of SEM that are variance–covariance-based method and components-based method. The components-based method (Partial Least Square or PLS) is used in this study as it is similar to regression, but simultaneously models the structural paths (i.e., theoretical relationships among latent variables) and measurement paths (i.e., relationships between a latent variable and its indicators). In addition, PLS estimates the structural model using an iterative ordinary least square (OLS) regression-like procedure, with a primary goal to explain and maximize variance of the dependent variables by minimizing the residual variance of all dependent variables, both latent and observed. Because of its iterative estimation algorithm that proceeds block by block,

Table 1. Descriptive Statistics for Indicators.

Variables	*N*	Min	Max	Mean	Std Dev
Learning and growth					
LG1	274	1	7	4.80	1.377
LG2	274	1	7	4.31	1.425
LG3	274	1	7	4.16	1.634
LG4	274	1	7	4.68	1.393
LG5	274	1	7	4.44	1.411
Internal					
Int1	274	1	7	5.07	2.023
Int2	274	1	7	4.83	1.977
Int3	274	1	7	4.04	2.017
Int4	274	1	7	4.26	2.053
Int5	274	1	7	4.87	1.876
Customer					
Cus1	274	1	7	4.17	1.903
Cus2	274	1	7	5.30	1.644
Cus3	274	1	7	5.27	1.631
Cus4	274	1	7	5.00	1.862
Cus5	274	1	7	5.59	1.579
Job relevant information					
JRI1	274	1	7	5.43	1.071
JRI2	274	1	7	5.07	1.204
JRI3	274	1	7	5.14	1.304
Role clarity					
RC1	274	1	7	5.46	1.284
RC2	274	1	7	5.15	1.275
RC3	274	1	7	4.78	1.184
RC4	274	1	7	5.81	1.078
RC5	274	1	7	5.50	1.282
RC6				5.58	1.230
Procedural fairness					
PF1	274	1	7	3.67	0.866
PF2	274	1	7	3.42	0.871
PF3	274	1	7	3.55	0.834
PF4	274	1	7	3.30	0.944

PLS requires less stringent assumptions about the distributional characteristics of the raw data and sample size (Hartmann & Slapnicar, 2009). Furthermore, PLS incorporates multiple dependent constructs and explicitly recognizes measurement error. As a result of these advantages, PLS has been frequently used in many accounting studies (e.g., Chenhall, 2004; Hall, 2008; Hartmann & Slapnicar, 2009). Accordingly, SmartPLS software version 2.00 is chosen for data analysis in this study.

Reliability and Validity Tests

Unless the reliability and validity of the measurement model have been examined, its application in testing structural relationships between latent variables may lead to equivocal conclusions (Chin & Todd, 1995). In order to ensure the reliability and validity of the measurement model, we assess the quality of the measurement models. This includes their construct reliability, convergent validity, and discriminant validity for the reflective constructs.

To assess the reliability of the latent variables estimated by PLS in this study, the composite reliabilities suggested by Fornell and Larcker (1981), Cronbach alpha as indicated earlier, as well as the average variance extracted (AVE) statistics are used to ensure acceptable reliability. The results of these three tests are presented in Table 2. As shown in the table, both the composite reliability and alpha score for each variable are above 0.70. This demonstrates acceptable reliability (Nunally, 1978). Furthermore, the AVE statistics of all the variables are above 0.5 (Chin, 1998a), thereby providing support for adequate reliability of the measurement models.

Convergent validity is obtained when a measure correlates well with other tests measuring the same construct (Kaplan, 1987). It is examined by using the AVE statistics. Table 3 shows that the AVEs for all the constructs are above 0.50, which is the benchmark for satisfactory convergent validity (Chin, 1998b; Hair, Anderson, Tatham, & Black, 1998). This indicates that the instruments adopted in this study have satisfactory convergent validity.

Table 2. Reliability Tests of Variables.

	AVE	Composite Reliability	Cronbach Alpha
Learning and growth	0.590	0.878	0.826
Internal	0.753	0.938	0.917
Customer	0.671	0.910	0.877
Job relevant info	0.688	0.869	0.772
Role clarity	0.726	0.914	0.874
Procedural fairness	0.714	0.937	0.919

Table 3. Discriminant Validity Coefficient of Models.

Learning and Growth				
	Learning and Growth	Job Relevant Information	Role Clarity	Procedural Fairness
Learning and growth	**0.768**			
Job relevant information	0.390	**0.830**		
Role clarity	0.396	0.760	**0.845**	
Procedural fairness	0.268	0.438	0.497	**0.852**

Internal Business				
	Internal Business	Job Relevant Information	Role Clarity	Procedural Fairness
Internal business	**0.868**			
Job relevant information	0.106	**0.830**		
Role clarity	0.073	0.438	**0.845**	
Procedural fairness	0.179	0.761	0.497	**0.852**

Customer				
	Customer	Job Relevant Information	Role Clarity	Procedural Fairness
Customer	0.819			
Job relevant information	0.204	**0.830**		
Role clarity	0.169	0.438	**0.845**	
Procedural fairness	0.282	0.761	0.497	**0.852**

There are two ways to assess discriminant validity of the measurement instruments. Similar to assessing reliability, it can be accessed through the AVE as well as by comparing the square root of the AVE statistics to the correlations among the latent variables (Chin, 1998b). The purpose of this comparison stems from the definition of discriminant validity that is satisfied when a construct shares more variance with its measures than it shares with other constructs (Fornell & Larcker, 1981). Table 3 shows that the square roots of the AVEs are all greater than the respective correlations between latent variables. Another way to test the discriminant validity of the measurement models is to check whether the factor loadings of each item are higher on the construct it intends to measure than any other construct (Chin, 1998b). The results in Tables 4, 5, and 6 support adequacy of discriminant validity of the models.

Table 4. PLS Factor Loadings for Learning and Growth Perspective Model.

	Learning and Growth	Job Relevant Information	Role Clarity	Procedural Fairness
L&G1	**0.744**	0.316	0.252	0.156
L&G2	**0.808**	0.301	0.325	0.267
L&G3	**0.802**	0.324	0.343	0.235
L&G4	**0.773**	0.267	0.301	0.180
L&G5	**0.705**	0.287	0.289	0.173
JRI1	0.283	**0.775**	0.674	0.315
JRI2	0.408	**0.874**	0.626	0.374
JRI3	0.272	**0.835**	0.588	0.401
RC1	0.250	0.654	**0.836**	0.441
RC2	0.385	0.682	**0.878**	0/492
RC3	0.317	0.600	**0.732**	0.265
RC4	0.312	0.630	**0.863**	0.375
RC5	0.431	0.686	**0.917**	0.489
RC6	0.296	0.591	**0.829**	0.421
PF1	0.219	0.423	0.435	**0.861**
PF2	0.220	0.365	0.434	**0.859**
PF3	0.264	0.361	0.421	**0.835**
PF4	0.207	0.338	0.397	**0.850**

Table 5. PLS Factor Loadings for Internal Business Perspective Model.

	Learning and Growth	Job Relevant Information	Role Clarity	Procedural Fairness
L&G1	**0.867**	0.081	0.192	0.032
L&G2	**0.855**	0.040	0.111	−0.012
L&G3	**0.935**	0.149	0.181	0.088
L&G4	**0.942**	0.092	0.159	0.069
L&G5	**0.722**	0.055	0.118	0.115
JRI1	0.056	**0.779**	0.674	0.316
JRI2	0.157	**0.870**	0.626	0.374
JRI3	0.048	**0.837**	0.588	0.401
RC1	0.094	0.656	**0.837**	0.441
RC2	0.154	0.682	**0.878**	0.492
RC3	0.079	0.602	**0.731**	0.265
RC4	0.162	0.631	**0.863**	0.376
RC5	0.225	0.687	**0.918**	0.489
RC6	0.198	0.592	**0.830**	0.421
PF1	0.001	0.424	0.436	**0.863**
PF2	0.062	0.365	0.435	**0.861**
PF3	0.106	0.361	0.422	**0.833**
PF4	0.088	0.338	0.398	**0.850**

Table 6. PLS Factor Loadings for Customer Perspective Model.

	Learning and Growth	Job Relevant Information	Role Clarity	Procedural Fairness
L&G1	**0.701**	0.218	0.252	0.128
L&G2	**0.899**	0.189	0.280	0.183
L&G3	**0.839**	0.106	0.163	0.085
L&G4	**0.766**	0.108	0.201	0.085
L&G5	**0.871**	0.164	0.211	0.169
JRI1	0.164	**0.780**	0.674	0.315
JRI2	0.246	**0.871**	0.626	0.374
JRI3	0.089	**0.834**	0.588	0.401
RC1	0.181	0.655	**0.836**	0.441
RC2	0.251	0.682	**0.877**	0.492
RC3	0.209	0.601	**0.731**	0.265
RC4	0.206	0.631	**0.863**	0.375
RC5	0.306	0.687	**0.917**	0.489
RC6	0.265	0.592	**0.830**	0.421
PF1	0.093	0.423	0.435	**0.861**
PF2	0.176	0.364	0.434	**0.860**
PF3	0.176	0.360	0.421	**0.834**
PF4	0.129	0.337	0.397	**0.850**

Results of Structural Models

The results of the structural models are presented in Table 7 and Figs. 2, 3, and 4. The path coefficients showing the strength between latent variables were checked by running PLS estimation algorithms. The statistical significance of each path coefficient is assessed using a resampling procedure – bootstrapping. Bootstrapping generates subsamples by randomly choosing a case from the original data set for the purpose of estimating the precision of the PLS coefficients (Chin 1998a). This study uses a set of 1,000 replacements (Hartmann & Slapnicar, 2009). The results generated from the bootstrapping procedures are reported in Table 7 and Figures 2, 3, and 4.

The standard t-statistic criterion to assess the significance level is > 1.645 significance level at 0.05; > 1.96 significance level at 0.01. Based on the criteria, the results indicate that the learning and growth perspective and the customer perspective of nonfinancial measurements produce significant positive effect on job relevant information (Hypothesis 1) at the significance level of 0.01. However, the internal perspective of nonfinancial measurements does not have a significant effect on job relevant information. Hypothesis 1 is therefore not supported for the internal business perspective model.

Table 7. Path Coefficients.

Dependent Variable	Independent Variables	Hypotheses	Path Coefficients	Significance Level	Results of Hypotheses
Learning and growth perspective model					
JRI	LG	H1	0.390	0.01	Supported
RC	LG		0.118	ns	
	JRI	H2	0.714	0.01	Supported
PF	LG		0.071	ns	
	JRI		0.129	ns	
	RC	H3	0.714	0.01	Supported
Internal business perspective model					
JRI	Int	H1	0.106	ns	Not supported
RC	Int		0.099	ns	
	JRI	H2	0.754	0.01	Supported
PF	Int		−0.015	ns	
	JRI		0.133	ns	
	RC	H3	0.428	0.01	Supported
Customer perspective model					
JRI	Cus	H1	0.204	0.01	Supported
RC	Cus		0.133	ns	
	JRI	H2	0.734	0.01	Supported
PF	Cus		0.033	ns	
	JRI		0.143	ns	
	RC	H3	0.378	0.05	Supported

ns = not significant

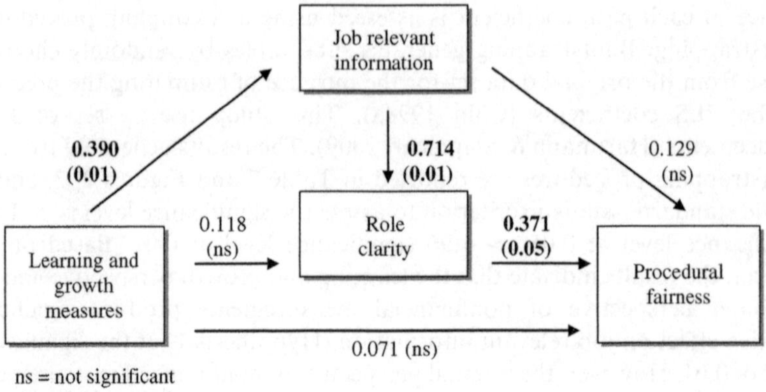

ns = not significant
parenthesis = p-value

Fig. 2. Learning and Growth Perspective Model: Path Coefficients.

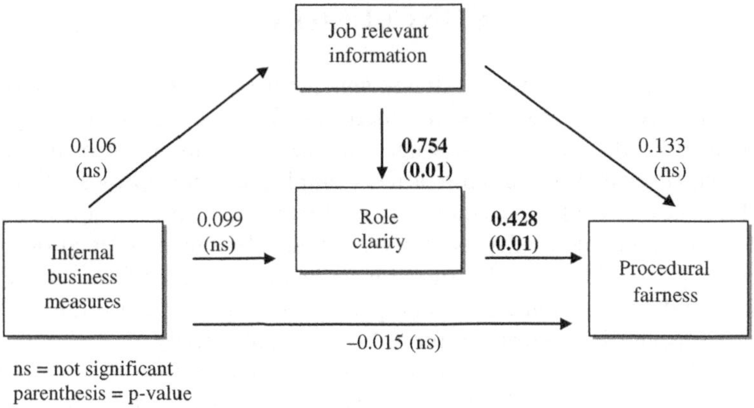

Fig. 3. Internal Business Perspective Model: Path Coefficients.

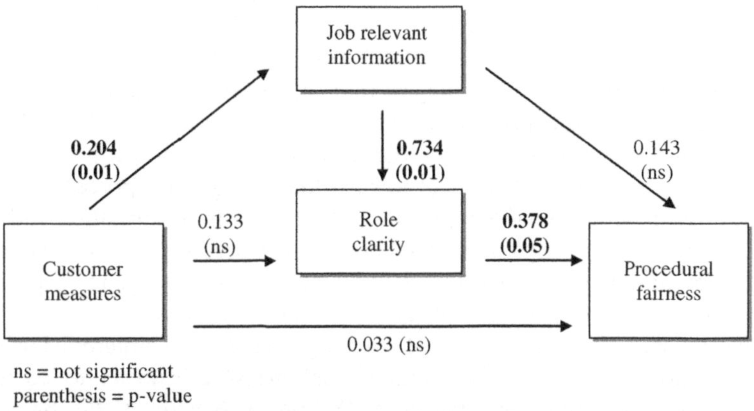

Fig. 4. Customer Perspective Model: Path Coefficients.

In addition, in the three models, job relevant information is positively associated with role clarity (Hypothesis 2) with significance level of 0.01. Hypothesis 3 is also supported by the result indicating that role clarity has a significant positive relationship with procedural fairness. Finally, Hypothesis 4 is supported as nonsignificant results indicate mediating effects. In summary, except for Hypothesis 1 for the internal business perspective, all other hypotheses are supported.

CONCLUSION

The aim of this study is to investigate *how* the use of nonfinancial measures affects employees' perceptions of procedural fairness. The focus is on the mechanism or the manner by which the use of nonfinancial measures as performance evaluation criteria affects employees' perceptions of procedural fairness. It also attempts to evaluate the roles of job relevant information and role clarity as intervening variables in these relationships. Specifically, it addresses the question of whether the effects of use of nonfinancial measure on employees' perceptions of procedural fairness are indirect through job relevant information and role clarity.

Using PLS, three structural models were tested to answer the research questions. The results indicate that they are generally consistent with the proposition that the relationships between the use of nonfinancial measures as performance evaluation criteria and procedural fairness is indirect through job relevant information and role clarity. These results are consistent with the arguments that cognitive mechanisms help to explain the effects of management control systems on behavioral consequences (Burney & Widener, 2007; Collins, 1982; Hall, 2008; Ilgen, Fisher, & Taylor, 1979). Specifically, the results suggest that the use of nonfinancial measures facilitates job relevant information acquisition by employees. Job relevant information, in turn, enhances role clarity, which, in turn, influences employees' perceptions of fairness toward their evaluation procedures.

These results underpin the importance of job relevant information and role clarity in improving employees' perceptions of procedural fairness. Overall, the use of nonfinancial measures as evaluation criteria engenders positive effects on procedural fairness. Analysis of the data is based on the three Balanced Scorecard perspectives of learning and growth, internal business process, and customer. Use of nonfinancial measure is likely to be most pervasive within these three perspectives (Kaplan & Norton, 1996). Based on the results, the learning and growth perspective of nonfinancial measures as performance evaluation method has the strongest impact on job relevant information among the three perspectives, as indicated by its highest standardized path coefficient. The customer perspective provides the second strongest significant influence on job relevant information. In contrast, the internal business process perspective has an insignificant effect on job relevant information. Thus, overall, the use of nonfinancial measures has positive effects on job relevant information. Using nonfinancial measures especially the measures of the learning and growth perspective and the customer perspectives will enhance job-related information to employees.

Our study contributes to prior research that examines the indirect effects of management control system components on employees' perceptions of procedural fairness. From a theoretical perspective, the study indicates that the development of theoretical models that include relevant cognitive variables, i.e., job relevant information and role clarity, can help to improve our understanding of how management control systems, i.e., nonfinancial measurement systems, affect employee fairness perceptions.

This study also extends the findings of several prior management accounting studies which have investigated the effects of comprehensive performance measurement systems involving a combination of financial and nonfinancial measures (e.g., Burney et al., 2009; Burney & Widener, 2007; Hall, 2008; Hartmann & Slapnicar, 2009). Apart from integrating the main variables investigated by different prior studies into a single comprehensive model, our study also focuses on the effects of nonfinancial measures *by themselves* without the confounding effects of financial measures. By isolating the effects of nonfinancial measures, our study is able to demonstrate that some of the effects found by prior studies may be derived from the use of nonfinancial measures by themselves. In other words, nonfinancial measures *by themselves* may be sufficient to promote positive behavioral consequences in their own rights. There may be practical significance in this finding. There is a belief that a combination of financial measures and nonfinancial measures is an essential requirement for effective performance measurement systems, a belief popularized by those who advocate comprehensive performance measurement systems such as the Balanced Scorecard. However, such comprehensive systems are costly and difficult to implement in practice. There is also the concern that the use of too many measures may cause information overload and confusion among employees who have to meet too many performance criteria. Our study provides empirical evidence to suggest that in some situations the use of nonfinancial measures alone may suffice.

While this study highlights the role of nonfinancial measures in promoting workplace fairness by recognizing the usefulness of cognitive mechanisms, the findings need to be interpreted in light of the limitations of the study. This study is based on survey data; therefore, the results are associated with the inherent limitation of the survey method. This research is also based on a sample of middle-level managers of large manufacturing organizations. Hence, the generalizability of the findings may not extend to other levels of employees and the nonmanufacturing sectors. Future research could be conducted in other industries such as the service industries to ascertain the generalizability of the findings of this study. Finally, this study has focused on the roles of cognitive variables, namely job relevant information and role clarity, in the relation between the use of nonfinancial measures and procedural fairness. It is likely that other variables such as organizational commitment, psychological empowerment, and interpersonal trust may also have important intervening effects. This provides opportunities for further research to investigate the possible mediating effects of these variables.

REFERENCES

Bernardin, H. J. (1979). The predictability of discrepancy measures of role constructs. *Personnel Psychology*, *32*(1), 139–153.

Burney, L. L., Henle, C. A., & Widener, S. (2009). A path model examining the relations among strategic performance measurement system characteristics, organizational justice, and extra- and in-role performance. *Accounting, Organizations and Society*, *34*, 2009.

Burney, L., & Widener, S. K. (2007). Strategic performance measurement systems, job-relevant information, and managerial behavioral responses-role stress and performance. *Behavioral Research in Accounting*, *19*, 43–69.

Chenhall, R. H. (2004). The role of cognitive and affective conflict in early implementation of activity-based cost management. *Behavioral Research in Accounting*, *16*, 19–44.

Chenhall, R. H., & Brownell, P. (1998). The effect of participative budgeting on job satisfaction and performance: role ambiguity as an intervening variable. *Accounting, Organizations and Society*, *13*(3), 225–233.

Chin, W.W. (1998a). Issues and opinion on structural equation modeling. *MIS Quarterly*, *22*(1), vii–xvi.

Chin, W. W. (1998b). The partial least squares approach for structural equation modeling. In G. A. Marcoulides (Ed.), *Modern methods for business research* (pp. 295–336). Mahwah, NJ: Lawrence Erlbaum Associates.

Chin, W. W., & Todd, P. A. (1995). On the use, usefulness, and ease of use of structural equation modeling in MIS research: A note of caution. *MIS Quarterly*, *19*(2), 237–246.

Chong, V. K., & Chong, K. M. (2002). Budget goal commitment and informational effects of budget participation on performance: A structural equation modeling approach. *Behavioral Research in Accounting*, *14*, 65–86.

Collins, F. (1982). Management accounting systems and organizational control: A role perspective. *Accounting, Organizations & Society*, *7*(2), 107–122.

Figler, H. R. (1980). Managing stress. *Management Accounting*, *62*, 22–28.

Fornell, C., & Larcker, D. F. (1981). Evaluating structural equation models with unobservable variables and measurement error. *Journal of Marketing Research*, *18*(1), 39–50.

Gregson, T., Wendell, J., & Aono, J. (1994). Role ambiguity, role conflict, and perceived environmental uncertainty: Are the scales measuring separate constructs for accountants. *Behavioral Research in Accounting*, *12*, 31–67.

Hair, J. F., Anderson, R. E., Tatham, R. L., & Black, W. C. (1998). *Multivariate data analysis*. Upper Saddle River, NJ: Prentice Hall.

Hall, M. (2008). The effect of comprehensive performance measurement systems on role clarity, psychological empowerment and managerial performance. *Accounting, Organizations & Society*, *33*(2/3), 141–163.

Hartmann, F., & Slapnicar, S. (2009). How formal performance evaluation affects trust between superior and subordinate managers. *Accounting, Organizations & Society*, *34*(6/7), 722–737.

Hopwood, A. G. (1972). An empirical study of the role of accounting data in performance evaluation. *Journal of Accounting Research*, *10*(3), 156–182.

House, R. J., & Rizzo, J. R. (1972). Role conflict and ambiguity as critical variables in a model of organizational behavior. *Organizational Behavior & Human Performance*, *7*(3), 467–505.

Ilgen, D. R., Fisher, C. D., & Taylor, M. S. (1979). Consequences of individual feedback on behavior in organizations. *Journal of Applied Psychology*, *64*(4), 349–371.

Johnson, T. H., & Kaplan, R. S. (1986). *Relevance lost: The rise and fall of management accounting*. Boston, MA: Harvard Business School Press.

Kahn, R. L., Wolfe, D. M., Quinn, R. P., Snoek, J. D., & Rosenthal, R. A. (1964). *Occupational stress: Studies in role conflict A and ambiguity*. New York, NY: Wiley.

Kaplan, R. M. (1987). *Basic statistics for the behavioral sciences*. Newton, MA: Allyn and Bacon, Inc.

Kaplan, R. S. (1984). The evolution of management accounting. *Accounting Review*, *59*(3), 390.

Kaplan, R. S., & Atkinson, A. (1998). *Advanced management accounting*. Upper Saddle River, NJ: Prentice Hall.

Kaplan, R. S., & Norton, D. P. (1996). *The balanced scorecard*. Boston, MA: Harvard Business School Press.

Konovsky, M. A., & Folger, R. (1989). Effects of procedural and distributive justice on reactions to pay raise decisions. *Academy of Management Journal*, *32*, 115–130.

Kren, L. (1992). Budgetary participation and managerial performance: The impact of information and environmental volatility. *Accounting Review*, *67*(3), 511–526.

Lau, C. M., & Lim, E. W. (2002). The effects of procedural justice and evaluative styles on the relationship between budgetary participation and performance. *Advances in Accounting*, *19*, 139–160.

Lau, C. M., & Moser, A. (2008). Behavioral effects of non-financial performance measures: The role of procedural fairness. *Behavioral Research in Accounting*, *20*(2), 55–71.

Lau, C. M., & Tan, L. C. (2006a). The effects of procedural fairness and interpersonal trust on job tension in budgeting. *Management Accounting Research*, *17*(2), 171–186.

Lau, C. M., & Tan, L. C. (2006b). The effects of participation and job relevant information on the relationship between evaluative style and job satisfaction. *Review of Quantitative Finance and Accounting*, *21*(1), 17–34.

Lau, C. M., Wong, K. M., & Eggleton, I. (2008). Fairness of performance evaluation procedures and job satisfaction: The role of outcome-based and non-outcome-based effects. *Accounting and Business Research*, *38*(2), 121–135.

Leventhal, G. S. (1980). What should be done with equity theory? New approaches to the study of fairness in social relationships'. In K. Gergen, M. Greenberg & R. Willis (Eds.), *Social exchange: Advances in theory and research*. New York, NY: Plenum Press.

Libby, T. (1999). The influence of voice and explanation on performance in a participative budgeting setting. *Accounting, Organizations and Society*, *24*, 125–137.

Lind, E. A., & Tyler, T. R. (1988). *The social psychology of procedural justice*. New York, NY: Plenum Press.

Lindquist, T. M. (1995). Fairness as antecedent to participative budgeting: Examining the effects of distributive justice, procedural justice and referent cognitions on satisfaction and performance. *Journal of Management Accounting Research*, *7*, 122–147.

Lipe, M. G., & Salterio, S. E. (2000). The balanced scorecard: Judgmental effects of common and unique performance measures. *Accounting Review*, *75*(3), 283.

Magner, N., Welker, R. B., & Campbell, T. L. (1996). Testing a model of cognitive budgetary participation processes in a latent variable structural equations framework. *Accounting & Business Research*, *27*(1), 41–50.

McFarlin, D. B., & Sweeney, P. D. (1992). Research notes: Distributive and procedural justice as predictors of satisfaction with personal and organisational outcomes. *Academy of Management Journal*, *35*(3), 626–637.

Nunally, J. C. (1978). *Psychometric theory*. New York, NY: McGraw-Hill.

Oppenheim, A. N. (1992). *Questionnaire design, interviewing and attitude measurement*. London: Pinter Publishers.

Otley, D. T. (1978). Budget use and managerial performance. *Journal of Accounting Research*, *16*(1), 122–149.

Parasuraman, S., & Alutto, J. A. (1981). An examination of the organizational antecedents of stressors at work. *Academy of Management Journal*, *24*(1), 48–67.

Renn, R. W. (2003). Moderation by goal commitment of the feedback-performance relationship: Theoretical explanation and preliminary study. *Human Resource Management Review*, *13*(4), 561–580.

Rizzo, J. R., House, R. J., & Lirtzman, S. (1970). Role conflict and ambiguity in complex organizations. *Administrative Science Quarterly*, *15*(2), 150–163.

Sholihin, M., & Pike, R. (2009). Fairness in performance evaluation and its behavioural consequences. *Accounting and Business Research*, *39*(4), 397–413.

Simon, H. A., Guetskow, H., Kozmetsky, G., & Tyndall, G. (1954). *Centralization vs. decentralization in organizing the controller's department*. New York, NY: Controllership Foundation, Inc.

Sprinkle, G. B. (2003). Perspectives on experimental research in managerial accounting. *Accounting, Organizations & Society*, *28*(2/3), 287–318.

Tiessen, P., & Waterhouse, J. H. (1983). Towards a descriptive theory of management accounting. *Accounting, Organizations & Society*, *8*(2/3), 251–267.

Wentzel, K. (2002). The influence of fairness perceptions and goal commitment on managers' performance in a budget setting. *Behavioral Research in Accounting*, *14*, 247–271.

PART VI
NEW TRENDS – GOVERNANCE AND NON-PROFIT ORGANIZATIONS

CONTROL OF DIRECTORS' COMPENSATION IN SPANISH COMPANIES: CORPORATE GOVERNANCE AND FIRM PERFORMANCE

Elena Merino, Montserrat Manzaneque and Regino Banegas

ABSTRACT

Purpose – *The purpose of this chapter is to examine the hypothesized effects of board characteristics and performance on directors' compensation in the Spanish corporations, whose corporate governance is a special example of a unitary board system.*

Methodology/approach – *In order to test the influence of a set of factors on directors' compensation levels, we have developed several models based on linear panel data regression. The sample included 76 listed companies on the Spanish computerized trading system or Continuous Market for the period 2004–2009.*

Performance Measurement and Management Control: Global Issues
Studies in Managerial and Financial Accounting, Volume 25, 391–425

ISSN: 1479-3512/doi:10.1108/S1479-3512(2012)0000025018

Findings – *The control mechanisms, like board characteristics and performance and their effect on the level of directors' compensation, depend on the types of director (executive, independent and proprietary).*

Research limitations/implications (if applicable) – *This study has certain limitations mainly related to problems associated with obtaining information. The methodology should be complemented by other types of analyses, such as the influence of the characteristics of the board on the remuneration structure in a greater level of disaggregation.*

Practical implications (if applicable) – *The results of this research chapter give reasons to regulators and investors to be aware of the importance of the board's characteristics as corporate control mechanisms over the directors' remuneration and the necessity of connection between directors' compensation and the firm's performance.*

Originality/value of paper – *Firstly, descriptive empirical evidence on the level of directors' compensation is provided within a unitary board system for different types of directors. Secondly, an ample panel data set enables the examination of a set of determinants using panel data methods which control for unobserved firm heterogeneity. Finally, the perspective is extended from executive director compensation to other types of directors, such as proprietary or independent, which are very important features of the Spanish board structure.*

INTRODUCTION

Agency problems have once again become relevant due to excesses in compensation committed by numerous boards of directors, as has recently been demonstrated. Some corporate scandals (Enron, WorldCom, Nortel Networks and AIG, among others) have questioned the behaviour of a number of boards of directors in relation to their performance in their function of supervision and control. In fact, in some of these cases, the salaries received by directors came from artificially obtained benefits at the same time that these corporations were facing financial and economic difficulties, or, worse still, were bankrupt (Ingley & Walt, 2005). Nowadays, this problem has intensified the lack of investor confidence in their advisers, whose compensation levels have been criticized, especially over the last few years.

From the agency theory point of view, the role of the board of directors is critical to supervision and control over management, preventing opportunistic behaviour of shareholders and controlling shareholders against minority shareholders. However, previous research (Fama & Jensen, 1983) suggests that excessive compensation awarded to directors leads to increased laxity in their monitoring and control functions. In this sense, the fast growth of directors' compensation in recent years casts doubt on the efficacy of the board.

In concrete terms, the board of directors is responsible for determining the amount of compensation received by directors. Because of this, the role of directors and the adequacy of their incentive schemes have been called into question. In fact, several researchers have argued that directors have too much discretion and are able to influence their compensation contracts (Bebchuk & Fried, 2004) and that powerful directors are even able to use their compensation to extract income out of the firm in order to satisfy their private benefits (Duffhues & Kabir, 2008). Therefore, an important issue in corporate governance is whether or not the failure of some mechanisms furnishes board members with the possibility of establishing a reward system that is more suitable to their personal interests.

Despite the importance of this issue, directors' compensation has only very recently started to receive attention. In contrast, management compensation has been widely discussed over the years (Andreas, Rapp, & Wolff, 2009; Brick, Palmon, & Wald, 2006).

In this sense, this work adopts an empirical approach to examine the hypothesized effects of board characteristics and performance on directors' compensation in Spain, whose corporate governance is a special example of a unitary board system, like the United Kingdom and Italy. In this context, a unique panel of data has been put together from 76 listed Spanish companies for the period 2004–2009, in order to test the influence of a set of factors on director compensation levels. The findings of this research work reveal that the average compensation per executive director has gone up around 42% from 2008 to 2009. These results are in stark contrast to the problems suffered by the rest of the economy over the past two years, hence the understandable interest that the analysis of the determinants of payments received by directors has aroused.

Drawing on agency theory arguments and previous literature, a comprehensive group of determinants is analysed, which are representative of corporate performance and board characteristics. It was found that the control mechanisms, like board characteristics and performance and their effect on the level of directors' compensation, depend on the type of director.

The work contributes to the existing literature in several ways. Firstly, descriptive empirical evidence on the level of directors' compensation is provided within a unitary board system for different types of directors, which is a very important feature of the Spanish board structure. So while there are several studies regarding executives' compensation, there is only scarce empirical evidence on the problem of independent or proprietary directors, which contribute to guaranteeing the level of independence of the board. Secondly, an ample panel data set enables the examination of a set of determinants using panel data methods which control for unobserved firm heterogeneity.

This work is organized accordingly: firstly, a review of previous literature on the subject is carried out; secondly, the research design is set out, defining the sample and the variables under study; thirdly, the application of relevant statistical techniques is dealt with and the main results are analysed; and finally, the main conclusions are discussed.

SPANISH DIRECTORS' COMPENSATION

Spanish Directors' Compensation Practices

In Spain, the typical board structure is composed of internal (*inside*) and external (*outside*) directors. In the second case, there are two categories: (a) independent: those who are in a position to perform their duties without being influenced by any connection within the company, its shareholders or its management and (b) proprietary: those who own equity state above or equal to the legally determined threshold for a significant holding, or otherwise appointed due to their status as shareholders or those who represent shareholders (*Unified Good Governance Code*, CNMV or Spanish Securities Markets Commission, 2006) (Fig. 1).

The size of the board of directors on average is 10.4 with a maximum of 24 members in 2009 (CNMV, 2009a). According to Heidrick & Struggles Report (2009), Spanish boards are situated in third place in relation to board size, behind countries like Germany and France, but the representatives of reference shareholders make up 43% of the total, which is way above these countries (Germany, 16% and France, 17%).

With regard to the presence of different types of directors on the board of directors, outside directors dominate (with an average of 81.7% in year 2009), whereas independent directors only comprise 30.5% (CNMV, 2009a).

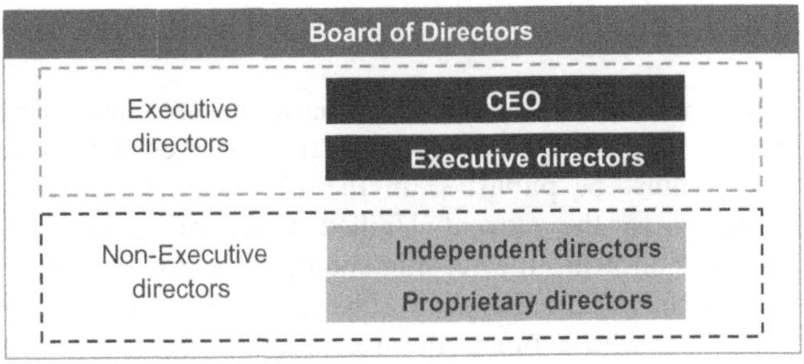

Fig. 1. Typical Board Structure in Spain.

On average, the compensation received by each member of the board of directors has gone up by 3.19% in the period 2008–2009.

In general terms, the structure of remuneration of directors may comprise the following concepts: fixed salary, attendance fees, variable salary and benefits in respect of bylaws, stock options and/or other financial instruments and other compensations.[1]

As in other European countries, the Spanish compensation system is characterized by fixed salary and attendance fees. In fact, in Europe, the percentage of cash pay comprises 83% of the total remuneration received by members of the board of directors (Heidrick & Struggles, 2009). In contrast, in countries like United States, salary only comprises just over one-quarter of the total remuneration and there is a big amount of compensation linked to the company performance.

In relation to the level of compensation, Spain is one of the European countries – along with Switzerland, Germany or the United Kingdom – with higher remuneration per director; in concrete terms, €108.000 per director, where fixed fees represent 79% of the total (Heidrick & Struggles, 2009). According to the *Corporate Governance Annual Report* (CNMV, 2009a), the compensation received by executives has gone up by 11.7% with respect to 2008. The same behaviour has occurred with independent director compensation, which has been increased, on average, by 6.5%. In contrast, the proprietary director compensation has gone down by around 6%.

To sum up, the main characteristics of the Spanish board of directors' compensation system are (a) it is mainly made up of fixed concepts and is insufficiently linked to the company's performance and (b) the level of compensation is considerable and has increased in particular for executives and independent members of the board.

Excesses in Compensation in Spain: Recent Normative Changes

In the Spanish normative context, the board of directors is responsible for determining the amount of compensation earned by its members. These amounts must be published in the *Corporate Governance Annual Report* according to the current Unified Code of Good Governance (CNMV, 2006). Following the recommendations of the previous Code, it is habitual that the companies have a special Nomination and Remuneration Committee, whose functions are, among others, to evaluate the balance between skills, knowledge and experience on the board and define the roles and capabilities required of the candidates to fill each vacancy; examine or organize the succession of the chairman and the chief executive officer (CEO); and, finally, to make proposals to the board of directors regarding the remuneration policy of directors and the individual remuneration and other contractual conditions of executive directors (CNMV, 2006). However, in spite of these normative requirements, the remuneration of directors is currently considered to be excessive in some cases.

To avoid this situation, recent regulatory attempts to control the opportunistic behaviour of directors have been oriented primarily towards increasing the transparency of information regarding the board's remuneration and the inclusion of recommendations about the structure and composition of this remuneration, as well as the strengthening of corporate control mechanisms based on the characteristics of the board of directors.

As such, the recent Law on Sustainable Economy (2011) includes the obligation to disclose the individual remuneration granted to each director, with the aim of ending the lack of transparency surrounding this matter. In addition, the Spanish Securities Markets Commission (CNMV) in its role as a supervisor of markets is considering the possibility of making certain modifications to the Unified Good Governance Code (CNMV, 2009b) in order to include the recommendations of the Green Paper published by the European Commission (2010) on corporate governance in financial institutions and remuneration policies. This chapter deals with issues relating to reducing the number of directors, the separation of the roles of the chairman and the CEO, external evaluation of the board or the determination of content and type of compensation, among others. The European Union (3177/2009/CE) proposals also make reference to the link between variable compensation and performance criteria, and compliance with this in the long term;

the establishment of limits on the compensation given for termination of contract; and the limitation of equity-based compensation awarded to executive directors, as a way of controlling the remuneration excesses of the board members (European Commission, 2009).

In summary, all of these normative advances highlight the importance of an empirical approach in order to examine the influence of corporate governance and performance on the remuneration level of board members in the Spanish context.

PREVIOUS RESEARCH ON THE CONTROL OF DIRECTORS' COMPENSATION

As in the case of top management, compensation is one of the most important incentive mechanisms to align interests between directors and stockholders. A number of previous studies, however, show that *excessive compensation* could contribute to a lack of independence and control of the board over the management and expropriation of stockholders' wealth through this compensation.

The literature has emphasized, inter alia, the risk of 'capture by the first executive' and the risk of 'collusion with other executives' that can be suffered by board members (Paz-Ares, 2008). As a matter of fact, the inclusion of top management among board members may give rise to an increase of these risks. For instance, the CEO usually controls the process by which directors are elected to the board (Berle, 1962; Galbraith, 1971; Herman, 1981) and the directors' compensation. In this sense, the directors could serve according to the wishes of the CEO (Crystal, 1991) in return for a seat on the board or more compensation.

In order to avoid these situations, the literature proffers two alternatives. The first one is to increase the grade of *board independence* through different mechanisms related to its structure (Andreas et al., 2009; Cheng & Firth, 2005; Conyon & He, 2008; Fahlenbrach, 2009; Ryan & Wiggins, 2004): (a) to increase the shareholding of board members, (b) to separate the roles of both chairman of the board (or president) and CEO, (c) to include a certain number of independent members and (d) to moderate the size of the board. In this sense, the empirical evidence shows that certain board structures reduce agency problems (Hermalin & Weisbach, 1991; Yermack, 1996) and, consequently, it is expected that they could be relevant in explaining the level and structure of directors' compensation.

From a theoretical point of view, the *managerial power approach* developed by Bebchuk and Fried (2004) confirms the previous idea. They argue that executives generally have powers in one form or another and can therefore extract a certain amount in terms of income. The particular characteristics of a firm, however, especially the board structure, furnish its executives with a greater or fewer powers. According to this theoretical perspective, the characteristics of board of directors are relevant in explaining the directors' compensation.

The second alternative is related to the question of how the members of the board of directors are compensated. Some theoretical perspectives argue that if the directors receive adequate compensation, the assumption is made that they work harder and contribute more to the increase in corporate performance (Duffhues & Kabir, 2008). In fact, the agency theory provides a positive relationship between corporate performance and the level of director compensation in order to align incentives between directors and shareholders. One way of aligning interests between directors and the rest of the shareholders is to make director compensation a function of firm performance (Fernandes, 2005). Thus, directors will be motivated to contribute more to the increase in corporate performance (Duffhues & Kabir, 2008), on the basis of which the link between directors' compensation and performance could limit excesses in remuneration, in particular, when the economic and financial situation of the company is not good. Such a mechanism could avoid certain situations that are currently occurring and that shareholders perceive as an expropriation of their wealth.

To sum up, previous documented research puts into evidence that director compensation is associated with corporate governance mechanisms and performance.

BOARD CHARACTERISTICS AND THEIR HYPOTHESIZED EFFECTS ON DIRECTORS' COMPENSATION

The measures that are frequently used to review the effectiveness of the board's governance and their independence are (Andreas et al., 2009; Cheng & Firth, 2005; Conyon & He, 2008; Fahlenbrach, 2009; Ryan & Wiggins, 2004) the participation of the board of directors in shareholding, CEO and board chair duality, the inclusion of independent members

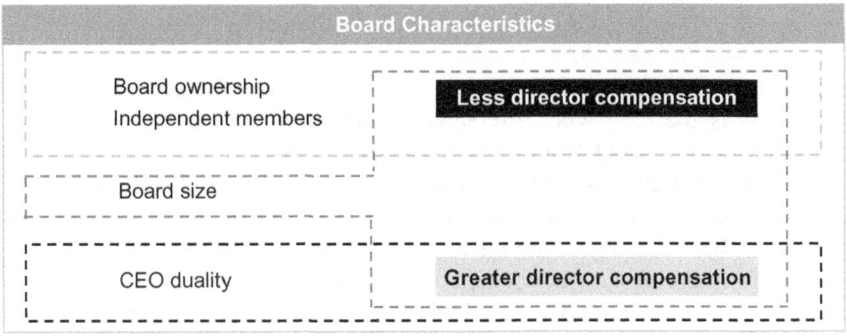

Fig. 2. Board Characteristics and Their Effects on Directors' Compensation.

and the size of the board. As mentioned above, these variables can significantly affect the capacity of the board of directors to act efficiently (Fig. 2).

Board Ownership

The Agency Theory supports the idea that the participation of the board of directors in shareholding is a powerful incentive to exert control (Shleifer & Vishny, 1997), achieving the alignment of directors' interests with those of other shareholders. In relation to director compensation, theoretically, in firms with a higher degree of board or management ownership, both directors and management have a greater incentive to maximize stock returns. This incentive downplays the importance of high compensation (Cordeiro, Veliyath, & Eramus, 2000; Cheng & Firth, 2005).

Empirically, some authors (Boyd, 1996; Bryan, Hwang, Klein, & Lilien, 2000; Cordeiro et al., 2000) found director compensation to be lower in firms in which a substantial proportion of shares is held by the board or by management. However, in Spain, Arrondo, Fernández, and Fernández (2008) did not find a significant relationship between the compensation of directors and shareholding.

In keeping with the most common results, the following hypothesis is proposed:

H1. There is a negative relationship between board ownership and director compensation.

CEO and Board Chair Duality

One of the relevant aspects of the board's structure is the accumulation of power, as is the case when the same person holds down the position of chairman of the board and CEO of the corporation.

Though theoretically the accumulation of power could facilitate the transmission of information and reduce the costs of coordination, this aspect has been the object of considerable criticism, given that it could lead to decision-making to the benefit of management over and above the company's shareholders (Coles, Mcwilliams, & Sen, 2001; Jensen, 1993). In fact, previous empirical studies suggest that agency problems are greater when the chairman of the board is, at the same time, the CEO (Wang, Chuang, & Lee, 2010; Yermack, 1996), which negatively affects the actual control of the board (Morck, Shleifer, & Vishny, 1988) and their effectiveness (Hermalin & Weisbach, 2003).

In the area of director remuneration, such inefficiency has been demonstrated by Brick et al. (2006) who found that the accumulation of power is positively related to director compensation.

Taking these arguments into consideration, we propose to test the following hypothesis:

H2. There is a positive relationship between CEO duality and directors' compensation.

Outside Directors

Authors like Conyon and He (2008) point out the role of the outsiders as a control mechanism; due to their experience in decision-making, they are less exposed to influences and have a reputation to maintain in the job market. This leads us to think that the outside members of the board are more effective in looking after the interests of shareholders. In the Spanish context, Sánchez and Lucas (2008) found a negative relationship between the number of outside directors (proprietary and independent) and director compensation. Similarly, Arrondo et al. (2008) found that the percentage of independent outside directors has a negative correlation with the level of compensation of directors. This is interpreted accordingly – that the presence of independent directors encourages moderate remuneration more in line with the interests of shareholders.

This suggests the following hypothesis:

H3. There is a negative relationship between outside directors and director compensation.

Board Size

In general, there is no consensus in relation to the size of the board of directors to optimize the efficiency of its independence. Some authors, like Pearce and Zahra (1992) and Lehn, Patro, and Zhao (2003), coincide, pointing out that a large number of members favour the diversity of criteria and the level of information about the factors that could affect the value of the corporation. Yet others, like Lipton and Lorsch (1992) and Jensen (1993), argue that a larger board implies less speed and efficiency in the decision-making process, due to the fact that some problems related to coordination and information could emerge.

In fact, the results of previous studies are not unanimous regarding the relationship between the size of the board and director compensation; so while Brick et al. (2006) found that the number of directors is not significantly related to director compensation, Ryan and Wiggins (2004) and Andreas et al. (2009) found a significant but inverse relationship between board size and director compensation. In the Spanish context, Sánchez and Lucas (2008) found a positive relationship between board size and board compensation.

As such, there is no consensus regarding the direction of this relationship in previous literature, thus the following hypothesis may be proposed:

H4. There is a positive or negative relationship between board size and director compensation.

PERFORMANCE AND THEIR HYPOTHESIZED EFFECTS ON DIRECTORS' COMPENSATION

At empirical level, there is no conclusive evidence of a large and positive pay–performance relationship as literature suggests. Some studies find positive but low pay–performance relationship (Banghoj, Gabrielsen, Peterson, & Plenborg, 2010; Brunello, Graziano, & Parigi, 2001; Buck, Bruce, Main, & Udueni, 2003; Kato, Kim, & Lee, 2004; Kato & Kubo, 2006; Linck, Netter, & Yang, 2008; Main, Bruce, & Buck, 1996). For example, Linck et al.

(2008) find that stock performance is positively correlated with total director pay for a sample studied in the period post Sarbanes–Oxley Act. For its part, Banghoj et al. (2010) also find a positive but weak pay–performance relationship. Lazarides, Dimpretas, and Dimitrios (2008) confirmed in their study the positive relationship between financial performance and remuneration. The study used three variables. Two of them are based on accounting measures (ROA and Total Assets) and the third is Tobin's *Q*, which is mainly based on stock market value. Tobin's *Q* is the only variable that was not statistically significant in the final model. Andreas et al. (2009) find a positive relationship between performance (dividend and accounting-based measures) and the level of directors' compensation.

In the Spanish context, Crespí and Gispert (2003) find a positive relationship between board remuneration and company performance, which is stronger for book values (ROA) than for stock market measures (stock return) for a sample of Spanish listed companies during the period 1990–1995.

On the contrary, other studies document non-existent relationship (Cordeiro et al., 2000; Fernandes, 2005; Haid & Yurtoglu, 2006) between performance and directors' compensation. For example, Cordeiro et al. (2000) find no relationship between stock return and total compensation per director. In similar terms, Fernandes (2005) find that not even variable compensation is related to performance, either market-based, or accounting-based for a large sample of Portuguese firms during the period 2002–2004. The work of Duffhues and Kabir (2008) fails to detect a positive pay–performance relationship for a large sample of Dutch listed firms during the period 1998–2001.

The following three hypotheses related to the performance measures and the level of director remuneration have therefore been tested (Fig. 3):

H5. There is a positive relationship between return on assets and director compensation.

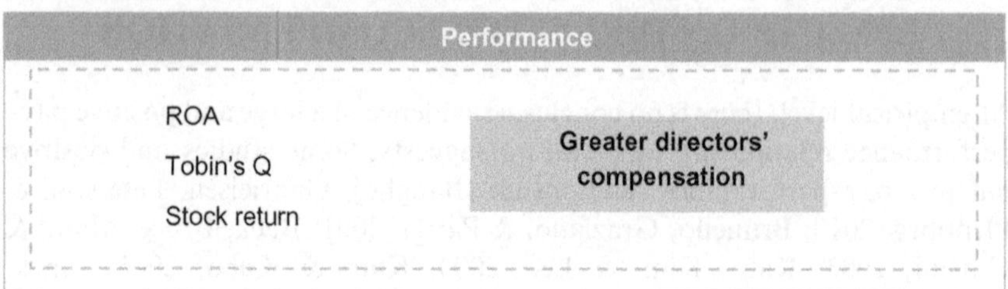

Fig. 3. Performance and Their Effects on Directors' Compensation.

H6. There is a positive relationship between stock return and director compensation.

H7. There is a positive relationship between the Tobin's Q and director compensation.

SAMPLE, VARIABLES AND METHODOLOGY

Sample

The sample included all listed companies on the Spanish computerized trading system (SIBE) or Continuous Market for the period 2004–2009. All finance-related firms (bank, insurance, unit trusts and utilities companies) are excluded from the sample due to their differences in terms of regulatory requirements, financial reporting standards and compliance, in keeping with Sulong and Mat Nor (2010). As such, the sample comprises 76 firms out of a total of 125 listed companies which represent a broad range of industry sectors (see Table 1). To verify the representative nature of the sample, the maximum allowable error[2] for a finite population was estimated. The maximum error is small, 7.07% to be exact, with a level of confidence of 95% ($p = 5\%$), leading to the consideration that the sample is representative of the population.

Table 1. Composition of the Population and Sample Firms According to the Industry Type.

	Listed Companies on the Spanish Computerized Trading System		Sample	
	N	%	*N*	%
1. Oil and energy	20	16.00	11	14.47
2. Basic materials, manufacturing and construction	40	32.00	23	30.26
3. Consumer goods	36	28.80	26	34.21
4. Consumer services	20	16.00	11	14.47
5. Technology and telecommunications	9	7.20	5	6.58
Total	125	100.00	76	100.00

Source: Spanish computerized trading system (SIBE) or Continuous Market. Retrieved from http://www.bolsamadrid.es.

Regarding the issue of excesses of compensation awarded to directors, the Spanish context is especially interesting, given the level of concentration of ownership and the characteristics of the board and the increasing political pressure to encourage the level of transparency and reasonable nature of remuneration systems. These are the main reasons why this geographic and normative context has been selected.

Therefore, to test the hypotheses, data was obtained on corporate governance characteristics, compensation given to directors and performance measures from these 76 Spanish firms. As sources of information, the CNMV database, the Annual Report regarding Corporative Governance and Annual Accounts of each corporation were used. On occasions, it has also been necessary to consult other sources of information, such as the web pages of the corporations, information gathered from the registry of the Madrid Stock Exchange, the Barcelona Stock Exchange or information from the SABI database.

Variables

The variables selected for the analysis are shown in Table 2.

As the dependent variable, the level of compensation given to board's members has been used. The total compensation includes the following concepts: fixed and variable remuneration, salary fees, attendance fees, travelling indemnity, stock options and/or other financial instruments (Crespí & Pascual-Fuster, 2008; Haid & Yurtoglu, 2006; Kato & Long, 2004). Following the standard approach in other studies (Boyd, 1994; Cheng & Firth, 2005; Finkelstein & Hambrick, 1989), a logarithmic transformation of the dependent variable was applied to reduce heteroscedasticity. In the same way, remuneration concepts like redundancy payment or insurance premiums were not included, due to their sporadic character and high amount, which could distort the results and make their interpretation difficult.

Furthermore, the dependent variable has been categorized into three different types of compensation, according to the types of director (executive, independent and proprietary).

As independent variables, several measures concerning the board of directors' characteristics and corporate performance indicators are proposed. The first four variables are related to *board of directors' characteristics*, which have been widely used in previous studies (Boyd, 1994, 1996; Bryan et al., 2000; Coles, Daniel, & Naveen, 2008; Conyon & He, 2008; Cordeiro et al.,

Table 2. Definition and Typology of the Variables.

Variables	Definition	Typology
Compensation variables		
LNCOMP	Natural log of total compensation by member of the board of directors	Numeric
LNEXECUTIVES	Natural log of total compensation by executive member of the board of directors	Numeric
LNINDEPENDENT	Natural log of total compensation by independent of the board of directors	Numeric
LNPROPRIETARY	Natural log of total compensation by proprietary of the board of directors	Numeric
Board characteristics		
OWNERDIRECT	Proportion of shares owned by the board of directors	Numeric
CEODUALITY	Dummy variable which takes value of 1 when both roles are held by the same person, and 0 when they are not	Dichotomic
OUTSIDERS	Proportion of outside directors on the board of directors	Numeric
BRDSIZE	Number of members in the board of directors	Numeric
Firm performance variables		
ROA	Return on assets, ratio of operating income to net assets	Numeric
TOBINQ	Tobin's Q measured as the sum of market value of common shares and book value of debt to book value of total assets according to Duffhues and Kabir's (2008) specifications.	Numeric
STOCKRET	Stock return measured as the sum of stock price and dividend per share over stock price in the year before	Numeric
Control variables		
CRPSIZE	Corporate size measured by the logarithm of total assets	Numeric
INDUSTRY	1. Oil and energy 2. Basic materials, manufacturing and construction 3. Consumer goods 4. Consumer services 5. Technology and telecommunications	Dichotomic

2000; Core, Holthausen, & Larcker, 1999; Dávila & Peñalva, 2004; Ertugrul & Hedge, 2008; Larcker, Richardson, & Tuna, 2007; Li, 1994; Mínguez & Martín, 2003; Sánchez & Lucas, 2008).

A second group of independent variables are related to the *corporate performance* indicator. Performance measures can be related to accounting performance (e.g. return on assets, return on equity) and to the capital market performance (e.g. total shareholder return, return on adjusted market). In order to take a wide range of performance variables, three different measures are used. As an accounting performance measure, the *return on assets* (ROA) is proposed, which has been widely used by previous studies (Andreas et al., 2009; Angbazo & Narayanan, 1997; Arrondo et al., 2008; Dávila & Peñalva, 2004). However, this kind of measure is subject to management manipulation (Kren & Kerr, 1997), so two other variables have been included. One of them represents capital market performance – the *annual stock return* – that is measured as the sum of stock price and dividend per share over stock price in the year before (Andreas et al., 2009; Brick et al., 2006; Cordeiro et al., 2000; Duffhues & Kabir, 2008; Ryan & Wiggins, 2004). Finally, the last variable is the *Tobin's Q* measure, which is a hybrid of accounting and capital market-based measures (Duffhues & Kabir, 2008). This variable is defined as the ratio of the sum of market value of common shares and book value of debt to book value of total assets.

The last group of variables proposed is 'control variables'. These variables are *firm size* and *industry*. In order to make the sector classification, the approach established for companies listed on any Spanish Stock Exchange and that came into effect on 1 January 2005 was adopted. This classification distinguishes five sectors, excluding financial services and the real estate sector.

Methodology

In order to test the hypothesis, several dependence models were developed based on linear panel data regression, which is applied to multiple phenomena observed over multiple periods for the same firms (Hsiao & Lightwood, 1994). Following this methodology, a data sample of 456 (76 companies × 6 years) was developed, which is a short ($T = 6$), lineal and strongly balanced panel.

More specifically, to evaluate the validity of the models, the application of fixed and random-effects methodologies was checked using the Hausman test (1978). This test verifies the absence of the correlation between the individual effects and the independent variables, so when the null hypotheses is not rejected it indicates that the higher degree of efficiency in the estimation leads to the use of the random-effects model (Andreas et al., 2009; Croissant & Millo, 2008).

This enables the unobservable but constant heterogeneity to be taken into consideration, due to it being a common problem in identifying the effect of corporate governance mechanisms and performance over board's compensation (Andreas et al., 2009; Elsas & Florysiak, 2008; Mayers & Smith, 2010).

In accordance with what has been put forward, different variants of the following model of panel data have been estimated[3]:

$$y_{it} = \alpha + \sum_{k=1}^{4} \beta X_{it} + \sum_{j=1}^{3} \beta P_{it} + \sum_{m=1}^{2} \beta CV_{it} + u_i + \varepsilon_{it} \quad (1)$$

where y_{it} is the endogenous variable (logarithm of remuneration by director); $\sum_{k=1}^{4} \beta X_{it}$ are independent variables representative of board characteristics, where $k =$ board ownership (OWNERDIRECT$_{it}$), duality of the chairman of the board and the CEO (CEODUALITY$_{it}$), proportion of external board members (OUTSIDERS$_{it}$), board size (BRDSIZE$_{it}$); $\sum_{j=1}^{3} \beta P_{it}$ are independent variables representative of corporate performance, where $j =$ return on assets (ROA$_{it}$), Tobin's Q ratio (TOBINQ_{it}) and stock return (STOCKRET$_{it}$); $\sum_{m=1}^{2} \beta CV_{it}$ are control variables, where $m =$ corporate size (CRPSIZE$_{it}$), and industry dummies (INDUSTRY$_{it}$), and u_{it} is the individual error component and ε_{it} is the idiosyncratic error.

Therefore, four variants of the model were estimated based on the type of director, called 'total compensation per director' (model 1), 'total compensation by executive director' (model 2), 'total compensation by independent director' (model 3) and 'total compensation by proprietary director' (model 4).

In terms of the hypotheses, it is therefore expected that $\beta_1 < 0$ (H1) will represent a positive relationship between the board ownership and directors' compensation. A negative relationship is also expected to exist between the proportion of external board members and compensation per director ($\beta_3 < 0$, H3) indicating that greater independence in the board has a moderate effect over the amount of compensation received by them. On the contrary, a positive relationship is foreseen between duality ($\beta_2 > 0$, H2) and compensation per director. As such, a specific relationship between board size and compensation is not expected (β_4>0 or β_4<0, H4). Finally, a positive relationship between corporate performance (returns on assets, stock return and Tobin's *Q*) and the level of compensation in order to align incentives between directors and shareholders is expected (β_5>0, H5; β_6>0, H6; and, β_7>0, H7).

In addition, all the models regarding the corrections over panel data proposed by Arellano and Bond (1991) and Blundell and Bond (1998) have been re-estimated in order to overcome the problems of heterogeneity and endogeneity in the model. The GMM system was therefore applied using a two-step estimator, so that the estimators of these models are efficient and asymptotically robust in the presence of heteroscedasticity.

RESULTS

Descriptive Statistics

The descriptive analysis for the payments received by members of the board is reported in Table 3.

According to the sample, the average compensation received by each member of the board of directors is €160,139.17 for 2004 and €225,749.99 for 2009, representing a rate of inter-annual growth of approximately 7.11% since 2004. (Graph 1).

This tendency is followed by remunerations received by each type of director. The compensation of executive members of the board has therefore gone up by 150.4% from 2004 to 2009, the compensation of independent members has increased around 38.93% and the compensation of proprietary members shows an inter-annual increase of 5.10%, from an average of €77,686.13 in 2004, to €99,629.78 in 2009. This behaviour shows that the level of compensation given to members of the board has increased significantly from 2004 to 2009.

The statistical behaviour of the independent variables is shown in Table 4.

The average percentage of control owned by board of directors is 24%. This result coincides with the average (of 23.64%) obtained by Arrondo et al. (2008) for the Spanish market in the year 2005 but exceeds the results of previous studies. This variable has grown over the years; in fact, Fernández, Gómez, and Fernández (1998) obtained a rate of 7% for the year 1993 and Mínguez and Martín (2003) obtained an average participation of 13.6% for 1999.

Table 3. Descriptive Statistics on Board of Directors Compensation.

Variable	2004	2005	2006	2007	2008	2009
Total compensation by member of the board of directors						
Mean	160,139.17	165,853.05	203,628.06	219,127.99	218,766.63	225,749.99
Rate of change		3.36	22.78	7.61	−0.16	3.19
Standard deviation	142,459.64	153,510.68	218,431.09	201,097.81	203,467.78	228,111.27
Minimum	12,333.33	1,875.00	11,636.36	17,142.86	14,000.00	15,888.89
Maximum	687,988.79	782,470.59	1,257,800.00	902,250.00	942,764.71	1,225,764.71
Total compensation by executive member of the board of directors						
Mean	588,697.35	608,658.15	882,797.47	911,935.50	1,036,260.71	1,474,074.28
Rate of change		3.39	45.04	3.30	13.63	42.25
Standard deviation	592,352.24	558,858.75	1,139,084.00	1,030,368.76	1,223,130.29	2,512,409.70
Minimum	15,000.00	15,000.00	15,000.00	7,000.00	68,000.00	84,000.00
Maximum	3,398,300.50	2,145,500.00	5,912,500.00	4,925,500.00	6,371,000.00	18,430,500.00
Total compensation by independent of the board of directors						
Mean	67,270.51	74,585.67	79,645.94	111,199.52	110,151.55	93,459.56
Rate of change		10.87	6.76	39.62	−0.94	−15.15
Standard deviation	54,623.20	62,174.34	67,591.29	171,077.19	151,374.97	84,963.83
Minimum	6,500.00	7,000.00	7,500.00	8,500.00	6,500.00	3,666.67
Maximum	257,749.33	325,000.00	325,285.71	1,344,000.00	905,000.00	494,875.00
Total compensation by proprietary of the board of directors						
Mean	77,686.13	85,261.03	83,110.88	92,707.99	95,073.27	99,629.78
Rate of change		9.75	−2.52	11.55	2.55	4.79
Standard deviation	64,965.85	82,126.93	61,619.88	81,927.69	78,838.51	88,958.01
Minimum	9,000.00	3,000.00	7,500.00	6,857.14	1,000.00	3,666.67
Maximum	277,222.22	353,333.33	298,500.00	383,000.00	316,000.00	329,000.00

This table shows a summary of statistics on director compensation measured in Euros. The compensation includes the following concepts: fixed and variable remuneration, salary fees, attendance fees, travelling indemnity, stock options and/or other financial instruments. Also, this table includes information distinguishing between executives, independent and proprietary directors' compensation.

In connection with the duality of chairman and CEO (as roles being performed by same person), the results showed this to be the case in 64% of the cases looked at.

In relation to the presence of independent outsiders on the board of directors, an average of 31% was obtained. This result is close to those obtained by Arrondo et al. (2008), which was also in reference to the Spanish market (35.35%), and those of Conyon and He (2008) for the Chinese market (34.4%). However, the results differ from those of Fahlenbrach (2009), with

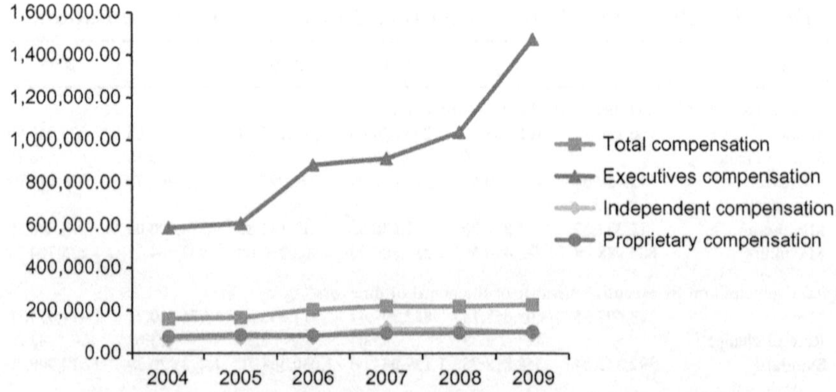

Graph 1. Evolution of Compensation by Director (2004–2009).

Table 4. Descriptive Summary Statistics on Panel Data Variables.

Variable	Mean	Standard Deviation	Minimum	Maximum
Dependent variables				
LNCOMP	14.11	1.21	9.62	16.85
LNEXECUTIVES	13.06	1.17	8.85	15.67
LNINDEPENDENT	10.99	0.90	8.21	14.11
LNPROPRIETARY	11.00	0.97	6.91	12.86
Board characteristics variables				
OWNERDIRECT	0.24	0.25	0	0.99
CEODUALITY	0.64	0.48	0	1
OUTSIDERS	0.31	0.18	0	0.86
BRDSIZE	11.25	3.98	3	24
Firm's performance				
ROA	0.03	0.12	−1.06	0.47
TOBINQ	2.58	2.88	0.20	24.22
STOCKRET	1.11	0.66	0.03	8.69
Control variables				
CRPSIZE	20.20	1.79	16.45	25.14
INDUSTRY	2.92	1.22	1	5

This table details a summary of statistics on the basis variables of interest that are defined in Table 2.

findings that obtained an average of 73.2% overall for outside directors within the American market, and Weir and Laing (2001), with an average of 47% for the UK market.

The size of the board of directors on average is 11.25 members, a result that is close to those obtained by other Spanish studies, such as Fernández et al. (1998), Sánchez and Lucas (2008) and Crespí and Pascual-Fuster (2008), who obtain averages of 10.75 (1993), 11 (2005 and 2006) and 10.9 (2004, 2005 and 2006), respectively. Along the same lines, recent studies relating to the US market, such as those of Coles et al. (2008) or Fahlenbrach (2009), obtained an average of 10.4 (1992–2001) and 9.8 (1993–2004) members on the board of directors. The results of this actual study therefore show that boards of directors in Spanish businesses tend to be greater in size. In fact, further analysis of this variable reveals that the size of the board can be very large indeed with the maximum of these running to 24 board members in size.

The binary correlation between all variables is reported in Table 5.

Board's Characteristics, Performance and Level of Directors' Compensation

The first set of regression results is reported in Table 6. In column (2), the predicted sign on each variable in the regression is indicated. Industry and yearly indicator variables are included in all models to capture potential impact in director payments across industries and years.

As a result of the Hausman test, a fixed effect model appears to be the most suitable methodology in all cases, so information regarding the results obtained through OLS and fixed effects methodologies are included.

In panel A of Table 6, the results of regression **LNCOMP**, the log of compensation by director, on board characteristics, performance and control variables are all shown.

We find a significant and positive relationship between the variable representing the concentration of powers of the chairman and the CEO (CEODUALITY) in the same person (coefficient 0.221) and the level of remuneration of board members, as previous studies show (Brick et al., 2006). These results highlight the importance of separating these two figures in order to moderate the remuneration of directors, an idea that is consistent with the recommendations contained in the reports and codes of good governance that have been enacted in Spain.

Also, the size of the board (BRDSIZE) (coefficient 0.045) has a significant positive effect on the compensation level. These results show that a large

Table 5. Correlation Matrix.

	LNCOMP	LNEXE-CUTIVES	LNINDE-PENDENT	LNPROPRIETARY	OWNER-DIRECT	CEO-DUALITY	OUT-SIDERS	BRDSIZE	ROA	TOBINQ	STOCKRET	CRPSIZE
Dependent variables												
LNEXECUTIVES	0.876***											
LNINDEPENDENT	0.719***	0.583***										
LNPROPRIETARY	0.765***	0.574***	0.729***									
Board characteristics variables												
OWNERDIRECT	−0.144***	−0.134***	−0.062**	−0.156***								
CEODUALITY	0.218***	0.051	0.200***	0.107**	−0.105**							
OUTSIDERS	0.066	0.123**	0.024	0.071	−0.114**	0.073						
BRDSIZE	0.647***	0.430***	0.386***	0.461***	−0.162***	0.119**	−0.048					
Firm's performance												
ROA	0.221***	0.119**	0.133***	0.239***	0.103**	0.064	−0.120**	0.105**				
TOBINQ	0.195***	0.192***	0.199***	0.167***	−0.137***	−0.055	0.114**	−0.011	0.172***			
STOCKRET	0.030	0.017	0.009	0.059	0.037	0.019	0.128**	−0.030	0.067	0.069		
Control variables												
CRPSIZE	0.678***	0.568***	0.542***	0.579***	−0.091**	0.171***	0.036	0.617***	0.117**	−0.182***	0.022	
INDUSTRY	−0.126**	−0.161***	−0.236***	−0.197***	0.013	−0.194***	0.009	−0.017	−0.042	−0.001	−0.045	−0.068

*The correlation is significant at 0.001 (bilateral).
**The correlation is significant at 0.05 (bilateral).
***The correlation is significant at 0.01 (bilateral).
Variables are defined in Table 2.

Table 6. Estimation: OLS, Fixed and Random Effects.

(1)	(2)	(3)		(4)		(5)		(6)	
	Expected Signs	Panel A: Compensation by Director LNCOMP (Model 1)		Panel B: Compensation by Executive Director LNEXECUTIVES (Model 2)		Panel C: Compensation by Independent Director LNINDEPENDENT (Model 3)		Panel D: Compensation by Proprietary Director LNPROPRIETARY (Model 4)	
		OLS	**Fixed effects**	**OLS**	**Fixed effects**	**OLS**	**Fixed effects**	**OLS**	**Fixed effects**
OWNERDIRECT	–	–0.079	0.184	–0.059	0.485	**0.098**	**–0.099**	–0.121	–0.188
		(0.143)	(0.207)	(0.184)	(0.338)	(0.144)	(0.228)	(0.149)	(0.208)
CEODUALITY	+	**0.282*****	**0.221*****	0.101	0.128	**0.184*****	**0.138****	0.017	**–0.133****
		(0.076)	**(0.072)**	(0.099)	(0.109)	**(0.074)**	**(0.079)**	(0.077)	**(0.078)**
OUTSIDERS	–	0.298	–0.158	0.304	**–0.599****	–0.017	**–1.392*****	–0.093	**0.601*****
		(0.210)	(0.209)	(0.273)	**(0.299)**	(0.227)	**(0.256)**	(0.224)	**(0.233)**
BRDSIZE	?	**0.099*****	**0.045*****	**0.034****	–0.007	–0.008	**–0.037*****	**0.021****	**–0.055*****
		(0.012)	**(0.014)**	**(0.015)**	(0.020)	(0.012)	**(0.015)**	**(0.012)**	**(0.014)**
ROA	+	**1.107*****	**0.643*****	0.583	0.196	0.244	0.106	**1.359*****	**0.971*****
		(0.327)	**(0.262)**	(0.438)	(0.395)	(0.365)	(0.316)	**(0.364)**	**(0.365)**
TOBINQ	+	**0.119*****	**0.013**	**0.116*****	**0.059*****	**0.101*****	0.016	**0.083*****	–0.021
		(0.013)	(0.015)	**(0.016)**	**(0.021)**	**(0.013)**	(0.017)	**(0.013)**	(0.015)
STOCKRET	+	–0.032	**–0.075*****	–0.045	**–0.077****	–0.073	**–0.088*****	–0.023	–0.061
		(0.054)	**(0.032)**	(0.066)	**(0.044)**	(0.051)	**(0.034)**	(0.064)	(0.037)
CRPSIZE	+	**0.331*****	**0.129*****	**0.330*****	**0.186*****	**0.270*****	**0.147*****	**0.257*****	**0.105****
		(0.028)	**(0.049)**	**(0.035)**	**(0.069)**	**(0.028)**	**(0.053)**	**(0.028)**	**(0.048)**
Intercept		**5.919*****	**10.882*****	**6.002*****	**9.293*****	**5.792*****	**8.883*****	**5.824*****	**9.537*****
		(0.573)	**(0.017)**	**(0.720)**	**(1.457)**	**(0.566)**	**(0.098)**	**(0.568)**	**(0.999)**
Industry (dummies)		Yes		Yes		Yes		Yes	
Hausman test			45.64***		17.23***		40.08***		58.05***
***F*-test**		71.14***	6.99***	27.72***	2.49***	29.44***	6.98***	31.87***	6.21***
		[12, 420]	[8, 350]	[12, 381]	[8, 313]	[12, 386]	[8, 319]	[12, 383]	[8, 316]
Wald Chi^2									
***R*-squared**		0.6703		0.4661		0.4778		0.4996	

Table 6. *(Continued)*

(1)	(2)	(3)		(4)		(5)		(6)	
	Expected Signs	Panel A: Compensation by Director LNCOMP (Model 1)		Panel B: Compensation by Executive Director LNEXECUTIVES (Model 2)		Panel C: Compensation by Independent Director LNINDEPENDENT (Model 3)		Panel D: Compensation by Proprietary Director LNPROPRIETARY (Model 4)	
		OLS	**Fixed effects**	**OLS**	**Fixed effects**	**OLS**	**Fixed effects**	**OLS**	**Fixed effects**
Within			0.1378		0.0598		0.1490		0.1358
Between			0.6378		0.3303		0.1122		0.0107
Overall			0.5693		0.2923		0.1268		0.0144
Number of firms		76	76	76	73	76	72	76	72
Number of observations		433	433	394	394	399	399	396	396

This table displays the impact of characteristics of the board and performance on the level of compensation.
Variables are defined in Table 2.
Random or fixed effects are shown, depending on the value obtained for Hausman test.
Industry dummies are included in all models.
The bold text shows significant coefficients.
Standard error is shown in brackets.
Symbols *, **, *** respectively indicate significance levels at 10%, 5% and 1%.

board increases the discretion of its members, resulting in greater levels of compensation per director, consistent with Sánchez and Lucas (2008) and Core et al. (1999). However, our finding contrasts with Ryan and Wiggins (2004) and Andreas et al. (2009) who find a negative relationship between board size and directors' compensation.

Regarding the variables related to performance, ROA has a significant and positive relationship with the directors' compensation (coefficient 0.643), which reveals that the compensation awarded to directors is, in this case, related to the good or bad performance of the company. However, if we analyse the result for the stock return variable (STOCKRET), the relation is just the opposite (coefficient −0.075). This suggests that despite the fall in market prices of the shares, directors' compensation has gone up. The coefficient of both variables implies that the relationship between board remuneration and company performance is stronger for book values than for stock market measures, consistent with Crespí and Gispert (2003).

The coefficient of TOBINQ, with a value of 0.013, is positive, although not statistically significant, suggesting that this performance measure does not explain directors' compensation.

To sum up, these results are consistent with the perspective that directors' compensation is less important in aligning the interests of directors and shareholders when the corporate governance mechanisms are stronger (Bryan et al., 2000). Also, all of the aforementioned results demonstrate that the majority of performance measures used (ROA and STOCKRET) are linked to the level of compensation, but STOCKRET does not exert the desirable motivation for the directors to work more assiduously on the shareholders' behalf rather than side with managers' or significant shareholders' interests. Only the accounting measure (ROA) is positively linked with the directors' compensation level, with the consequent management manipulation risks that this variable sometimes suffers (Cordeiro et al., 2000; Kren & Kerr, 1997).

In addition, this study reviews the directors' compensation by type of director, thus obtaining different results.

For the model developed on compensation by executive director, LNEXECUTIVES (Table 6, panel B), the results are different. Regarding the characteristics of the board, only the OUTSIDERS variable is significant and negatively related to executive directors' compensation (coefficient –0.599), showing the importance of this kind of director. These findings are consistent with the previous empirical evidence, which suggest that a greater number of outsiders mean stronger corporate governance and lower executive compensation (Brick et al., 2006).

The coefficient of TOBINQ is positive and significant (coefficient 0.059). Economically, executive pay is 5.9% greater when the ratio goes up 1%. Finally, STOCKRET has a negative relationship in relation to executive compensation (coefficient −0.077).

These results show that in relation to executive directors only the number of outsiders has a moderate effect over their compensation. Thus, and contrary to what one would expect, this research work did not discover the desired effect of corporate governance on control of executive directors' compensation.

Panel C of Table 6 shows the results for the compensation of independent directors, **LNINDEPENDENT**. The main coefficient change, in relation to panel B, is for BRDSIZE (coefficient –0.037), which is now statistically significant and negative, as predicted. There are also more statistically significant variables in the expected direction. So the OUTSIDERS variable (coefficient −1.392) is statistically significant and negative. Finally, the CEODUALITY variable (coefficient 0.138) shows a positive relationship with the compensation of independent directors. These results are consistent with the perspective that independent directors' compensation is less important in aligning the interests of directors and shareholders when the corporate governance mechanisms are stronger (Bryan et al., 2000), as in case of the executive directors' compensation.

Regarding performance measures, the STOCKRET variable shows a negative relationship with independent directors' compensation (coefficient −0.088), at the same level as Model 2.

The main concern here is the independence of directors. The results show that when the corporate governance mechanisms are stronger, the level of independent directors' compensation is lower. However, performance measures do not show the expected relationship.

The more interesting results are with the proprietary directors' compensation (**LNPROPRIETARY**) and the independent variables. In this case, CEODUALITY (coefficient –0.133), OUTSIDERS (coefficient 0.601), BRDSIZE (coefficient –0.055) and ROA (coefficient 0.971) are statistically significant.

These findings might indicate that more outsider members and the separation of power are linked to greater levels of proprietary directors' compensation, possibly because they have more power over their remuneration. This finding is in contrast to the expected relationship; thus, it could be misinterpreted by shareholders as they tend to think that a higher level of outsiders and the separation of power are linked with a lower level of director compensation. Specifically, this is important in the context of this work because, as has been previously mentioned, the directors who are representatives of majority shareholders (proprietary) reach an average of 43% of total directors. Thus, the problems of lack of shareholder confidence in the directors may be exacerbated.

Robustness Checks

When analysing the coefficients for independent variables in the linear models that are estimated with System-GMM methodology (Table 7), the coefficients for the total compensation per director (Model 1) confirm the aforementioned relationships except for OUTSIDERS and STOCKRET which change to significant, revealing the importance of simultaneous causality among variables under analysis. Thus, the proposed hypotheses related to CEODUALITY ($\beta_2 > 0$), ROA ($\beta_5 > 0$) and TOBINQ ($\beta_7 > 0$) are accepted. However, the hypotheses proposed for STOCKRET ($\beta_6 > 0$) and OUTSIDERS ($\beta_3 < 0$) are rejected.

Furthermore, the results of Models 2, 3 and 4 confirm what has been previously discussed – that only the percentage of outside directors (OUTSIDERS, $\beta_3 < 0$) has a negative influence on the level of independents' compensation and, contrary to expectations, the relationship with the level of proprietary and executive directors' compensation (OUTSIDERS, $\beta_3 > 0$).

In general terms, the results are also consistent with the hypothesis that the level of directors' compensation depends on the level for the previous year (LnCompensation_1).

Regarding performance, a significant relationship was found in the expected direction between ROA ($\beta_5 > 0$) and proprietary and executive directors' compensation, and between TOBINQ ($\beta_7 > 0$) and all types of directors' compensation. Contrary to expectations, STOCKRET ($\beta_6 < 0$) shows a negative relationship with the directors' compensation level.

Finally, we find a significant and negative relationship between the percentage of shares owned by members of the board (OWNERDIRECT) and the compensation received by the different types of directors. This is consistent with the results obtained by previous studies (Boyd, 1996; Bryan et al., 2000; Cordeiro et al., 2000). These results support the point of view that the ownership of shares downplays motivation exerted through compensation, since they already receive cash payments though dividends from the shares they hold (Cheng & Firth, 2005). It is therefore noted that the ownership of shares is a proper corporate governance control device

Table 7. Estimation: System-GMM in Two Steps.

(1)	(2) Expected Signs	(3) Panel A: Compensation by Director LNCOMP (Model 1)	(4) Panel B: Compensation by Executive Director LNEXECUTIVES (Model 2)	(5) Panel C: Compensation by Independent Director LNINDEPENDENT (Model 3)	(6) Panel D: Compensation by Proprietary Director LNPROPRIETARY (Model 4)
LnCompensation_1	+	**0.640*****	**0.518*****	**0.637*****	**0.558*****
		(0.015)	**(0.022)**	**(0.018)**	**(0.020)**
OWNERDIRECT	–	−0.067	**−0.211*****	**−0.155*****	**−0.614*****
		(0.094)	**(0.065)**	**(0.076)**	**(0.076)**
CEODUALITY	+	**0.018*****	0.011	**0.029*****	0.008
		(0.005)	(0.009)	**(0.007)**	(0.006)
OUTSIDERS	–	**0.677*****	**0.602*****	**−0.706*****	**0.854*****
		(0.084)	**(0.124)**	**(0.148)**	**(0.134)**
BRDSIZE	?	**0.042*****	**0.019*****	**0.009*****	**−0.046*****
		(0.006)	**(0.005)**	**(0.005)**	**(0.008)**
ROA	+	**1.063*****	**0.449*****	0.061	**0.914*****
		(0.212)	**(0.185)**	(0.191)	**(0.187)**
TOBINQ	+	**0.043*****	**0.059*****	**0.029*****	**0.024*****
		(0.003)	**(0.004)**	**(0.002)**	**(0.004)**
STOCKRET	+	**−0.034*****	**−0.046*****	**−0.020**	**−0.069*****
		(0.008)	**(0.013)**	**(0.027)**	**(0.018)**
CRPSIZE	+	**0.147*****	**0.212*****	**0.106*****	**0.155*****
		(0.010)	**(0.016)**	**(0.014)**	**(0.014)**
Intercept		**1.223*****	**1.443*****	**1.967*****	**2.155*****
		(2.62)	**(0.365)**	**(0.374)**	**(0.473)**
Test of joint significance					
Explanatory variables		741.26***	375.74***	551.15***	215.94***
		(8, 74)	(8, 71)	(8, 71)	(8, 70)
Dummy year variables		22.47*	4.17***	7.59***	4.98***
		(4, 74)	(4, 71)	(4, 71)	(4, 70)
Overidentifying test					
Hansen		47.24	50.33	54.15	51.93
		(96)	(97)	(97)	(97)
Autocorrelation test					
AR(1)		−3.25***	−3.06***	−2.95***	−3.12***
AR(2)		−0.82	0.28	−0.96	−0.23

This table displays the impact of characteristics of the board on the level of compensation.
Variables are defined in Table 2.
Models are run with the System-GMM methods.
The bold text shows significant coefficients.
Standard error is shown in brackets.
*, **, *** respectively indicate significance levels at 10%, 5% and 1%.

Table 8. Summary of Findings.

Impact	Total Compensation by Director	Total Compensation by Executive Director	Total Compensation by Independent Director	Total Compensation by Proprietary Director
+	CEODUALITY OUTSIDERS[b] BRDSIZE ROA TOBINQ[b] CRPSIZE	OUTSIDERS[b] BRDSIZE[b] ROA[b] TOBINQ CRPSIZE	CEODUALITY CRPSIZE BRDSIZE[b] TOBINQ[b]	OUTSIDERS ROA CRPSIZE TOBINQ[b]
–	STOCKRET	OWNERDIRECT[b] OUTSIDERS STOCKRET	OWNERDIRECT[b] OUTSIDERS BRDSIZE STOCKRET	OWNERDIRECT[b] CEODUALITY[a] BRDSIZE STOCKRET[b]

Nature of the impact of board of directors' characteristics and performance measures on the board compensation.
This table summarizes the findings obtained about the influence of board characteristic and performance measures on level of compensation received by them.
[a]Non-significant results when System-GMM method are used.
[b]Significant results when System-GMM method are used.

over compensation received by board members for Spanish listed companies, in line with the theoretical approaches developed by the Agency Theory which advocates it as a way to align board directors' interests with those of shareholders.

A summary of the findings is shown in Table 8.

CONCLUSIONS

Currently, the lack of investor confidence in the directors, with respect to their independence and the extraction of wealth of minority shareholders, has not been alleviated due to the fact that the compensation levels of directors are still an object of criticism, especially over the past few years. In this sense, the data of this work revealed significant increases in remuneration in recent years despite the current financial and economic difficulties.

The absence therefore or failure of certain mechanisms of corporate governance has furnished board members with the possibility of establishing a reward system that is more in keeping with their own personal interests rather than those of the investors. In the current climate, this problem has intensified the lack of investor confidence in their advisers.

From a theoretical point of view, the composition and structure of the board of directors and the performance of the firms are important control mechanisms to align the directors' and shareholders' interest. Consequently, aspects such as the size of the board, the addition of independent directors, separation of the roles of CEO and chairman of the board, the participation of the directors in the capital of the company, or the link of compensation to performance measures are necessary mechanisms to ensure compliance with the control of compensation assigned to board members.

Our results provide empirical evidence of how board characteristics and performance affect the level of compensation given to a director in a unitary board system, as is the case in Spain.

Using a sample of 76 Spanish listed firms for the period 2004–2009, this work has provided strong evidence to support the hypothesized determinants. It has been found that the directors' compensation is less important to align the interests of directors and shareholders when the corporate governance mechanisms are stronger and the performance measures show the desirable relationship with directors' compensation in order to motivate directors to work more assiduously on the shareholders' behalf. In concrete terms, the results show that the level of directors' compensation is positively related to board size, return on asset, CEO duality, number of outsiders, Tobin's Q and corporate size, and negatively associated with stock return. In comparison with executive directors' compensation, we find that the level of compensation is negatively related to the percentage of outsiders on the board and board ownership.

Regarding outsider directors' compensation, the main conclusions are (a) stronger corporate governance is important to control the level of compensation of independent directors and (b) the risk of extraction of wealth by proprietary directors from shareholders through their remuneration is greater when there are more outsider members on the board, contrary to the expected.

To sum up, our results suggest that Spanish firms with weaker corporate governance have a greater probability of suffering from problems concerning the level of directors' compensation. They also show that the desirable relationship between performance and remuneration of directors does not exist and that it is important to control the extraction of wealth that could be carried out by the proprietary directors. It, therefore, appears that the Spanish directors' compensation level is not designed around director-shareholders agency problems control.

As such, Spanish regulators have been especially active in this context, by using a Unified Code of Good Governance which included recommendations regarding this issue. The results of this research work, however, give reason to regulators and investors, within a concentrated ownership context, to be aware of the importance of the board's characteristics as corporate control mechanisms over the directors' remuneration and the poor connection between directors' compensation and the firm's performance. This raises the question whether it is necessary to establish other mechanisms to exercise control over the board of directors – for example a supervisory board as is the case in Germany – and the connection between directors' compensation and firm's performance and, more importantly, whether these mechanisms should be enforced, in order to control the level of remuneration received by board members.

Finally, it is noted that this study has certain limitations mainly related to problems associated with obtaining information. The methodology should be complemented by other types of analyses, such as the influence of the characteristics of the board on the remuneration structure in a greater level of disaggregation which can provide regulators and investors with basic and primary information regarding the board's discretion level concerning their own remuneration. The breakdown of individualized payments for each member of the board of directors proposed by the new legislative context in Spain could therefore be an important source of information to extend the results of this study.

NOTES

1. Fixed salary only includes the amounts paid to directors for their participation as board directors. This pay remains constant and is independent of the level of activity or any potential impact that management may have on the company's results. Attendance fees are the expenses incurred by directors for attending meetings of the board and its delegated committees. The aim of this type of compensation is to encourage the active participation of directors.

Variable salary includes amounts received in the form of profit sharing or any other formula that links these perceptions to the achievement of specific measurable objectives (company's net profit or other financial management indicators).

Other potential benefits that directors may obtain include advances, loans, funds and pension plans, insurance premiums and guarantees provided for directors.

2. Maximum allowable error:

$$\varepsilon = Z_{1-(\alpha/2)}\sqrt{\frac{N-n}{N-1}\frac{pq}{n}}$$

where

$Z_{1-\alpha/2} = z$ value associated with the degree of confidence $1-\alpha$

$N =$ size of the population

n = size of the sample
p = proportion
$q = (1{-}p)$

3. We assumed parameter homogeneity, which means that $\alpha_{it} = \alpha$ for all i,t and $\beta_{it} = \beta$ for all i,t.

REFERENCES

Andreas, J. M., Rapp, M. S., & Wolff, M. (2009). Determinants of director compensation in two-tier systems: Evidence from German panel data. *Review of Managerial Science*, *6*(6), 33–79.

Angbazo, L., & Narayanan, R. (1997). Top management compensation and the structure of the board of directors in commercial banks. *European Finance Review*, *1*, 239–259.

Arellano, M., & Bond, S. (1991). Some tests of specification for panel data: Monte Carlo evidence and an application to employment equations. *Review of Economic Studies*, *58*(2), 277–297.

Arrondo, R., Fernández, C., & Fernández, E. (2008). Influencia de la estructura de gobierno corporativo sobre la remuneración de los consejeros en el mercado español. *Información Comercial Española (ICE): Revista de Economía*, *844*, 187–203.

Banghoj, J., Gabrielsen, G., Peterson, C., & Plenborg, T. (2010). Determinants of executive compensation in privately held firms. *Accounting and Finance*, *50*, 481–510.

Bebchuk, L. A., & Fried, J. (2004). *Pay without performance*. Cambridge: Harvard University Press.

Berle, A. (1962). Modern functions of the corporate system. *Columbia Law Review*, *62*(3), 433–449.

Blundell, R., & Bond, S. (1998). Initial conditions and moment restrictions in dynamic panel data models. *Journal of Econometrics*, *87*(1), 115–143.

Boyd, B. K. (1994). Board control and CEO compensation. *Strategic Management Journal*, *15*, 335–344.

Boyd, B. K. (1996). Determinants of US outside director compensation. *Corporate Governance*, *4*, 202–211.

Brick, I. E., Palmon, O., & Wald, J. K. (2006). CEO compensation, director compensation and firm performance. *Journal of Corporate Finance*, *3*, 403–423.

Brunello, G., Graziano, C., & Parigi, B. (2001). Executive compensation and firm performance in Italy. *International Journal of Industrial Organization*, *19*, 133–161.

Bryan, S., Hwang, L. S., Klein, A., & Lilien, S. (2000). *Compensation of outside directors: An empirical analysis of economic determinants*. Retrieved from http://ssrn.com/abstract=244540. Accessed on 10 February 2012.

Buck, T., Bruce, A., Main, B., & Udueni, H. (2003). Long term incentive plans, executive pay and UK company performance. *Journal of Management Studies*, *40*, 1709–1721.

Cheng, S., & Firth, M. (2005). Ownership, corporate governance and top management pay in Hong Kong. *Corporate Governance*, *13*(2), 291–302.

CNMV. Spanish Securities Markets Commission. (2006, May). Unified Good Governance Code. Madrid.

CNMV. Spanish Securities Markets Commission. (2009a). *Corporate Governance Annual Report*. Retrieved from http://www.cnmv.es/DocPortal/Publicaciones/Informes/IAGC2009.pdf. Accessed on 3 October 2011.

CNMV. Spanish Securities Markets Commission. (2009b). Propuesta de actualización del Código Unificado de Buen Gobierno de las Sociedades Cotizadas. Diciembre. Madrid. 2005).

Coles, J. W., Daniel, N., & Naveen, L. (2008). Boards: Does one size fit all? *Journal of Management*, *27*(1), 23–50.

Coles, J. W., Mcwilliams, V. B., & Sen, N. (2001). An examination of the relationship of governance mechanisms to performance. *Journal of Management*, *27*(1), 23–50.

Conyon, M. J., & He, L. (2008). *Executive compensation and CEO equity incentives in China's listed firms*. Retrieved from http://ssrn.com/abstract=1261911. Accessed on 20 January 2012.

Cordeiro, J., Veliyath, R., & Eramus, E. J. (2000). An empirical investigation of the determinants of outside director compensation. *Corporate Governance*, *8*(3), 268–279.

Core, J. E., Holthausen, R. W., & Larcker, D. F. (1999). Corporate governance, chief executive officer compensation, and firm performance. *Journal of Financial Economics*, *51*(3), 269–305.

Crespí, R., & Gispert, C. (2003). Total board compensation, governance and performance of Spanish listed companies. *Labour*, *17*, 103–126.

Crespí, R., & Pascual-Fuster, B. (2008). *Executive directors pay and networks in Spanish listed companies*. Retrieved from http://ssrn.com/abstract=1287100. Accessed on 10 February 2012.

Croissant, Y., & Millo, G. (2008). Panel data econometrics in R: The plm package. *Journal of Statistical Software*, *27*(2). Retrieved from http://www.jstatsoft.org/v27/i02/paper. Accessed on 10 February 2012.

Crystal, G. S. (1991). Why CEO compensation is so high. *California Management Review*, *34*(1), 9–29.

Dávila, A., & Peñalva, F. (2004). *Corporate governance and the weighting of performance measures in CEO compensation*. IESE Research Paper No. 556. IESE Business School, University of Navarra, Spain.

Duffhues, P., & Kabir, R. (2008). *Is the pay-performance relationship always positive? Evidence from the Netherlands*. Retrieved from http://ssrn.com/abstract=962030. Accessed on 10 February 2012.

Elsas, R., & Florysiak, D. (2008). *Empirical capital structure research: New ideas, recent evidence, and methodological issues*. Discussion Paper 2008–10. Munich School of Management, Munich.

Ertugrul, M., & Hedge, S. (2008). Board compensation practices and agency cost of debt. *Journal of Corporate Finance*, *14*, 512–531.

European Commission. (2009, April). Commission recommendation complementing Recommendations 2004/913/EC and 2005/162/EC as regards the regime for the remuneration of directors of listed companies (3177/2009/EC). European Commission, Brussels.

European Commission. (2010, June). Green Paper: Corporate governance in financial institutions and remuneration policies. European Commission, Brussels.

Fahlenbrach, R. (2009). Shareholder rights, boards, and CEO compensation. *Review of Finance*, *13*, 81–113.

Fama, E., & Jensen, M. C. (1983). Separation of ownership and control. *Journal of Law and Economics*, *26*, 301–325.

Fernandes, N. (2005). *Board compensation and firm performance: The role of "independent" board members*. Working Paper No. 104/2005. European Corporate Governance Institute (ECGI). Brussels, Belgium.

Fernández, A. I., Gómez, S., & Fernández, C. (1998). El papel supervisor del consejo de administración sobre la actuación gerencial. Evidencia para el caso español. *Investigaciones Económicas*, *22*, 501–516.

Finkelstein, S., & Hambrick, D. C. (1989). Chief executive compensation: A study of the intersection of markets and political processes. *Strategic Management Journal*, *10*, 121–134.

Galbraith, J. K. (1971). *The new industrial state* (2nd revised ed.). Boston: Houghton-Mifflin.

Haid, A., & Yurtoglu, B. B. (2006). *Ownership structure and executive compensation in Germany*. Retrieved from http://ssrn.com/abstract = 948926

Hausman, J. A. (1978). Specification tests in econometrics. *Econometrica*, *46*(6), 1251–1271.

Heidrick & Struggles. (2009). *Boards in turbulent time*. Corporate Governance Report. Heidrick & Struggles International Inc. Retrieved from http://www.heidrick.com/PublicationsReports/PublicationsReports/CorpGovEurope2009.pdf. Accessed on 10 February 2012.

Hermalin, B., & Weisbach, M. S. (1991). The effect of board composition and direct incentives on corporate performance. *Financial Management*, *20*, 101–112.

Hermalin, B., & Weisbach, M. S. (2003). Board of directors as an endogenously determined institution: A survey of the economic literature. *Economic Policy Review*, *9*(1), 7–26.

Herman, E. S. (1981). *Corporate control, control power*. Cambridge: Harvard University Press.

Hsiao, C., & Lightwood, J. (1994). Análisis de especificación para datos panel. *Cuadernos Económicos del ICE*, *56*, 6–27.

Ingley, C., & Walt, N. (2005). Do board processes influence director and board performance? Statutory and performance implications. *Corporate Governance: An International Review*, *13*, 632–653.

Jensen, MC. (1993). The modern industrial revolution, exit, and the failure of internal control systems. *The Journal of Finance*, *48*(3), 831–880.

Kato, T., Kim, W., & Lee, J. H. (2004). *Executive compensation and firm performance in Korea*. Retrieved from http://ssrn.com/abstract=480544. Accessed on 20 January 2012.

Kato, T., & Kubo, K. (2006). CEO compensation and firm performance in Japan: Evidence from new panel data on individual CEO pay. *Journal of Japanese and International Economies*, *20*, 1–19.

Kato, T., & Long, C. (2004). *Executive compensation, firm performance, and ownership structure: An empirical study of listed firms in China*. Retrieved from http://www.wesleyan.edu/econ/ChinaStockMarket.pdf. Accessed on 10 February 2012.

Kren, L., & Kerr, J. L. (1997). The effects of outside directors and board shareholdings on the relation between chief executive compensation and firm performance. *Accounting and Business Research*, *27*(4), 297–309.

Larcker, D. F., Richardson, S. A., & Tuna, I. (2007). Corporate governance, accounting outcomes and organizational performance. *The Accounting Review*, *82*(4), 963–1008.

Law 2/2011 on Sustainable Economy (2011). BOE on March 5.

Lazarides, T., Dimpretas, E., & Dimitrios, K. (2008). Executive Board members' remuneration: A longitudinal study. *Corporate Ownership & Control*, *6*, 94–103.

Lehn, K., Patro, S., & Zhao, M. (2003). *Determinants of the size and structure of corporate boards: 1935–2000*. Retrieved from http://ssrn.com/abstract=470675. Accessed on 10 February 2012.

Li, J. (1994). Ownership structure and board composition: a multi-country test of agency theory predictions. *Managerial and Decision Economics*, *15*, 359–368.

Linck, J. S., Netter, J. M., & Yang, T. (2008). The determinants of board structure. *Journal of Financial Economics*, *87*(2), 308–328.

Lipton, M., & Lorsch, J. W. (1992). A modest proposal for improved corporate governance. *Business Lawyer*, *48*, 59–77.

Main, B. G., Bruce, A., & Buck, T. (1996). Total board remuneration and company performance. *The Economist Journal*, *106*, 1627–1644.

Mayers, D., & Smith, C. W. (2010). Compensation and board structure. *Evidence for the Insurance Industry*. *The Journal of Risk and Insurance*, *77*(2), 297–327.

Mínguez, A., & Martín, J. F. (2003). *El Consejo de Administración como mecanismo de control: Evidencia para el mercado español*. Working Paper WP-EC 2003-02. Instituto Valenciano de Investigaciones Económicas, Valencia.

Morck, R., Shleifer, A., & Vishny, R. W. (1988). Management ownership and market valuation: An empirical analysis. *Journal of Financial Economics*, *20*, 293–315.

Paz-Ares, C. (2008). El enigma de la retribución de los consejeros ejecutivos. *Revista para el análisis del derecho*, *1*, 1–74.

Pearce, J. A., & Zahra, S. A. (1992). Board compensation from a strategic contingency perspective. *Journal of Management Studies*, *29*, 411–438.

Ryan, H. E., & Wiggins, R. A. (2004). Who is in whose pocket? Director compensation, board independence, and barriers to effective monitoring. *Journal of Financial Economics*, *73*, 497–524.

Sánchez, G., & Lucas, M. E. (2008). *El nivel retributivo de los altos directivos de las empresas cotizadas españolas: Influencia del tamaño y composición del Consejo de Administración*. 22° Congreso Anual de la AEDEM, Salamanca.

Shleifer, A., & Vishny, R. W. (1997). A survey of corporate governance. *The Journal of Finance*, *2*, 737–783.

Sulong, Z., & Mat Nor, F. (2010). Corporate governance mechanisms and firm valuation in Malaysian listed firms: A panel data analysis. *Journal of Modern Accounting and Auditing*, *6*(1), 1–18.

Wang, Y., Chuang, C., & Lee, S. (2010). Impact of compositions and characteristics of board of directors and earnings management on fraud. *African Journal of Business Management*, *4*(4), 496–511.

Weir, C., & Laing, D. (2001). Governance structures, director independence and corporate performance in the UK. [illegible]

Yermack, D. (1996). Higher market valuation of companies with a small board of directors. *Journal of Financial Economics*, *40*, 185–211.

CHANGES IN PERFORMANCE MEASUREMENT AND EVALUATION SYSTEMS AS INSTITUTIONAL PROCESSES: THE CASE OF AN ITALIAN TEACHING HOSPITAL

Antonio Leotta and Daniela Ruggeri

ABSTRACT

Purpose – *In the last decades, Italian healthcare organisations have been subject to important normative changes, aimed at increasing their efficiency. As a response, performance measurement and evaluation (PME) systems have been introduced. The present study attempts to examine PME system changes as institutional processes. In studying such processes the healthcare literature acknowledges the presence of two logics: managerial and professional, as peculiar to healthcare settings, whose convergence or divergence can explain the success of any institutional process.*

Design/methodology/approach – *We adopt Busco et al.'s (2007) framework as an approach for unbundling PME system change into four relevant coordinates, namely: (1) the object (PME system), (2) the*

Performance Measurement and Management Control: Global Issues
Studies in Managerial and Financial Accounting, Volume 25, 427–463

ISSN: 1479-3512/doi:10.1108/S1479-3512(2012)0000025019

subjects (institutional forces), (3) the place and time of change (the managerial and professional logics) and (4) the how and why change happens (change as an institutional process). We conducted a longitudinal case study at a large teaching hospital in Southern Italy, directed to interpret PME system changes during the period from 1998 until 2009.

Findings – *Our observation distinguishes episodes of successful institutional processes, where the introduced innovations are transformed into* objectivated *practices, from episodes of missed institutionalisation, where new procedures were rapidly abandoned.*

Research and social implications – *This theoretical framework can be useful for interpreting the PME system changes in different institutional contexts.*

Originality – *The Busco et al.'s framework allows us to understand PME system changes by integrating the perspectives from Neo-Institutional Sociology, representing healthcare organisational responses to external institutional pressures, and Old-Institutional Economics, conceptualising PME system changes as an institutionalisation process.*

INTRODUCTION

In the last decades, as in other European countries, Italian healthcare organisations have been subject to important normative changes, aimed at increasing their efficiency. As a response, performance measurement and evaluation (PME) systems have been introduced and redesigned. Throughout this process, several difficulties have arisen. Firstly, in terms of continuous changes, due to the diversity of the institutional constituencies, being public administrations (State and Region) and social actors (trade unions), all exercising pressure on healthcare organisations. Secondly, in terms of impediments mainly due to organisational frictions that hindered the introduction of new procedures. Finally, as failures of the introduced procedures, which have been rapidly abandoned to be substituted with new ones. Thus, the process of modernisation of the Italian healthcare sector has been and is still unaccomplished. In line with the accounting nature of this study, we focus on the dynamics related to PME system changes and attempt to understand the driving forces underlying these dynamics. More specifically, drawing on healthcare literature, we acknowledge the co-presence of two logics within healthcare organisations: the managerial logic, underpinning managerial and administrative behaviours, and the professional one, behind the physicians' point of view. We consider these two logics very relevant for understanding the dynamics of PME system changes, and posit the question of how such logics have contributed to the observed changes.

To address this question we adopt Busco, Quattrone, and Riccaboni's (2007) framework as an approach for unbundling PME system changes into four relevant coordinates, namely: (1) the object, (2) the subjects, (3) the place and time of change, and (4) the how and why change happens. Having defined PME systems as the object of change, we identify the subjects of change with the institutional forces, which require healthcare organisational responses, and the place of change with the different organisational logics, that is managerial and professional. Hence, we attempt to understand the mechanisms underlying the how and why of change, studying change as a process of institutionalisation. Accordingly, the Busco et al.'s framework allows us to understand PME system changes integrating the perspectives from Neo-Institutional Sociology (NIS) and Old-Institutional Economics (OIE). The former is useful for representing healthcare organisational responses to external institutional pressures, thus explaining changes through episodes on isomorphism and decoupling (DiMaggio & Powell, 1991); the latter is aimed at conceptualising PME system changes as an institutionalisation process, thus highlighting the mechanisms behind how the introduced PME systems are translated into practices and how these practices are reproduced (Burns & Scapens, 2000) and may become 'objectivated' (Berger & Luckmann, 1966). Both the NIS and OIE perspectives are consistent in showing that managerial and professional logics explain the abandoning or the institutionalisation of PME systems where they respectively diverge or converge.

Two critical assumptions underlie the present study: first, since the process of change is the result of divergence or convergence of the managerial and professional logics, change is viewed through a teleological model, according to which 'Influences in the external environment or within the entity itself may create instabilities that push it towards a new developmental path' (Van de Ven & Poole, 1995, p. 517). This new path here refers to the introduction of new PME systems, which become reproduced and *objectivated* practice when the different interest groups are convergent. The second assumption refers to the nature of PME system as a configuration of elements not necessarily consistent with each other. This potential inconsistency of PME system elements is both the cause and effect of PME system changes. Accordingly, we use the term PME 'package'.

The main contribution of this study is to offer a meaningful understanding of change and stability in PME systems within healthcare settings. To do so, it integrates NIS and OIE within Busco et al.'s framework and applies this integrative perspective to interpret how the PME system changes occurred within a teaching hospital in Southern Italy over the period from 1998 to 2009. Field data offers evidence for our theoretical interpretation, according to which the convergence between the managerial and professional logics is viewed as convergence between coercive and normative isomorphism, leading to institutionalised PME systems. Conversely, divergence between the two logics is interpreted as divergence between the two types of isomorphism, which entails decoupling behaviours leading to the abandoning of the introduced PME systems.

The rest of the paper is structured as follows. Second section provides a theoretical interpretation of the change phenomenon in PME systems within healthcare settings. After reviewing the extant literature that has addressed the topic, and defining the object of change in PME package, this section illustrates our interpretive framework, embracing NIS and OIE within the approach suggested by Busco et al. (2007). Third section describes the research method that guided the explanatory case study. Fourth section reports the field evidence and offers an interpretation of it based on the theoretical framework previously developed. Fifth section concludes the paper.

INTERPRETING CHANGES IN PERFORMANCE MEASUREMENT AND EVALUATION SYSTEMS WITHIN HEALTHCARE SETTINGS

Changes in Performance Measurement and Evaluation Systems within Healthcare Organisations

Management accounting literature on healthcare organisations has considered accounting change processes: in some contributions on specific topics, such as performance measures (Modell, 2001), and Diagnosis-Related-Groups (DRG) and case-mix accounting systems (Lehtonen, 2007); in others on the broader management accounting practices such as a set of transferable tools or techniques (Kurunmäki 2004, p. 329), and control mix or control systems package (Abernethy & Chua, 1996, p. 573). In the present study, we focus on PME systems, which we consider as not necessarily consistent with each other, thus referring to them as PME package. More specifically, we use the concept of 'package' because it 'points to the fact that different systems are often introduced by different interest groups at different times' (Malmi & Brown, 2008, p. 291). We see this argument as very appropriate to the healthcare field, where new systems are often required by different institutional constituencies, such as normative institutions, that is State and Regions, social actors, that is trade unions, patient groups, and so on, which exercise institutional pressure on the renovation and redesign of the healthcare service and of its management systems. However, one of the main concerns in the management literature on healthcare organisations has been the co-presence of two main logics underpinning organisational behaviour, that is managerial and professional logics, often seen as causing resistance and difficulties in new systems implementation (Chua & Degeling, 1993; Covaleski, Dirsmith, & Michelman, 1993; Doolin, 1999; Jacobs, Marcon, & Witt, 2004; Lehtonen, 2007; Lowe & Doolin, 1999; Weiner, Maxwell, Sapolsky, Dunn, & Hsiao, 1987). Power struggles between managers and clinicians have been considered as one of the main arguments in explaining why attempts to implement new management accounting control systems have failed to meet expectations (Abernethy & Stoenwinder, 1995; Chua & Degeling, 1993; Covaleski et al., 1993; Doolin, 1999; Jones, 1999; Lapsley, 1994; Lehtonen, 2007; Pettersen, 1995, 2001; Weiner et al., 1987). For these reasons we refer to institutional, managerial and professional constituencies as the most relevant interest groups whose convergence/divergence mainly explains success/failure in the implementation of new PME systems as elements of a control package.

As argued by Burns and Scapens (2000), the enactment of systems (rules) and practices (routines) 'may be subject to resistance, especially if the rules and routines challenge existing meanings and values, and actors have sufficient resources of power to intervene in this process' (Burns & Scapens, 2000, p. 10). Consistent with the arguments developed above, which refer to the different interest groups in healthcare settings representing institutional, managerial and professional constituencies, we see sources of resistance in the possible divergences among these groups. In this sense, if on the one hand, interest divergence is a source of resistance to translate new PME systems into practices, on the other, interest convergence seems to favour the retention of the introduced systems and their success in becoming *objectivated* practices. So, interest convergence being the main motor of change, we represent change through a teleological model, according to which: 'Goals are socially reconstructed and enacted based on past actions

(Weick, 1979). Influences in the external environment or within the entity itself may create instabilities that push it towards a new developmental path' (Van de Ven & Poole, 1995, p. 517). This new path here refers to the introduction of new PME systems, which become *objectivated* practices when the organisation acts as a singular, discrete entity, that is when the different interest groups are convergent.

Change from the Institutional Perspective: Organisational Responses to Institutional Pressures as an Institutionalisation Process

The purpose of this section is to define a framework for the analysis of PME system changes as organisational responses to institutional pressures. In doing so, we draw on Busco et al. (2007) in order to define coordinates for change interpretation. Relying on this framework, we examine PME system changes (object of change) assuming institutional forces as the main subject of change, organisational responses, based on the managerial and professional logics, as the place of change, and the process of institutionalisation as when and how change happens.

Accordingly, we adopt the perspective of NIS to analyse the organisational responses to institutional pressures. This theory is also in line with the teleological model of change, mentioned in the previous section, for it identifies the influences that have created stability or instability that pushes the organisation towards a new developmental path. Then, we draw on OIE, relying on Burns and Scapens' (2000) framework, to interpret PME system changes as a process of institutionalisation.

Unbundling Management Accounting Change: Coordinates for Interpretation

Many management accounting studies focused on healthcare settings have examined 'change' by longitudinal case studies based on NIS (Abernethy & Chua, 1996; Lehtonen, 2007; Modell, 2001). Coherently with the previous contributions, we adopt, as abovementioned, the institutional perspective for explaining accounting changes in healthcare context. To better understand the process of accounting change, we need to factorise management accounting change, identifying the key points for interpreting this phenomenon.

In analysing the change process we rely on the framework proposed by Busco et al. (2007) which, in line with the previous management accounting literature, investigates the nature of management accounting systems and practices along some key dimensions that represent 'coordinates of change', offering several reflections on the process of change.

The first coordinate refers to what or who 'drives management accounting change and whether this is some hidden structural force, some clear individual agency, an opaque combination of the two or even something else' (Busco et al., 2007, p. 126).

The second coordinate concerns how management accounting techniques are able to spread and become practice (Quattrone, 2006), referring to the process through which new accounting innovation travels and becomes successful (Lapsley & Wright, 2004). Busco et al. (2007) argue that understanding this process can help interpret why this happens. This coordinate invites reflection on the ontology of change, understanding how management accounting changes occur and why.

The third coordinate refers to 'something else, something which is contextual to the practice that is object of change or related to the action of specific individuals' (Busco et al., 2007, p. 128). Analysing the process of change is important in order to understand what and who has changed, in other terms the object of change and those people involved in change. In this case the focus is on the effects on 'who', those individuals, subject to change and not on the active role of the subjects of change.

Finally, the forth coordinate concerns where and when change happens and who is there to observe it. It means referring to specific spatial and temporal concepts of change.

All the four coordinates are strictly interrelated, for instance, the what, how and why are differently interpreted according to who and where. This is shown in Fig. 1 by the bidirectional arrows that link the four coordinates, acting on 'change'.

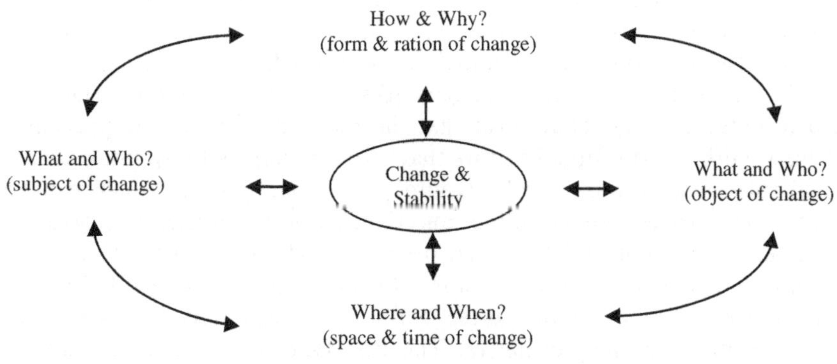

Fig. 1. Coordinates of Management Accounting Change and Stability.
Source: Busco et al. (2007, p. 127).

Following this framework, we seek to understand how accounting change in healthcare settings is driven by organisational responses under the influence of institutional pressures.

We also try to understand how change occurs and spreads within the healthcare organisation, explaining the reasons 'why' change is made. This reflection is supported by the identification of 'the object of change' that might affect some aspects of management accounting systems and the practices people follow. Finally, we ask in what period the phenomenon of change occurs and how it is perceived in different organisational areas.

This analysis suggests a holistic approach to the interpretation of accounting changes that stem from a combination of the coordinates described above. The analysis of change along these four coordinates allows us to observe not only the dynamics of accounting systems and practices, and those who promoted it, but also the processes of convergence/divergence between managerial and professional logics. Hence, we see this framework as useful for integrating NIS and OIE perspectives, in order to offer a more in-depth understanding of respectively: (1) the institutional forces and the organisational responses leading to the introduction of new PME systems; (2) the process by which the new PME systems are put into practice, which may or not be reproduced and *objectivated.*

Organisational Responses to Institutional Pressures: The Perspective of Neo-Institutional Sociology

Management accounting literature on healthcare organisations has adopted the institutional theoretical lenses to better understand accounting change. The successful/unsuccessful implementation of new accounting and control systems has been investigated justifying how different organisational responses to institutional pressures can determine the final implementation results. Neo-Institutional sociologists view organisations as embedded within larger interorganisational networks and cultural systems (DiMaggio & Powell, 1991). They posit that, in addition to economic pressures, there are other institutional factors that affect managerial responses, that is legal rules, regulations, standards and professional values, which can be viewed as other influences on the objectivation of reproduced practices.

A central concept of NIS is that organisations are pressured to become isomorphic, that is to conform to a set of institutionalised beliefs. DiMaggio and Powell (1991) identify three forms of institutional isomorphism: coercive, mimetic and normative. The first type seems to better explain the adherence of the organisational processes to external pressure exerted by other organisations, cultural expectations, government requests, etc. The second type occurs when organisations face uncertainty and model themselves on other organisations. Hence, the coercive and mimetic isomorphism seems to better explain the adherence of organisations to external pressures. The normative isomorphism, instead, arises when professionals, operating in organisations, are pressured to comply with a set of norms and rules developed by professional groups, as 'the experience of a specialised education, and the involvement in professional networks, influences how professional personnel undertake their activities within the organisation' (Carruthers, 1995, p. 317). This type of isomorphism seems to explain the adherence of organisational personnel to professional logic, coherent to their professional status.

Some studies have highlighted that in a complex setting, such as a hospital organisation, there co-exist two logics: managerial and professional. They converge, for example, when medical professionals encounter calculative practices of managerial accounting, adopting management accounting techniques, thus giving rise to a hybridisation phenomenon (Kurunmäki, 2004). Another contribution underlines that the effects of new reforms on hospital management control (i.e. DRG-based perspective pricing and case-mix accounting systems) are successful when accounting information penetrates deeply into the organisation, making healthcare professionals cost conscious and increasingly aware of their limited financial resources (Lehtonen, 2007). Otherwise, the managerial logic, expression of a bureaucratic control, may conflict with professional control, deriving from standards and values shared by professional networks, so implying a failure of accounting systems implementation (Abernethy & Stoenwinder, 1995).

Within the NIS we refer to Oliver's (1991) conceptual framework in order to better understand the organisational responses to institutional pressures. In particular, Oliver identifies a number of possible responses ranking from acquiescence choice to high avoidance choice. Other previous studies use this framework (Abernethy & Chua, 1996; Modell, 2001) finding support to Oliver's hypotheses.

More specifically, Abernethy and Chua (1996) illustrate 'how change in control mixes (in healthcare context) are both a function of an organisation's institutional environment and the strategic agendas of their dominant coalitions' (p. 599). Some other studies indicate that organisational personnel (senior managers and staff specialists) are not necessarily confined to passively comply with institutional pressures, but they may show proactive behaviours as responses to them (Beckert, 1999; DiMaggio & Powell, 1991; Modell, 2001).

Within this theoretical framework, the present study focuses on the decoupling effect, that is when the professional logic shows a refusal behaviour in accepting institutional practices promoted by managerial logic.

We retain that the observation at intra-organisational level highlights how the presence of two logics (managerial and professional) could be adopted, respectively, by administrative staff and physicians.

We argue that if there are different types of isomorphism co-present in a given organisational context, then it is possible that they would not be convergent, because they tend to different goals. The degree of convergence of objectives between coercive isomorphism and the normative one may be related to the degree of convergence between the managerial and the health professional logics, here interpreted as 'influences (…) within the entity itself (that) may create instabilities that push it toward a new developmental path' (Van de Ven & Poole, 1995, p. 517). Any difference between the two logics, and between the two types of isomorphism, may explain decoupling behaviour.

Change as a Process of Institutionalisation: The Perspective of Old-Institutional Economics

Given different institutional external forces leading to the introduction of a new PME system, the convergence or divergence between the managerial and professional logics within the organisation can explain the institutionalisation of the introduced PME system or its abandoning. We now need to better understand how this occurs by examining change as a process.

An extensive and growing literature exists on management accounting change (e.g. Burns & Scapens, 2000; Burns & Vaivio, 2001; Busco et al., 2007; Dambrin, Lambert, & Sponem, 2007; Johansson & Siverbo, 2009; Lukka, 2007) in which the institutional theory has played an important role. The basic assumption is that economic behaviour is influenced by structure and in particular, by social institutions (Scapens, 1994): organisations compete not just for resources and customers, but for political power and institutional legitimacy, for social as well as economic fitness (DiMaggio & Powell, 1983). Hence, organisations are shaped according to the rules imposed on them, the expectations of the environment and the cognitive patterns of the actors (Scott, 2001), on the basis of objectified collective beliefs (institutional logics).

In the institutional perspective Burns and Scapens (2000) analyse change as a process of institutionalisation. In line with Scapens (1994), they focus on rules (formalised statements of procedures, i.e. management accounting systems) and routines (informal practices actually in use, i.e. management accounting practices in use), pointing out that 'rules are normally changed only at discrete intervals; but routines have the potential to be in a cumulative process of change as they continue to be reproduced' (Barley & Tolbert, 1997, p. 96).

In describing the process of institutionalisation, Burns and Scapens distinguish two realms: institutional and action, as shown in Fig. 2. Both of them are characterised by synchronic elements when the institutional principles are *encoded* into rules and routines and the actors are involved in *enacting* new rules and routines (these two steps are represented, respectively, by 'a' and 'b' solid lines in Fig. 2). The institutional and action realms are also characterised by diachronic elements when actions are *reproduced* into routines and routines are *objectivated* into institutions, giving rise to a cumulative process of change through time (represented, respectively, by 'c' and 'd' solid lines).

The central part of Fig. 2 shows how rules and routines act as modalities that link the institutional realm and the realm of action (Burns & Scapens, 2000, p. 10). The movement from the first box to the second one underlines that new rules can be implemented and new routines can emerge during the *encoding* and *reproduction* phases, shaping the process of change. This is due to various sources of resistance to change, intentional or unintentional, that highlight a discrepancy between prescribed behaviour and actual behaviour,

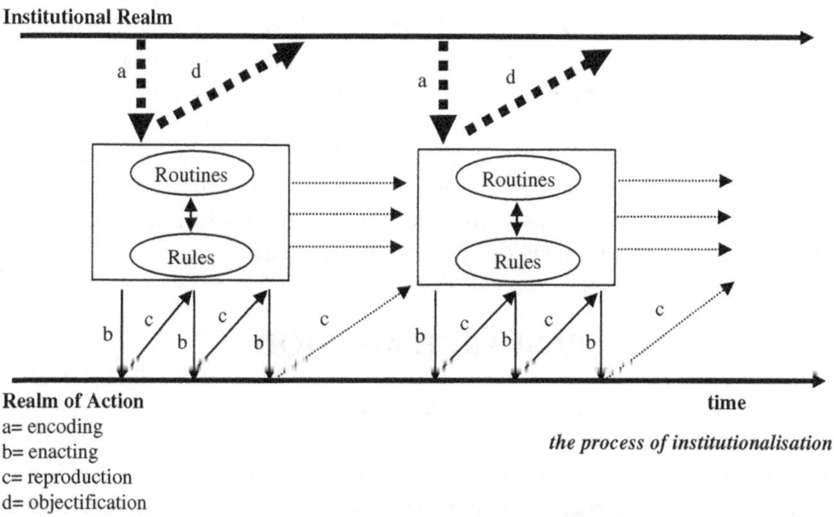

Fig. 2. Revision of 'The Process of Institutionalisation'. *Source:* Burns and Scapens (2000, p. 9).

occurring when new management accounting rules meet an existing set of routines (Johansson & Siverbo, 2009).

As suggested by Englund and Gerdin (2008) the Burns and Scapens' framework refers to rules and routines in two alternative senses: as 'interpretation of the actions which can be observed' (Burns & Scapens, 2000, p. 14) or as a representation of 'empirically observable practice as such' (Burns & Scapens, 2000, p. 9). In our study, we interpret rules and routines in the former way being the implied concepts related to the ideas of 'modalities' (Giddens, 1984, p. 28). We view rules and routines 'as the non-situated and subject-less norms of conduct (formal or informal) which actors draw upon in the production and reproduction of practice' (Englund & Gerdin, 2008, p. 1128). In this way, rules and routines help us understand why individuals interact and act in a management accounting context.

Accordingly, in examining PME system changes, we distinguish, within their package, PME systems and PME practices, which we consider as rules and routines, respectively (Burns & Scapens, 2000). We see this distinction as very relevant for it provides explanations of the PME system changes as an institutionalisation process, where system reproduction stems from practices *objectivation* (Berger & Luckmann, 1966). In fact if, on the one hand, PME systems stem from reproduced PME practices through their cumulative influence over time, that is diachronically, on the other, PME systems shape PME practices at a specific point in time, that is synchronically (Burns & Scapens, 2000). Therefore, revising Burns and Scapens's framework coherently to the original model of Barley and Tolbert (1997), we embrace the concept of objectification including it in the place of institutionalisation phase (represented by solid line 'd' in Fig. 2).

In this paper, we will explain changes of PME package in a teaching hospital over a given period of time investigating how these changes were influenced by intra-organisational responses to institutional pressures, distinguishing them in terms of adaptation and resistance.

RESEARCH METHOD

Case Research Site and Approach

In line with our research objective, directed to understand how intra-organisational influences explain PME system changes within healthcare organisations, we conducted a longitudinal case study at a large teaching hospital in the Southern Italy, aimed at interpreting PME system changes throughout the period from 1998 to 2009. We focused on this period of time because it begins from the year when the process of corporatisation was introduced into Italian hospitals.

The case was chosen because it appears as a confusing situation of frequent PME system changes. The immediate challenge was to explain the process of frequent changes by making sense of it. Thus, after selecting the theoretical perspective that seems to offer a meaningful understanding of the phenomenon, we performed a more in-depth field research, with the aim of providing a deeper and more concrete description of change process and of showing how the theoretical lenses of NIS and OIE can explain the phenomenon. In line with this role, the theories presented above are used to explain the specifics of PME system change in healthcare settings, rather than to provide generalisations (Scapens, 2004). To do so, after observing episodes on isomorphism, we show how the concept of decoupling is useful to examine intra-organisational resistance to new PME systems, as resulting from divergence of managerial and professional logics. Then we adopt Burns and Scapens' framework to show its usefulness for representing the PME system changes we observed. The case study approach is thus explanatory.

Moreover, case research allows us to follow an in-depth qualitative and interpretive approach intended to grasp individuals' perceptions of the appropriateness of new PME systems to meet expectations, and so to be translated into practices, by showing that these perceptions depend on intra-organisational factors, such as organisational department, hierarchical level, and professional role of the interviewees. The investigation at an intra-organisational level allows us to take account of the multifaceted nature of change process, which must be examined by departing from a linear conception of time and space and assuming the multiple viewpoints of the different agents involved in it (Busco et al., 2007). In this way, the field research method followed here allows an analysis of managerial ambiguities, tensions and contradictions 'since it permits an analysis of suggestive themes and counterpoints, interpretations and counter-interpretations, and different voices around accounting in organisations' (Ahrens & Dent, 1998, cited in: Lukka, 2007, p. 78).

Data Sources

The longitudinal nature of the study regards the time to which data refers, while data collection was concentrated during the period from November 2010 to May 2011. We collected archival and field data.

Archival data comes from: (a) internal documents, that is budget forms, cost and performance reports, and internal letters; (b) institutional sources, that is protocols between the hospital and other public administrations, protocols and agreements with trade unions; (c) public sources, that is hospital website, newspaper articles. All this data was collected during a first-round visit, in the period from November 2010 to February 2011, in close collaboration with the Internal Control Manager.

Field data comes from interviews conducted with various subjects, who were selected from the personnel working at the hospital from 1998 to 2009, classified as administrative managers and staff, medical managers, physicians and clinical staff, from medical units such as Cardiology, Surgical Oncology and Digestive System Surgery, which were selected for their task and organisational complexity, and cost level. Among medical personnel, we further considered the hierarchical level, that is whether the doctor was a top clinician or not, and the institutional role, that is whether the interviewee was unionist or not (this was the case for the manager of Digestive System Surgery unit). We considered this selection criterion consistent with the need to map individual perceptions along various coordinates, such as organisational area, hierarchical level, professional characteristics and institutional or social role. All interviews were conducted during the period from April to May 2011. They mostly aimed at gathering individual perceptions on the appropriateness of the various PME systems, which we referred to as 'procedures', in line with the terminology used in hospital documents. All interviews were preceded by a phone call followed by an email to the selected subjects. In the email we enclosed a sheet containing a brief chronological description of the PME procedures we drew from documentary analyses, after reviewing it with the Internal Control Manager. In this brief description we relied strictly on hospital documents so as to avoid any influence on interviewees' judgments and perceptions. The interviews, which lasted an hour on average, focused on: (a) the perceived objectives of each of the past procedures; (b) the extent to which each procedure is perceived to be translated into practice; (c) the perceived advantages of each procedure and its main limits; (d) the extent to which it is perceived that each past procedure has influenced current practice.

Data Analysis

Archival data allowed us to recognise both PME systems, that is procedures, as 'the ways things should be done' according to protocols and agreements, and PME practices, as 'the ways things are actually done', drawing on internal letters, and budget forms and reports. We thus analysed data by chronologically summarising the content of the documents over each year.

Further data was gathered from newspaper articles and the hospital website, which were coupled with the annual summaries mentioned above, so as to improve our understanding of each procedure and practice.

Field data was drawn from the interviews, which were not type-recorded because of the confidential nature of the issues discussed (Kajüter & Kulmala, 2005), and were conducted in an informal style; all were written down in a draft report and emailed to the interviewees.

Data triangulation derives from relating archival and field data to the same object (change of PME systems and practices). Furthermore, within archival data, we examined each PME procedure relying on two different sources: hospital and public. These multiple sources, at least partially, cope with the limit of relying on retrospective approach: we asked managers to describe, explain and reflect on events they had experienced in the past (Kasim Nor-Aziah & Scapens, 2007). A retrospective bias has been noticed, for example, on studies of voting choices, which have shown that respondents misreport their past choices in order to appear more consistent with their current choice (Shachar & Eckstein, 2007). We do not see such a bias as a limit to the validity of this study: firstly, for we contain it through data triangulation, as explained above; secondly, we see retrospective data as opportune, since only retrospectively can interviewees develop a perception of the extent to which past procedures influence the current practices.

Changes in PME 'Package' in an Italian Teaching Hospital

The teaching hospital under observation is one of the largest university hospitals in Southern Italy, with a very articulated organisational structure, currently composed of five departments and thirty-three medical units. This organisational complexity increased during the corporatisation process, which was enacted in 1998 in response to a reform in the public healthcare sector. For the sake of clarity, it has to be specified that Italian healthcare system is funded by the Regional Government. This means that, as well as normative changes and reforms, changes in hospital management systems can also be due to the dynamics of regional finance.

Within the hospital under study, the abovementioned institutional phenomena stimulated the introduction of different PME procedures. In particular, referring to the period from 1998 up to 2009, five main procedures were identified, which were labelled as: (1) 'Budget negotiation', in 1998; (2) 'Pool competition', in 2001; (3) 'Department projects', in 2003; (4) 'Drug budgeting', in 2004; (5) 'Current procedure', in 2009.After a 'selective' description of PME system changes interpreted from the managerial and the professional points of view, we complement the two views of change to finally interpret them with the lenses of the integrative framework discussed earlier.

Interpreting Changes in PME Package from the Managerial View

As the Control Manager specified:

> 'Performance measurement and evaluation procedures are aimed at accounting for individual contribution to organisational productivity and, thereby, at increasing productivity of each individual and the medical unit where he/she works'.

In 1998, process of corporatisation was accompanied by the introduction of a new procedure, called 'Budget negotiation', focused on efficiency and effectiveness, which allocated a large amount (70%) of resources through a new budgeting procedure. Compared to the previous more bureaucratic logic, the new procedure stressed the following points:

(a) the adoption of budgeting process in defining specific organisational targets;
(b) the introduction of a responsibility accounting system in order to clearly identify the results each subject was made responsible for, and how each medical manager evaluated his/her staff;
(c) the institution of an evaluation committee, with the aim of assuring the reliability of the whole evaluation process.

The new procedure was enacted in June, when budgets from single units were assembled as outputs of various meetings in which medical managers interacted with the general manager and the controller in order to set specific targets. Budget forms were prepared by an external consultant and were structured as follows: form A comprised efficiency parameters in terms of targets as well as actual values, referring to the previous year and external sources, in order to make comparison with other hospitals; form B comprised comments to form A, while form C showed organisational corrective actions. The two latter forms were meeting tools, and had to be filled in by medical unit managers in order to gather their proposals.

To support management control, a management accounting system was implemented, based on cost centre accounting and performance indicators. The latter were intended as parameters to be used by medical unit managers in evaluating the individual productivity within their team. More specifically, these parameters were articulated in the following sets:

(a) quality targets (in terms of commitment, professionalism, working days, etc.), which accounted for 50% of the individual productivity bonus;
(b) extra-hours, as percentage of the monthly budget, which accounted for 50% of the individual productivity bonus;
(c) sum of the previous two parameters as a percentage of the budget reached by the subject.

As can be noticed, the above parameters were mostly qualitative and subjective.

Thus, bonuses were distributed within the medical units according to the contribution of each individual.

At the end of the evaluation process, the evaluation committee emphasised in its report that managerial culture was still underdeveloped within the hospital, whence the need to encourage training for medical managers and staff. Moreover, it suggested revising performance indicators in order to avoid the influence of factors outside an individual's control.

In 2001, a new procedure was introduced to refine bonus distribution. To do so, it took into account the activities within the units, which differed in terms of care-services and required resources. Thus, medical units were pooled according to the homogeneity of their activities, and 'pools' were considered as the organisational level intermediate between the general manager and medical unit managers. The 'Pool competition' procedure was illustrated by the general manager in a letter addressed to each medical unit manager, where quoting the trade union agreement, he specified that this agreement:

> 'has established that some medical units performing similar activities should pool resources that would normally be allocated to single units. This system allows the evaluation of the activities of single units by comparing them with those of other units within the same pool'

And referring medical unit managers themselves, he continued:

> 'The comparison among different units gives each single unit participating in a specific pool the possibility to improve its initial budget with respect to the following performance indicators: (…)'

The 'Pool competition' procedure was thus intended to improve individual productivity and its measurement through a better consideration of the peculiarities of the unit the individual belonged to, and an increase in the competition among the units within the same pool.

Performance indicators were selected by the general manager with respect to each unit, assigning target values, in terms of quantitative parameters, namely revenues per bed, appropriateness of hospitalisation based on most frequent DRG, direct cost per bed, total direct cost, waiting lists, occupation rate, internal customer satisfaction index, etc.

According to the incentive system based on the 'Pool competition' procedure, each unit should have received budget resources on the basis of its own performance.

This procedure was appreciated by the evaluation committee, which noticed its consistency with the previous procedure, in terms of both budget forms and incentive systems. The latter tended to stimulate an increase in the operational volumes, taking account of cost containment. Particularly referring to the 'Pool competition' procedure, the evaluation committee specified in its report:

> 'The specifics of the incentive mechanisms adopted for every pool (...) are to induce competition among the medical units within each pool in order to improve performance in terms of volume of revenues, accounting for direct costs and care-service complexity'.

However, the evaluation committee also noticed:

> 'the possibility that, within a pool, one unit may receive an amount of resources higher than the theoretical amount, because of restrictions afforded by other units within the same pool, whose performance did not reach the predefined threshold. Similar mechanism is not prescribed for medical units which, for organisational reasons, do not participate in any pool. Such an unequal treatment may have negative consequences'.

Moreover, the evaluation committee observed that the stimulus to overcome predefined thresholds with respect to revenue volume could entail an increase in direct costs higher than the incremental benefits.

In 2003, a new procedure was introduced with the aim of bringing the hospital's incentive system into line with the new organisational structure and economic constraints, emulating the business structure of departments, as required by a public reform. Moreover, this change was consistent with the regional financial constraints, acknowledging the concept of 'economic equilibrium' as a focal point for department management.

The departments came from the institutionalisation of the previous pools of medical units, in line with the criterion of homogeneity of the activities within each pool.

The principles of project management were thus introduced. The general manager invited each department manager to formalise a project, specifying: (1) objectives, (2) economic analysis and (3) performance measurement criteria. Budget resources were allocated to departments according to the number of medical managers and healthcare staff. Residual resources had to be allocated to departments according to a set of rules focused on economic equilibrium. This concept brought with it a set of performance measures, based on objective parameters, that is revenue/cost ratio, which was ranked so as to assess the percentage of residual resources to be allocated to each department.

In 2004, further financial constraints within the Region brought about the alignment of hospitals' incentive systems with the goal of drug budget containment. So, within the teaching hospital under observation, each department was required to contain their costs for drugs within a budget level, based on the lowest cost for each drug over the previous three years. Formal justification had to be provided by managers whose department exceeded the budget level. A formal procedure was introduced to authorise pharmacy units to deliver drugs only after a formal request.

As the controller pointed out:

> 'One year after its introduction, the described procedure was no longer adopted, probably because of various changes in the corporate governance of the hospital, and the related substitution of the general manager. In fact, a new agreement was signed by the Region and the three Universities, regarding new rules for the general manager nomination and the composition of the hospital boards. During the period between 2005 and 2008 no regular evaluation procedure was observed, mainly due to the organisational reassessment of the hospital'.

The procedure observed in 2009 maintained both the budget negotiation and a department-based structure. The budgeting process started at the end of March, where the general manager required each department manager to formalise department plan. In so doing, the general manager specified the guidelines each department manager should follow in accordance with the Regional Plan of health spending containment. Many managers took part in the budgeting process, where an informal worksheet was adopted by the controller so as to take notes of department managers' proposals. The department plans had to define targets for responsibility centres. So, departmental plans were the result of internal meetings where the various medical unit managers contributed with their proposals. The budgeting process was repeated quarterly and at the end of each quarter, a performance report was sent to each department manager to communicate performance level and variances.

At the end of July 2010, the general manager required each department manager to fulfil a performance evaluation of each medical manager and healthcare staff working at his/her department. The set of parameters to be used for this evaluation had to be drawn from two forms: one prepared by the controller, comprising data on organisational activities and performance at the department and medical unit levels; another one prepared by the human resource manager, referring to monthly working hours of medical managers and staff at the department and medical unit levels.

In summary, the dynamics of PME systems over the period from 1998 up to 2009 reveals that, within the teaching hospital under study:

- Firstly, there was a progressive refinement of organisational structure of control: in fact, as unit-specific factors could interact with individual productivity, it was controlled for first pooling homogeneous units, then converting pools of medical units into organisational departments;
- Secondly, there was an evolution of the accuracy of performance measurement, with an increase in the number of parameters used, thus reducing the subjective bias in the individual performance evaluation.

Moreover, evaluation procedures and organisational structures, as two main elements of the evaluation package, have interacted with each other. On the one hand, the introduction of the 'Pool competition' procedure for evaluation purposes constituted an experimental practice for the following organisational change, introducing departments. On the other hand, department structure stimulated project management orientation, firstly (2003), and a higher participation in the budgeting process of medical unit and the department managers, secondly (2009).

Interpreting Changes in PME Package from the Professional View

The main changes identified in the observed period were perceived in different ways by physicians and medical unit managers.

Over the years, the teaching hospital was contaminated by many influences of institutional settings such as local government and political pressure resulting from the links between the teaching hospital and the university. This aspect was also confirmed at management level by frequent changes in the hospital board that indirectly conditioned the performance management and evaluation system. Before 1998, the mission of the hospital was teaching, scientific research and healthcare services. While after the corporatisation process, the hospital was required to give precedence to healthcare service within its mission. Consequently, the organisational performance was evaluated focusing on healthcare service without considering its teaching and research orientation. As described by a physician of Surgical Oncology unit:

> 'within the teaching hospital academic personnel must devote 60% of their activities to providing healthcare services, thereby sacrificing teaching and research time. This penalises those who still dedicate more time to research and teaching, actually accomplishing the principal mission of the teaching hospital which, over the years, has been forgotten'.

In 1998, the corporatisation process initially generated enthusiasm amongst professionals. Physicians and medical staff recognised the importance of the procedures introduced that emphasised efficiency as the main principle. They were aware that corporatisation process would lead to a differentiation between units and individuals on the basis of productivity levels. The efficiency principle led to the introduction of new, more objective evaluation criteria. Objective performance measures stressed the differences between units and individuals as the manager of Digestive System Surgery unit described:

> 'the differentiation between units increased, especially when objective criteria were adopted for incentive purposes. One year some departments had zero incentives and other departments obtained 200% of the incentives'.

Even though the objective evaluation criteria had improved the incentive allocation, some performance targets were considered unreachable because they included the effect of uncontrollable factors. Objective performance indicators introduced in 1998 were accepted by medical staff. However, later on some physicians expressed dissatisfaction arising from the awareness that performance dimensions were not 'perfectly' measurable. Referring to this aspect, the manager of Digestive System Surgery unit, after pointing out his role as unionist, argued that:

> 'Objectivity of performance measures can lead to satisfactory results in terms of individual evaluation ... but if the general manager keeps using these measures, the current performance evaluation system should include ex-post adjustments to take account for uncontrollable factors'.

Even if the objective evaluation criteria involved all physicians who sometimes actively participated in the evaluation process in their unit, a physician of Cardiology unit commented as follows:

> 'A limitation of objective evaluation was that not all medical managers put it into practice in all units. This lack of uniformity in the evaluation process represented a weakness of the system'.

The response on the part of some medical managers was to avoid the use of objective evaluation criteria because it was impossible to correctly differentiate between the individual contributions of physicians at their unit, as described by the manager of Surgical Oncology unit:

> 'I never make distinctions between members working in my team, because it is difficult to identify individual contributions with respect to team performance. I think that individuals may be differentiated by their organisational position, in this case they can be objectively evaluated on the basis of responsibility assigned to their position'.

Difficulties in applying objective criteria were explained by the constraint of 'structural resources'. Health professionals were made responsible for keeping within their budgets and also for increasing unit performance on the basis of estimated productivity levels from previous years. Dissatisfaction spread through the hospital organisation when structural resources did not fit performance objectives set by the general manager. The manager of Surgical Oncology unit discussed this problem highlighting:

> 'the problem of structural resources generated the "Alitalia care" phenomenon: namely a migratory flow of physicians towards the North of Italy where hospital structures and resources were not strictly limited. The general manager requires each medical unit to attract an increasing number of patients from outside the area. Given that the structural resources were limited, physicians attract new patients relying on their fame rather than on the hospital's reputation'.

The evaluation procedure called 'Pool competition' adopted in 2001 involved all hospital units. The physicians agreed on the criterion of the pool competition, which consisted of grouping medical units on the basis of homogeneous activities, thus stimulating team behaviours. However, the pool competition generated dissatisfaction between physicians because they perceived unequal treatment between physicians of the same organisational position in different pools. Physicians working in pools with better performance were more advantaged in incentive allocation than others working in less successful pools. The pool performance depended not only on physicians' productivity but also on the DRG associated with the care pathologies. Physicians, who processed low levels of healthcare services, with a high DRG, could appear more productive than others who processed higher levels of service with lower DRG. This aspect was clarified by the metaphor used by the manager of Digestive System Surgery unit:

> '... it's like in athletics races ... considering two individuals, A and B, running in different heats. Subject A, racing in the first heat, arrives first while B, racing in the second one, reaches third position. If the race ended at this point of the competition, we would say that subject A is more powerful than subject B. In real athletic competitions the overall outcome is not decided until the final, where the best performer wins ... in 2001 we missed the final race!'

Pool competition was hindered by a delay in the budgeting process. The general manager decided to introduce new evaluation criteria in the middle of the year, that is so late that the physicians did not have time to reach the targets. A physician from Cardiology unit argued:

> 'the pool evaluation procedure was a competition between pools but this competition was seriously affected by a seasonal effect. It means that our productivity, for example, is concentrated in the first part of the year ... when we were told about the pool procedure, we could not do anything to enhance our department performance!'

In 2003, 'the department projects' was introduced and new performance indicators were adopted. Medical managers perceived some of these performance indicators as being unfair since they did not take into consideration important aspects of unit activities that affected the performance achieved. In some cases, the general manager overly emphasised revenue/costs ratio making the incentive allocation between units unfair. Echoing this aspect, the manager of the Digestive System Surgery unit described his unit activities in the following negative terms:

> 'In the past, hospital general manager gave too much importance to costs/performance ratio, resulting in a disparity of assigned incentive between units. In the Digestive System Surgery unit, my medical staff spends a relevant number of back-office hours necessary to prepare the correct therapy for our patients. Back-office activities generated increased costs which were out of balance when compared to the achieved performance. Performance indicators such as revenue/cost ratio penalised my unit because we did not receive any incentive for an important part of our work. The general manager did not take into consideration the complexity of tasks carried out within my unit, ignoring aspects that did not have a clear financial manifestation, thus penalising my medical staff engaged in intense back-office activities'.

In 2004, the drug budget was introduced in order to limit the consumption of drugs. In particular, the consumption target was considered the minimum consumption of the three previous years. In doing so, it was assumed that higher consumption could be considered a waste of resources. The drug budget was perceived by physicians to be important but the consumption constrains generate deontological problems as the manager of Digestive System Surgery unit suggested:

> 'The drug budget neglected, however, the preventive function of many drugs, for example in cases of infections. So, the limitation of drug consumption resulted in a cost increase of care, which could have been avoided. Moreover, we risk forgetting the true nature of the health service, that is the guarantee of the quality of healthcare! … I cannot forget once when the director of hospital pharmacy told me that the drugs administered to terminal patients were considered a waste … It's unbelievable that for saving 500 € a person should die!'

Finally, the 'Current procedure' was introduced in 2009. According to the suggestions drawn from form 'B' in budget negotiation process, filled in by the manager of Digestive System Surgery unit, the general manager decided to associate subjective performance measures to objective performance indicators. In this regard, the above-mentioned manager told us:

> 'I think it is right to find a certain combination of objective and subjective parameters, especially because some parameters, like the DRG-based, are subject to continuous revision due to scientific innovations in surgery and the field of medicine'.

To sum up, the interviews with medical managers and staff reveal their perceptions of the different procedures adopted by the teaching hospital during the observed period. In particular, the interviews offer the following insights:

- Firstly, the introduction of organisational departments was facilitated by the previous pools, which were mainly acknowledged by medical managers and staff as homogeneous aggregations of medical units.
- Secondly, even if physicians recognised the benefits of pools, they highlighted that 'Pool competition' procedure suffered from the use of DRG-based parameters associated to healthcare services. More specifically, the difficulty related to task complexity and continuous changes in DRG-based parameters induced some physician to suggest integrating subjective performance measures with the other parameters.

Evidence of Isomorphism and Decoupling in Explaining Convergence/Divergence Between Managerial and Professional Logics in the Process of Institutionalisation

The dynamics of PME system changes is here explained drawing on Busco et al.'s (2007) framework as an approach embracing the perspectives of NIS and OIE, in order to offer a meaningful understanding of the phenomenon.

Unbundling the Changes in PME 'Package'

In what follows we examine the case study using four coordinates, as suggested by Busco et al.'s (2007) framework (see Fig. 3). In doing so, we identify the subject of change (*what and who*) and, in particular, we note that, at the beginning of the period under observation, 'what' stimulated a change was the healthcare reform of the corporatisation process. We consider this reform as an external force (institutional pressure) that drove the change into the teaching hospital. Carrying on our analysis, we also identify social actors, mainly trade unions, as 'who' exercised pressure on some PME system changes.

The object of change was different over the period we observed. In describing and analysing our case study we proceed following a chronological path, as shown in Fig. 3 by the chronological sequence below 'what and who?'.

Regarding the coordinates 'how, why and when', we modify the original Busco et al.'s framework by combining the coordinates 'how and why' with 'when' and leaving aside 'where', which we shall consider separately. This reasoning comes from the need to integrate this framework with that of Burns and Scapens, which helps us to understand better the ratio of PME system changes. In particular, we explain the coordinates 'how and why' by examining changes as a process of institutionalisation. In doing so, 'when' is

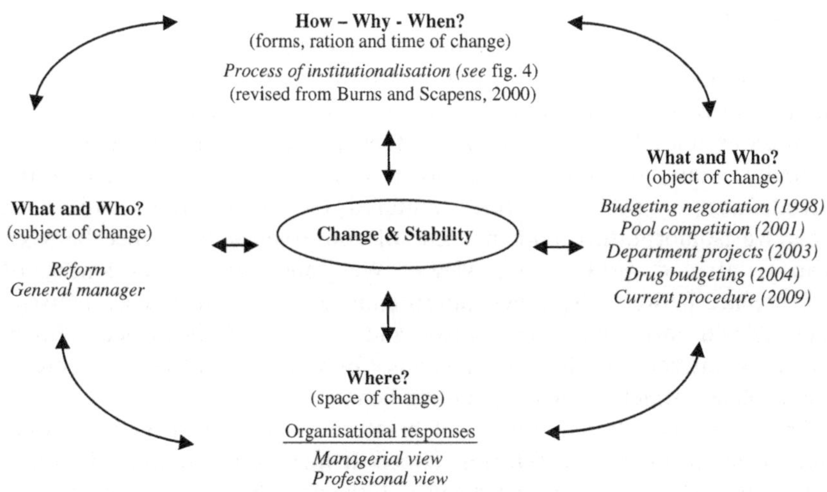

Fig. 3. Our Revision of Busco et al.'s (2007) framework, p. 9.

a fundamental coordinate that permits us to focus chronologically on the different procedures and to understand the transition from one to another.

The coordinate 'where' is instead explained by the two different organisational points of view described above: the managerial and professional. In particular, in order to have a more articulated view, in our description we considered the different organisational levels relating to managerial and professional roles, and distinguished three organisational levels within the managerial view (general manager, evaluation committee and controller) and two organisational levels within the professional view (manager of medical unit and physician).

Finally, we represent in the central circle of Fig. 3 'change and stability' that we interpret and explain using the four coordinates. According to NIS, stability is the result of the convergence between coercive and normative isomorphism, which represents the convergence between managerial and professional logics. Coherently with Burns and Scapens's framework, this convergence is the explanation of the 'objectification' of reproduced PME practices. Differently, divergence between the two types of isomorphism represents the case when the two logics diverge. This situation generates instability that leads the teaching hospital towards new PME system changes.

Evidence of Isomorphism and Decoupling from the Neo-Institutional Perspective

Adopting the Neo-Institutional lenses we describe PME system changes as organisational responses to institutional pressure, highlighting evidence of isomorphism and decoupling. We do so, firstly underlying the external institutions and the sort of pressure they put on the teaching hospital; secondly, examining the organisational responses both in terms of the introduction of new procedures required by the external institutions (coercive isomorphism), and in terms of convergence/divergence between managerial and professional views. We thus observe evidences of convergence between the views, interpreting them as convergence between coercive and normative isomorphism, and evidence of divergence between the views, interpreting them as divergence between normative and coercive isomorphism, which entails decoupling behaviours.

The main institutional sources of pressure are identified with National and Regional norms and reforms, aimed at increasing the individual and organisational productivity, both through the corporatisation of healthcare organisations, and through an introduction and a continuous refinement of

PME. Other institutional sources are related to social institutions, such as trade unions.

Looking at the organisational responses to institutional pressures, we identify, among the aforementioned PME procedures, 'Budget negotiation' in 1998, 'Department projects', in 2003, 'Drug budgeting', in 2004, 'Current procedure', in 2009, as the ones mainly due to normative changes; 'Pool competition', in 2003, was instead introduced by trade union agreement, explicitly quoted by the general manager in his letter to medical managers. We qualify these organisational responses to external institutional pressures as episodes of coercive isomorphism.

We now attempt to explain why elements of the introduced procedures were maintained or abandoned. In the organisational responses related to some elements of 'Budget negotiation' and 'Pool competition' we notice episodes of convergence between managerial and professional views, which we interpret in terms of convergence between coercive and normative isomorphism. In fact, a point of convergence between coercive and normative isomorphism can be recognised in the efficiency principle, leading the introduction of the 'Budget negotiation' procedure, which generated professional enthusiasm. This can partly explain why meetings on budget negotiation were maintained throughout the whole period of study, as noticed from the managerial view. Similar arguments relate to 'Pool competition' procedure, where the departments, instituted in response to a national public reform, maintained the same aggregation criterion as the one followed within the pools. This sort of stability can be explained by a convergence between the coercive isomorphism, related to the introduction of pools and departments, and normative isomorphism, mainly related to the logic of team that characterises medical culture.

Episodes of divergence between managerial and professional views are also noticed and here interpreted as divergence between coercive and normative isomorphism, leading to decoupling. Such episodes relate to PME criteria, which have been subject to continuous change. Referring to 1998 procedure, from the managerial viewpoint the main purpose was to improve the individual performance evaluation, so as to objectively define individual contribution to organisational productivity. From the professional view instead, objective indicators could not correctly differentiate the individual contributions of physicians. As argued by the manager of Surgical Oncology unit: 'I never make distinctions between members working in my team, because it is difficult to identify individual contributions with respect to team performance. I think that individuals may be differentiated by their organisational position, in this case they can be objectively evaluated on the basis of responsibility assigned to their position'. The quoted argument well explains decoupling through a divergence between coercive and normative isomorphism: the former is mostly represented by the managerial need to objectively measure individual performance, while the latter is related to team logic followed by physicians, which does not match the former need. Moreover, talking about objective parameters, the discussed episodes of decoupling were perceived as a cause of further limits by a physician of Cardiology unit, who observed: 'A limitation of objective evaluation was that not all medical managers put it into practice in all units. This lack of uniformity in the evaluation process represented a weakness of the system'.

Pool competition was also subject to decoupling due to different perceptions of appropriateness of the evaluation pool-based criteria. In fact, on the one hand the general manager, referring to the pool procedure, specified: 'The specifics of the incentive mechanisms adopted for every pool (…) are to induce competition among the medical units within each pool in order to improve performance in terms of volume of revenues, accounting for direct costs and care-service complexity'. On the other hand, during an interview with the manager of Digestive System Surgery unit, it was argued that physicians working in a better performing pool were advantaged in incentive allocation compared to those in a more poorly performing pool. The pool performance depended not only on physicians' productivity but also on the DRG associated with the care pathologies. Physicians, who processed low level of healthcare services, with a high DRG, could appear more productive than others who processed high level of service with lower DRG.

While the aforementioned episodes are related to some elements of the 1998 and 2001 procedures, other forms of decoupling can be related to the whole procedure as was the case for the 'Drug budgeting'. From the managerial perspective, the drug cost contention was consistent with the efficiency principle and had to be respected according to regional financial constraints. From their own point of view, instead, physicians did not comply with such a procedure because, as explained by the manager of Digestive System Surgery unit: 'we risk forgetting the true nature of the health service, that is the guarantee of the quality of healthcare!(…)'. In this argument we see another example of decoupling explained by divergence between coercive and normative isomorphism.

The NIS offers meaningful insights into PME system changes or stability, for it identifies the influences that have created stability or instabilities that have pushed the organisation towards a new developmental path, in line with the teleological model we refer to. A process view is also necessary so as to highlight the mechanisms that underlie the reproduction of PME practices and their objectification.

Evidence of Institutionalisation Process from the Old-Institutional Perspective
To complement NIS interpretation of the observed changes, we adopt OIE, relying on Burns and Scapens' framework, in order to describe the process of institutionalisation within the teaching hospital. In particular, in analysing PME system changes, we distinguish two package elements: PME systems (rules) and PME practices (routines). The identification of these two elements permits us to better explain the institutionalisation process, where system reproduction stems from practice objectification.

The aforementioned PME system changes described from the managerial and professional viewpoints have been systematised and illustrated as shown in Fig. 4.

Following Burns and Scapen's framework, we combine both synchronic and diachronic elements. Specifically, institutions constrain and shape action synchronically at a specific point in time. At the top of the figure we indicate the observed period from 1998 up to 2009, considering the five main changes: 'Budget negotiation', 'Pool competition', 'Department projects', 'Drug budgeting' and 'Current procedure'. The diachronic elements are due to actions that produce and reproduce institutions through their cumulative influence over the observed period.

Fig. 4 illustrates two important aspects of the phenomenon under observation: the 'continuity' and 'discontinuity' aspects. The continuity aspect is represented by the institutional realm and the realm of action: both are ongoing in a cumulative process of change through time, even if the first one occurs over a longer period of time than does change in the realm of action. Both realms are illustrated by the solid lines at the top and bottom of the figure.

The discontinuity aspect, instead, is provided by the central boxes that contain rules and routines. As the careful reader can note, rules and routines are each contained in circles of varying size depending on how articulated rules and routines are. Rules and routines were introduced as result of: (1) the encoding phase (arrow 'a'); (2) the transition from one box to the following one (dotted horizontal arrow); (3) the reproduction of previous practices, produced and reproduced under previous PME systems.

Following the first cause of discontinuity, in the 1998 the corporatisation process, which was introduced by a reform in healthcare sector, shaped a new rule called 'Budget negotiation' (encoding phase) that in turn led to the formation of the ongoing routines such as: (a) 'Budget negotiation meeting',

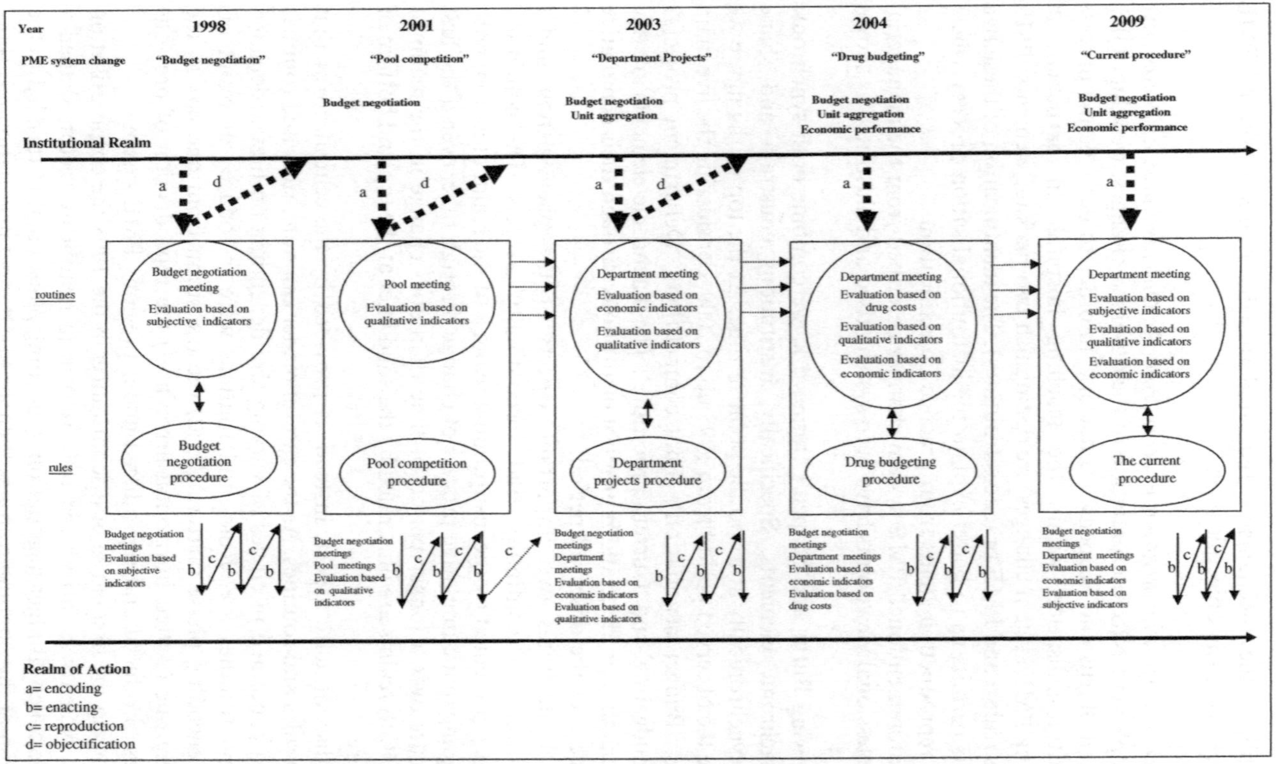

Fig. 4. Revision of 'The Process of Institutionalisation' Applied to the Teaching Hospital Case. *Source*: Burns and Scapens (2000, p. 9).

where medical managers interacted with the general manager and controller in order to fill in the budget forms specifying targets, comments and organisational corrective actions; and (b) 'Evaluation based on subjective indicators'.

An example of the second cause of discontinuity is given by the transition from 2003 to 2004 boxes. The department projects procedure led to the formation of routines such as: 'Department meetings'; 'Evaluation based on economic indicators'; 'Evaluation based on qualitative indicators'. In particular, we have to highlight that the economic indicators were also used in the previous years but they held less importance in the evaluation process, that is the reason why the 'Evaluation based on economic indicators' does not appear in 1998 and 2001 boxes. Regarding the use of economic and qualitative indicators, we notice that it persisted in the following years. This phenomenon is explained by the adoption of 'economic' and 'qualitative' categories of indicators in the evaluation process although in the following years (i.e. 2004, 2009) they were based on different principles and performance dimensions. It means that even if their use is a common factor in three of the observed years, economic and qualitative indicators were reproduced by routines (dotted horizontal arrows) without generating their objectification. We have to specify that an example of discontinuity shown by dotted horizontal arrows is also given by the transition from pool competition to department projects procedures, that is from 2001 to 2003, when the routine 'Evaluation based on qualitative indicators' came up again.

Finally, regarding the third cause of discontinuity, we observed that in only one case the reproduction of practices, derived from previous PME system changes, was objectified giving rise to the new PME system changes. An example of this phenomenon is given by the transition from 'Pool competition' procedure to 'Department project' procedure. In this case, the latter procedure emerged from the *realm of action* (established routines). In particular, the medical managers and staff 'enacted' the pool meeting (and the pool competition procedure) that encoded the institutional principles. The pool meeting stimulated new values and new ways of working modifying 'how things are done'. Specifically, the 'unit aggregation feeling' spread into the organisation, leading to the objectification of the 'unit aggregation' (arrow 'd'). The institutionalisation of routines (in this case), which had been reproduced through the behaviour of the individual actors, means that routines became simply 'the way things are'. This aspect was reflected in the department projects procedure and was formally introduced with the new organisational articulation in departments.

Paying attention to the top of Fig. 4, we indicate in the 'institutional realm' all rules and routines that were objectified. They become taken-for-granted assumptions of 'how things are'. The institutionalisation process of 'Budget negotiation' was the result of the convergence of managerial and professional logics. Before 1998, the teaching hospital was characterised by the dominance of medical professionals while the role of managerial personnel was limited. After the corporatisation process, the PME system was revised in order to better assist the managerial tasks that the general manager required medical managers to accomplish. The budget negotiation procedure was accepted by medical managers who showed enthusiasm, recognising the importance of efficiency as the main principle. In the period between 1998 and 2001, objective indicators slowly replaced subjective measures. Initially, objective indicators were successfully introduced although they did not become objectified because medical managers perceived them as the cause of excessive differentiation between both units and individuals. In particular, some of these indicators were linked to targets, considered unobtainable for the influence of uncontrollable factors. That was why objective indicators appeared to be insufficient for evaluation by medical managers. In fact, not all medical units followed the evaluation criteria indicated by the general manager, generating a lack of uniformity in the evaluation process between medical units. Some medical managers refused to apply objective indicators because they were unable to correctly differentiate the individual contribution of physicians in their unit. This was probably due to the way physicians worked in their unit. The evaluation process did not fit a team work model and this aspect generated dissatisfaction in medical managers.

In the following years, the incumbent general managers tried to improve the evaluation process by including qualitative indicators in 2003, when the 'Department projects' procedure was adopted. Only in 2009 did the general manager grant the request of some medical managers by associating subjective indicators to objective ones. The introduction of subjective indicators can be interpreted as an effort towards the convergence of professional and managerial logics.

Not all procedures introduced were objectified. For example, 'Pool competition' was abandoned after one year. This procedure redefined bonus distribution pooling medical units on the basis of their homogeneity. According to the managerial logic, the pool competition was intended to improve individual productivity. Moreover, the pool competition procedure considered the peculiarities of the units improving the individual productivity measurement. The evaluation committee appreciated this procedure because it improves '*performance in terms of volume of revenue, accounting for direct costs and care-service complexity*'; however, the committee underlined that this pool mechanism was not prescribed for all units and could generate negative consequences. What happened from professional view was a failure of pool competition procedure. The general manager and physicians espoused diverging interpretations of the usefulness of pools. This divide became even more obvious in the interview with the manager of Digestive System Surgery unit and a physician of Cardiology unit. The former stressed that medical managers perceived unequal treatment, disapproving of a procedure that nourished dissatisfaction. Basically, it was impossible to compare performance of physicians, who had the same organisational position, because they 'missed the final race'. Besides, the pool competition procedure did not take into account the seasonal effect, forgetting one important aspect of medical professional activities.

As well as the pool competition procedure, 'Drug budgeting' was abandoned. The aim of this procedure was to reduce the consumption of drugs. The introduction of this procedure was forced by the Region requiring the teaching hospital's incentive system to be linked to the goal of budget containment. The most important reason for abandoning this procedure was linked to deontological problems. The physicians felt bound by the drug containment not only from financial aspect but also from professional one. In many cases, when the mission of physicians was to guarantee the quality of healthcare by preventing infections, they did not respect the drug containment because it was in conflict with their therapy decisions.

Overall, pool competition and drug budgeting procedures represent an explanation of the decoupling effect. That is, the professionals expressed avoidance as a response to PME system changes. They attempted to preclude the necessity of conformity (isomorphism), altering their goals and activities as prescribed by the general manager. Thus, decoupling is due to the divergence of managerial and professional logics in the sense of conflicting interests and objectives.

CONCLUSIONS

This study has examined the dynamics of PME system changes as organisational responses to institutional pressures. In doing so, we have defined an integrative perspective from which analysing this phenomenon, drawing on Busco et al.'s (2007) framework as an interpretative scheme that embraces the NIS and OIE perspectives.

The adoption of Busco et al.'s framework has pushed us towards making some adjustment to the original contribution of the researchers, decomposing the coordinates 'how and why' and 'where and when'. This adjustment was due to our integration of Burns and Scapens' framework within that of Busco et al. In particular, understanding how change happens, in our view, means analysing change as a process of institutionalisation. The latter requires the period when change happened to be identified in order to better understand the transition from a procedure to another. That was the reason why we included 'when' in the coordinates 'how and why', leaving aside the coordinate 'where', which instead has allowed us to better focus on organisational responses to changes, and to understand when the managerial and professional logics converge or diverge. Thus, the main contribution of this study is the integration of NIS perspective and OIE one. The former has helped us to represent organisational responses to the external institutional pressures and, in particular, to interpret change through isomorphism and decoupling. The latter has allowed us to explain change as a process of institutionalisation.

Another contribution of our study is given by the adoption of Burns and Scapens' framework in our case analysis. In particular, we also have modified the original contribution of the two authors substituting the arrow 'd' labelled as 'institutionalisation' with the concept of objectification, according to the previous contribution of Barley and Tolbert (1997). Given this adjustment, our case findings have permitted us to show PME as a package composed of rules and routines 'as the non-situated and subject-less norms of conduct (formal or informal) that actors draw upon in the production and reproduction of practice' (Englund & Gerdin, 2008, p. 1128). So, the elements of the teaching hospital's PME package, rules and routines, have allowed us to understand how PME system shaped PME practice at a specific point in time. PME practice, through their cumulative influence over time, might be reproduced and objectified when the managerial and professional logics converged. Drawing on the NIS perspective, we interpret the convergence between the two logics as the convergence between coercive and normative isomorphism.

Otherwise, in some cases our field-based findings have made us aware of a further effect of change phenomenon like decoupling. In particular, we have observed that there are considerable differences between the adopted rules and the actual routines. This is explained by the divergence between the two logics, managerial and professional, which we interpreted as a divergence between the two types of isomorphism.

Future research in healthcare settings may investigate more deeply the effects of the specifics of medical units on physicians' perceptions of PME system changes. This point can shed light on further aspects that can facilitate or obstacle the introduction and institutionalisation of new PME systems, so enriching the Busco et al.'s framework.

ACKNOWLEDGMENTS

This paper results from the authors being involved into the national research project 'Strategic cost and performance management for the challenges of competitiveness and sustainability'- Progetto di RIcerca Nazionale (P.R.I.N.), financially supported by the Italian Government, directed by Lino Cinquini from Scuola Superiore Sant'Anna di Pisa and coordinated by Riccardo Giannetti from the University of Pisa and Rosa Alba Miraglia from the University of Catania.

REFERENCES

Abernethy, M. A., & Chua, W. F. (1996). A field study of control systems "redesign": The impact of institutional processes on strategic choice. *Contemporary Accounting Research*, *13*(2), 569–606.

Abernethy, M. A., & Stoenwinder, J. U. (1995). The role of professional control in the management of complex organisations. *Accounting, Organisations and Society*, *20*(1), 1–17.

Barley, S. R., & Tolbert, P. S. (1997). Institutionalisation and structuration: Studying the link between action and institution. *Organisation Studies*, *18*(1), 93–117.

Beckert, J. (1999). Agency, entrepreneurs, and institutional change. The role of strategic choice and institutionalized practice in organizations. *Organization Studies*, *20*, 777–799.

Berger, P. L., & Luckmann, T. (1966). *The social construction of reality: A treatise in the sociology of knowledge*. Garden City, NY: Anchor Books.

Burns, J., & Scapens, R. W. (2000). Conceptualizing management accounting change: An institutional framework. *Management Accounting Research*, *11*, 3–25.

Burns, J., & Vaivio, J. (2001). Management accounting change. *Management Accounting Research*, *12*, 389–402.

Busco, C., Quattrone, P., & Riccaboni, A. (2007). Management accounting issues in interpreting its nature and change. *Management Accounting Research*, *18*, 125–149.

Carruthers, B. M. (1995). Accounting, ambiguity, and the new institutionalism. *Accounting, Organisations and Society*, *20*(4), 313–328.

Chua, W. F., & Degeling, P. (1993). Interrogating an accounting-based intervention on three axes: Instrumental, moral and aesthetic. *Accounting, Organisations and Society*, *18*(4), 291–318.

Covaleski, M. A., Dirsmith, M. W., & Michelman, J. E. (1993). An institutional theory perspective on the DRG framework, case-mix accounting systems and health-care organisations. *Accounting, Organisations and Society*, *18*(1), 65–80.

Dambrin, C., Lambert, C., & Sponem, S. (2007). Control and change – Analysing the process of institutionalisation. *Management Accounting Research*, *18*, 172–208.

DiMaggio, P. J., & Powell, W. W. (1983). The iron cage revisited: Institutional isomorphism in organisational fields. *American Sociological Review*, *48*, 147–160.

DiMaggio, P. J., & Powell, W. W. (1991). The iron cage revisited: Institutional isomorphism and collective rationality in organisational fields. In W. W. Powell & P. J. DiMaggio (Eds.), *The neo institutionalism in organisational analysis*. Chicago, IL: The University of Chicago Press.

Doolin, B. (1999). Casemix management in a New Zealand hospital: Rationalisation and resistance. *Financial Accounting Management*, *14*(3–4), 397–417.

Englund, H., & Gerdin, J. (2008). Structuration theory and mediating concepts: Pitfalls and implications for management accounting research. *Critical Perspectives on Accounting*, *19*, 1122–1134.

Giddens, A. (1984). *The constitution of society*. Cambridge: Polity Press.

Jacobs, K., Marcon, G., & Witt, D. (2004). Cost and performance information for doctors: An internal comparison. *Management Accounting Research*, *15*, 337–354.

Johansson, T., & Siverbo, S. (2009). Why is research on management accounting change not explicitly evolutionary? Taking the next step in the conceptualisation of management accounting change. *Management Accounting Research*, *20*, 146–162.

Jones, C. S. (1999). Developing financial accountability in British acute hospitals. *Financial Accounting Management*, *15*(1), 1–20.

Kajüter, P., & Kulmala, H. I. (2005). Open-book accounting in networks: Potential achievements and reasons for failures. *Management Accounting Research*, *16*(2), 179–204.

Kasim Nor-Aziah, A., & Scapens, R. W. (2007). Corporatisation and accounting change. The role of accounting and accountants in a Malaysian public utility. *Management Accounting Research*, *18*, 209–247.

Kurunmäki, L. (2004). A hybrid profession – The acquisition of management accounting expertise by medical professionals. *Accounting, Organisations and Society*, *29*, 327–347.

Lapsley, I. (1994). Responsibility accounting revived? Market reforms and budgetary control in health care. *Management Accounting Research*, *5*, 337–352.

Lapsley, I., & Wright, E. (2004). The diffusion of management accounting innovations in the public sector: A research agenda. *Management Accounting Research*, *15*(3), 355–374.

Lehtonen, T. (2007). DRG-based prospective pricing and case-mix accounting – Exploring the mechanisms of successful implementation. *Management Accounting Research*, *18*, 367–395.

Lowe, A., & Doolin, B. (1999). Casemix accounting systems: New spaces for action. *Management Accounting Research*, *10*(3), 181–201.

Lukka, K. (2007). Management accounting change and stability: Loosely coupled rules and routines in action. *Management Accounting Research*, *18*, 76–101.

Malmi, T., & Brown, D. A. (2008). Management control systems as a package – Opportunities, challenges and research directions. *Management Accounting Research, 19*, 287–300.
Modell, S. (2001). Performance measurement and institutional process: A study of managerial responses to public sector reform. *Management Accounting Research, 12*, 437–464.
Oliver, C. (1991). Strategic responses to institutional processes. *The Academy of Management Review, 16*(1), 145–179.
Pettersen, I. J. (1995). Budgetary control of hospitals—Ritual rhetorics and rationalized myths? *Financial Accounting Management, 11*(3), 207–221.
Pettersen, I. J. (2001). Hesitation and rapid action: The new public management reforms in the Norwegian hospital sector. *Scandinavian Journal of Management, 17*, 19–39.
Quattrone, P. (2006). The possibility of the testimony. A case for case study research. *Organisation, 13*(1), 143–157.
Scapens, R. W. (1994). Never mind the gap: Towards an institutional perspective of management accounting practices. *Management Accounting Research, 5*(3–4), 301–321.
Scapens, R. W. (2004). Doing case study research. In E. Christopher & B. Lee (Eds.), *The real life guide to accounting research*. Oxford, UK: Elsevier Ltd.
Scott, W. R. (2001). *Institution and organisations*. Thousand Oaks, CA: Sage Publications, Inc.
Shachar, R., & Eckstein, Z. (2007). Correcting for bias in retrospective data. *Journal of Applied Econometrics, 22*, 657–675.
Van de Ven, A. H., & Poole, M. S. (1995). Explaining development and change in organisations. *Academy of Management Review, 20*(3), 510–540.
Weick, K. E. (1979). *The social psychology of organizing* (2nd ed.). Reading, MA: Addison-Wesley.
Weiner, S. L., Maxwell, J. H., Sapolsky, H. M., Dunn, D. L., & Hsiao, W. C. (1987). Economic incentives and organisational realities: Managing hospitals under DRGs. *Milbank Quarterly, 65*(4), 463–487.

ACCOUNTING INFORMATION AND PERFORMANCE MEASUREMENT IN A NONPROFIT ORGANIZATION

Luís Francisco and Maria-Ceu Alves

ABSTRACT

Purpose – *Accountability and performance measurement have become an important and urgent subject for nonprofit organizations (NPOs) as they encounter increasing competition from others agencies, all competing for governmental funding. This case study investigates how multiple stakeholder interests are incorporated into information systems to provide stakeholders with useful information, which contribute and make easier the assessment of the organization performance.*

Methodology/approach – *The paper draws on a combination of in-depth interviews with NPO managers, extensive documentary analysis, and participant observation in order to understand and explain how managers use accounting information. The study was conducted as a qualitative single case study of a large Portuguese NPO. Managers were inquired about their information uses and needs to evaluate the NPO performance and make decisions. The result of these interviews was complemented with the analysis of several documents of the NPO.*

Performance Measurement and Management Control: Global Issues
Studies in Managerial and Financial Accounting, Volume 25, 465–487

ISSN: 1479-3512/doi:10.1108/S1479-3512(2012)0000025020

Findings – *The subject of performance measurement for nonprofit organizations is extensive but generally inconclusive, because NPO lack the simple elegance of a financial measure, such as profitability, used by for-profit organizations to assess their performance. In such a context, this research provides empirical insights, and allows us to propose a framework that complements financial indicators with nonfinancial indicators.*

Research limitations/implications – *Because of the chosen research approach, our results may not be generalized and therefore researchers are encouraged to test the proposed framework.*

INTRODUCTION

Nonprofit organizations (NPOs) have grown both in number and importance over the past decades (Carvalho, 2010; Franco, Sokolowski, Hairel, & Salamon, 2005; Salamon, Sokolowski, & List, 2003). They are the result of citizens' initiatives to meet needs that are not covered by business organizations (for profit) or by governmental organizations (Weisbrod, 1975), but they also come from an entrepreneurial spirit based on ideological, religious, altruistic, or charitable reasons (Young, 1981). Each NPO relates itself and interacts with a range of people/organizations interested in it – the stakeholders – who are affected or may affect the achievement of objectives (Freeman & McVea, 2001; Freeman, Harrison, Wicks, Parmar, & De Colle, 2010) and have different perspectives on the organization (Barrett, 2001; Clarkson, 1995; Herman & Renz, 1997, 1998, 2004; Mitchell, Agle, & Wood, 1997; Wisniewski & Stewart, 2004).

For decision-making purposes, stakeholders need to assess the performance of the NPO with whom they relate, and it is relevant that performance evaluation might be done through financial and nonfinancial indicators (Connolly & Hyndman, 2000, 2004; Eccles, 1991; Hyndman, 1990; Hyndman & Anderson, 1995; Ittner & Larcker, 1998a; Martin & Kettner, 1997; Moxham & Boaden, 2007; Zimmermann & Stevens, 2006). In the literature review it appears that this issue is not clearly recognized and is treated in an inconsistent manner. According to Kaplan (2001), the literature on performance measurement in NPOs is extensive but generally inconclusive. This study intends to fill this gap by addressing the questions: What processes are used by stakeholders to evaluate NPO performance? What kind of information is used?

The remainder of the paper is structured as follows. Second section reviews the literature highlighting some contributions from the stakeholder theory, in the study of NPO information systems and performance evaluation. Third section contains a description of the methodology applied to the empirical study. An exploratory case study in a Portuguese Scout Organization – *Corpo Nacional de Escutas* (CNE) – was done to achieve our goals and answer the research questions. The last two sections explain and discuss the results obtained, present the conclusions drawn from the research, identify the limitations of the study, and provide directions for future research.

LITERATURE REVIEW

NPO and Performance Evaluation

Most researchers that work within the NPO sector assume that the deepest reason for an organization's existence lies in its mission, and this is what leads NPOs in strategic terms (Drucker, 1989; Niven, 2003, 2005; Weisbrod, 1998). NPOs will never resemble businesses organizations that can measure their success in purely economics terms (Sawhill & Williamson, 2001). In for-profit organizations the earnings, as return on capital invested, can justify its existence and thus be an indicator of their best or worst performance. But, even these organizations have recently recognized that financial measurements alone are inadequate for measuring and managing performance (Kaplan, 2001). To NPOs, the existence of profit does not have the same meaning and may even be an indicator of poor performance. NPOs are driven by their mission, not by the profit. "Success for nonprofits should be measured by how effectively and efficiently they meet the needs of their constituencies. Financial considerations can play an enabling or constraining role but can rarely be the primary objective" (Kaplan, 2001, p. 353).

Several authors (Barman, 2007; Zimmermann & Stevens, 2006) claim that NPOs should ensure effective organizational performance management and accountability to their stakeholders, using financial and nonfinancial indicators. Use of financial and nonfinancial indicators by NPOs to evaluate performance is crucial, from both the management and stakeholders perspectives. If business organizations evaluate performance with financial information, presented in the financial statements, NPO statements only inform about the financial position, the costs and revenues related to the activities and its comparison with budget (Charity Commission, 2005). So, this information alone is insufficient to understand the actual NPOs performance, since these organizations usually have specific purposes that do not fit in a straightforward way in the economical and financial scope. Still, although nonfinancial performance reporting should have a prominent role in the overall reporting practices of NPOs, financial reporting cannot be discarded as being totally irrelevant for their functioning, not only in terms of accountability, but also because of empirically observed effects on the amount of funds raised, these funds being necessary to enable NPO to pursue its objectives (Jegers, 2010).

To the extent that NPOs gain dimension, their activity increases and their impact grows, the assessment of the economy, efficiency, and effectiveness acquire relevance and become more complex to execute (Sawhill & Williamson, 2001). However, if there isn't a defined strategy to achieve the mission of the organization, with clear and communicated objectives, if there isn't a planning, it is difficult to achieve alignment of the organization as a whole and assess its performance (Crittenden, Crittenden, Stone, & Robertson, 2004; Poister, 2003). Therefore, as Cutt (1998) argues, systems for measuring NPOs performance must be used not only for a retrospective report but also prospectively, supporting the continuity in management, in a strategic way. Thus, to measure NPO performance it is required to take into account a multiplicity of components to provide information for internal and external requests, satisfying the managers' information needs as well as all the stakeholders of the organization (Cutt, 1998).

Recently, Salamon, Geller, and Mengel (2010) suggested that NPOs have firmly imbibed the mantra of performance measurement, being recognized for its relevance in the management and provision of information to stakeholders. At the same time, NPOs are struggling to implement performance measurement in the midst of competing pressures for limited resources of time and money. These authors (Salamon et al., 2010) highlight the difficulty in measuring and quantifying the NPO outcomes, often intangible. And, they also indicate as main requirements for improving the measurement of NPO performance: the availability of tools able to measure the qualitative outcomes, the less time consuming in using these tools, and the existence of features that support the function of measurement. Overall, the NPOs performance evaluation remains a contentious issue with little consensus (Kasperskaya, 2006).

In more recent research, the *Balanced Scorecard* (Kaplan & Norton, 1992, 1996) stands out as the performance evaluation tool more valued and mentioned in literature (Marr & Schiuma, 2003). This system has evolved from its early use as a simple performance measurement framework to a full strategic system, being used as a communication tool with stakeholders. The *balanced scorecard* uses the logic of causality, through chains of cause and effect, making the performance more measurable, observable, reportable, and generating referrals for the management of NPOs (Kasperskaya, 2006).

Stakeholders Theory

To study performance evaluation and reporting to NPO stakeholders, it becomes relevant to outline the theoretical framework within the scope of stakeholder theory and its developments.

Stakeholder theory has been developed from the mid eighties of the last century, in view of the strategic management of organizations. Freeman (1984) is widely recognized as being the founder of stakeholder theory; even though he points out that many of the key concepts go back to the 1960s. According to stakeholder theory, when organizations formulate their strategy, they should take into account not only the satisfaction of the shareholders but also the endorsement of all groups or individuals who are affected or may affect the achievement of the organization goals (*stakeholders*) (Friedman & Miles, 2006). "Stakeholders are recognized as being of particular importance in public and nonprofit organizations, which commonly have a more diverse group of stakeholders than private for-profit organizations making it more difficult to identify strategic issues" (Fletcher, Guthrie, Steane, Roos, & Pike, 2003, p. 509).

According to Donaldson and Preston (1995), stakeholder theory differs from other theories of the firm because it views the organization as an entity through which numerous and diverse participants accomplish multiple, and yet not always entirely congruent purposes. So, the stakeholder theory cannot be considered a single theory, but rather a set of theories for the management of stakeholders (Mainardes, Alves, & Raposo, 2011).

From performance evaluation and reporting on corporate social responsibility perspectives, stakeholder theory has served as reference to several works in the accounting area (Friedman & Miles, 2006) including

- the model proposed by Atkinson, Waterhouse, and Wells (1997) that suggests the monitoring of contractual relationships with different stakeholders;
- the "*performance prism*" (Neely, Adams, & Kennerley, 2002);
- some studies on performance evaluation in the public sector, were authors seek to align the performance of institutions with the stakeholders requests (McAdam, Hazlett, & Casey, 2005; Moullin et al., 2007; Wisniewski & Stewart, 2004)
- stakeholders accountability relatively to corporate social responsibility and sustainability, as a voluntary initiative or as required by national and international regulations, with the use of nonfinancial indicators (Belal, 2002; Eccles, Herz, Keegan, & Philips, 2001; Ittner & Larcker, 1998b; Perrini & Tencati, 2006);
- the use of integrated systems of accountability to stakeholders (Farkas & Molnár, 2005).

From an instrumental perspective (Donaldson & Preston, 1995), through accounting and management tools such as the *balanced scorecard* (Kaplan & Norton, 2001a, 2001b), the *performance prism* (Neely et al., 2002), the *tableau de bord* (Bourguignon, Malleret, & Nørreklit, 2004; Lebas, 1994) the consideration of the stakeholder interest and satisfaction has been a major issue. The instrumental perspective "explores how the stakeholder model may be used to strategic decision making, where certain results derive from enacting certain behaviours" (Mainardes et al., 2011, p. 234). In search of good results, the management of relationships with various stakeholders and the inherent need of information for proper decision making – by providing inside information – has led to the fact that stakeholders theory is at least implicit in processes such as "total quality management," "management of customer satisfaction," "participative management," "management of multifunctional teams," "management partnerships," and other (Friedman & Miles, 2006).

Information Disclosure and Accountability in NPO

For decision-making purposes, stakeholders need to regularly assess NPO performance and, to achieve that they need proper information. Nowadays, business organizations are recognizing the relevance of nonfinancial indicators to provide information to stakeholders. These complement the financial information and provide data that facilitate a better measurement of the value and importance of the organization (Eccles, 1991; Ittner & Larcker, 1998a).

While traditionally accounting is designed to express the ability to generate assets for shareholders, some international organizations that regulate the accounting principles and practices have given guidance on accounting information to serve the stakeholders interests. Proof of this is the support of organizations like the *International Federation of Accountants* (IFAC), *American Institute of Certified Public Accountants* (AICPA), *Association of Chartered Certified Accountants* (ACCA), *Canadian Institute of Chartered Accountants* (CICA) to documents that envisage the disclosure of accounting information to stakeholders (CIM, 2001; ACCA, 2005).

Recently, the IASB published its revised conceptual Framework (IASB, 2010) that was preceded by a debate between two objectives of financial reporting: decision usefulness and accountability. According to the IASB the primary users of financial reporting are present and potential investors, lenders, and other creditors, who use that information to make decisions about buying, selling or holding equity or debt instruments, and providing or settling loans or other forms of credit. However, the framework also notes that other parties, including market regulators, may find financial reports useful.

In Portugal, a recent Decree-Law (No. 36-A/2011 of March 9) allowed the creation of accounting rules to the NPO sector. This document argues that the creation of these rules is justified for two main reasons: first, due to the spread that these entities have come to know the important role they play in economy, which justifies strengthening disclosure requirements for the activities they undertake and the resources they use, including the obligation to provide reliable information on the management of resources entrusted to them, as well as on their achievements in developing their activities. Secondly, because the entities that comprise the nonprofit sector meet objectives of general interest that transcend the productive and sales activity of products or services.

The disclosure of information to outside organizations has evolved to reports that serve the interests and needs of different stakeholders (Friedman & Miles, 2006), going beyond the traditional economic and financial perspective, and advancing to contexts such as usually framed by the names as "environmental accounting," "social accounting," "sustainable development," "corporate and social responsibility," among others.

Concerned with the need for public organizations to provide performance information to stakeholders, Wisniewski and Stewart (2004) developed an instrument similar to the "performance prism." The "performance information portfolio" is a set of performance information about a particular public service provided by it, and necessary to enable the key stakeholders to assess it. The "performance information portfolio" as means of reporting can serve simultaneously not only for the stakeholders' performance evaluation but also for the traditional performance evaluation carried out by managers within the organization. The same need for information is posed for NPOs. "Just as for-profit organizations are accountable to the people that own them, nonprofit organizations ought to be accountable to the communities that support them" (Barrett, 2001). However, these organizations lack a common denominator, as is the profit for business organizations that could facilitate the task of developing models applicable to the generality of NPOs. However, there are some studies (Ebrahim, 2010; Molnár, 2008) that while recognizing the need for more research, advocate standards of providing information that can serve the generality of NPO stakeholders.

In NPO reports, the financial statements are interesting but seem to be far from satisfying their stakeholders, not least because the financial accounting and auditing are not very popular among people related to NPOs (Jegers, 2002). To meet the information needs of NPO stakeholders, additional information is required besides the one stated in the financial statements. So, management accounting information, which encompasses both financial and nonfinancial indicators, may be useful in providing information to NPO stakeholders.

This gap is recognized in recent studies like Salamon et al. (2010) and Ebrahim (2010). In the first study, it was pointed out that the main difficulties in measuring NPOs performance are the lack of: knowledge about appropriate tools for measuring performance; staff experience and expertise; and consensus on the best technique for assessing performance. In the second study, it is suggested that the difficulties in accountability are: the impossibility of satisfying all stakeholders, with all the information they want; the existence of multiple objectives/goals that are expected to meet accountability; the need to show the completion of the NPO mission; the influence of accountability in misrepresenting the NPO mission; and the choice of mechanisms of accountability that adequately serve the organization and its stakeholders.

Resolving Multiple Stakeholder Interests

Some features of NPOs make the assessment of their performance difficult.

- As it is recognized in the literature (Herman & Renz, 1998; Weisbrod, 1975, 1988), NPOs in general have different purposes, different operational modes, and are guided by their specific mission.
- Each NPO relates to and interacts with a set of stakeholders who are affected or may affect the achievement of the organization goals (Freeman & McVea, 2001). These stakeholders have different perspectives, interests, legitimacy, power, and influence over the various NPOs (Sowa, Selden, & Sandfort, 2004).

These difficulties are also recognized by Andreaus, Costa, and Ramus (2009) that point out that the "*mission-oriented nature*" of NPOs and their "*multi-stakeholdership*" are the two characteristics that make difficult the definition and development of models of accountability. So, each NPO turns out to be a particular case and the relationship with each stakeholder has its own characteristics.

In NPO accountability literature (Ebrahim, 2003a, 2003b, 2005; Fry, 1995; Kearns, 1994, 1996), the public relevance of performance reporting to NPO stakeholders is recognized. However, in more recent studies (Ebrahim, 2010; Molnár, 2008) the definition of standards for NPO reporting is still open, in particular for performance evaluation and decision making. It is within this area that we intend to contribute with this research.

RESEARCH AND METHODS

A case study research design was adopted to develop a better understanding on how stakeholders evaluate NPO performance and what accounting and management information is used.

When one wants to explain a social reality still little studied, complex, trying to understand it and, whose study requires detail, conducting a qualitative research appears to be the most recommended since the qualitative research can explore in a broader social context the phenomenon under study, aiming to understand the players, their reasons, needs, desired outcomes, and also the ways how the social processes, institutions and relationships between the parties work (Mason, 2002).

The case study is recommended when the phenomenon in question is not clearly distinguishable from the context where it belongs, when issues of "how" of "why" or even "which" emerge, when there is an advantage or necessity to use multiple sources of evidence, when it comes to a contemporary phenomenon with a real context identified (Dul & Hak, 2008; Yin, 2003). So, the case study seems to be the appropriate option and the kind of case study that seemed to fit best is the exploratory (Ryan, Scapens, & Theobald, 2002).

The organization chosen for this study is the catholic Scout association of Portugal (Corpo Nacional de Escutas – CNE) that was established in 1923. With nearly 70,000 Scout members, it is the largest youth association in Portugal. Its action is supported by 52 regional and sub-regional structures and one national based in Lisbon. The CNE is a founding member of the World Organization of the Scout Movement, the International Catholic Conference of Scouting, as well as of the Portuguese Scout Federation (http://www.cne-escutismo.pt, accessed on February 10, 2012).

It is a nonprofit association of young people integrated into the world organization of the scout movement for the integral education of young people, based on the method created by Baden-Powell[1] and in the volunteerism of its members. CNE is organized in 20 Regions, each one having its own Regional Office. Some Regions have many groups, each of them is self organized into areas called "núcleo." Each group (at local level) includes four units corresponding to the scout sections – Cub Scouts, Explorers, Pioneers, and Rovers.

For data collection, based on Mason (2002), Yin (2003), Denscombe (2003), and Gillham (2005) the selected techniques were "desk research," "interviews," and "participant observation."

The fieldwork to be carried out with the CNE can be divided into two phases. A first phase, already completed, in which we analyzed documentation and interviewed people in charge belonging to different executive bodies – from a national and regional level – within this association in order to characterize the accounting and management information system of the CNE. A second phase, still underway, in which we want, through the documentation review and interviews with various stakeholders of the CNE, to identify criteria and indicators that enable them to assess the performance of the CNE. The first phase of fieldwork is presented here. The ongoing second phase will be disclosed later. As stated in Yin (2003), case studies, especially the exploratory ones keep on evolving and their advancement determines and conditions the options for research into new phases of the study.

In this first phase of the study 40 leaders of the CNE were interviewed. We selected the main executive officers belonging to the different territorial levels – National Head, Regional Heads, Heads of "Núcleo," and Heads of Group – as it was felt that this selection would get an overall view of the need and use of accounting information in the CNE and a vision, more specifically, for each regional level. Respondents were thus two leaders at the national level, a volunteer (the National Scout Chief), a professional (responsible for the administrative/financial area), and 10 representatives from local leaders. The rest were representatives of regional or core (Núcleo) levels, of which two were professionals.

A total of 40 interviews were conducted. Each of them lasted, on average, one hour. All interviews were tape-recorded and transcribed into written form. It was important to have open-ended questions to ensure that interesting empirical aspects were not missed and to adapt the interview to the expertise of each interviewee without losing the overall direction (Irvine & Gaffikin, 2006).

The empirical data is based on interviews and paperwork in the form of financial and nonfinancial reports, manuals, and minutes of meetings. Paperwork cannot be ignored since "documents construct a particular view of reality, sometimes designed to put forward a case, justify a position, or present an image" (Irvine & Gaffikin, 2006, p. 131). To ignore this sort of documentation would be to distort the context in which organizational behavior occurred.

The researcher, due to the fact of knowing and engaging in the activities of the association under study, is himself a source of evidence, and permits Yin (2003) and Mason (2002) a "participant observation," relevant in the selection and interpretation of other sources of evidence. In conducting interviews, it is an advantage to be a participant observer first, in order to support the challenge of creating useful questions. According to Irvine and Gaffikin (2006), participant observation involves the collection of data in its natural setting and permits the researcher to understand a situation to an extent not entirely possible using only the insights of others obtained through interviews.

DISCUSSION OF RESULTS

Based on interviews and documents analysis, it was possible to collect a set of information about the information system used in the CNE.

Information System Characteristics

- All respondents claim that the information received is reliable.
- Information overload does not exist and the information received is considered by respondents as relevant and useful.

- The majority of respondents considers most useful and relevant periodic information and focuses on cash, including cash balances and banks movements. Some of the respondents consider as being useful comparisons between monthly budgeted costs and actual costs of different departments.
- However, not everyone gets this information so it is precisely here that gaps are identified by most respondents. Many believe that these comparisons between costs incurred and budgeted by departments and activities – namely, cost centers, activity centers, or centers of responsibility – represent the main information that they do not receive and would like to receive.
- Another gap identified concerns the Scout Shops activity. It would be important to get information on products or product type, such as "stock rotation" and "breakeven point."
- Manager involvement in planning and designing the accounting system is not very large, especially by those who have not mastered accounting and financial knowledge.
- Given the small size of many of the organizational units – local, "núcleo" and regional – most managers obtain the necessary accounting information directly, or ask for it to superiors. Moreover, the small number of financial movements helps to obtain the necessary information.
- Interviewees tell us that due to the small size of most of the activities and direct involvement of the people in charge, they can easily access the information they need. To do that, as some of the interviewees said: "meetings – even informal – are chosen over papers."

Methods and Means of Information Processing

- As one moves from local to national level, an increasing sophistication in the methods and means used in the treatment of accounting and management information can be seen.
- However, this sophistication of methods mainly depends on the capabilities and skills of those involved in treatment, at each level.
- In general, accounting work is done by computer and, the system used varies ranging from simple spreadsheets to modular systems.
- The staff dealing with accounting information is almost entirely staffed by volunteers who present different levels of competence. The National Office and some Regional Offices have professional staff attached to the finance and accounting area (usually part-time or retainer), but the oversight and executive responsibility is always voluntary.

- In terms of *accounting information system* there are some differences in the uses of the diverse territorial levels. Local groups have a "cash basis" accounting and current control of cash and banks accounts. The Regional Office financial accounting system is based on "accrual basis." The National Office has an organized financial accounting system and a management accounting system, which allows budgetary control and the management of revenue and costs by responsibility centers.

Users of Accounting, Financial and Management Information

- CNE's own leaders are the key stakeholders in accountability, management reports, plans, and budgets.
- There are other users of accounting information from the CNE. Among them are the donors who are the most representative and public organizations that financially support the CNE – such as the Portuguese Youth Institute and the municipalities.

Main Sources of Revenue and Expenses

- The main CNE sources of financial and material resources are the parents of young scouts, the government, the leaders, several Sales, sponsorship of businesses, and public entities for activities/projects; campaigns.
- The main expenses of the CNE are related to activities and structure.

Critical Success Factors and Management Indicators

- The quality of the board of directors was the key factor referred to as the most crucial to the success of the CNE. Respondents believe that it comes as the result of a careful recruitment and training.
- The proposed educational program and its implementation represent another key factor for the success of the CNE.
- The third factor is the quality of activities proposed, including the ability to leave these references in the young people who experienced them and in the sense of values perpetrated by the Scouts.
- The "Scout spirit" lived in meetings and activities, in the relationship between Scouts, and even in relation with the outside world and was mentioned as another key factor.

- In the management of the CNE the most useful indicators are: expenses by each Scout involved; revenues by Scout involved; the number of hours of activity; number of scouts per chief; the average number of hours that each leader gives to the CNE; the percentage of financial participation of young people (parents); the percentage of support received (for entity); the number of scouts per section, number of incoming and outgoing scouts per section and per year; the distribution of scouts in each stage of the progress system; the leaders' hours of training in total and per course, the percentage of leaders receiving training per year, the ratio number of scouts/number of young people; the average training cost per leader; Scout Shop average sales per scout and per group.

Strategic Planning and Performance Evaluation

- Evaluation and planning are part of the practice and a Scout tradition. Almost all respondents believe that among scouts reflection on what has been done, on the needs and how to meet those needs is something common. However in some organs and in certain circumstances the routine works, the same things are done, there is no perspective of new goals and the lack of time, that comes from the fact that the people involved are volunteers, is the main factor to work within routine parameters and only the most important things get to be done. It is also recognized the lack of systematization, organization and training in this area.
- The Plan and Budget, required for all levels of the CNE, have been instrumental in planning the activities of the CNE. However, in many cases, Plan and Budget are designed to meet rules more than as a result of effective strategic thinking.
- Performance evaluation of departments and their managers, such as the definition of objective is usually done in an informal way and through close monitoring of the evaluator.
- Just some Regional Offices and the National Office have a cost accounting system which allows cost control and analysis by "cost center" or "responsibility center".

Management Control and Regulatory Agencies

- The Audit Committees established at the CNE for the national and regional levels, have been working based on trust in managers, following the decisions and accomplishments. However, for the majority of respondents at regional level the Audit Committees should have a more active role in aspects such as regular monitoring of accounts, monitoring the action and advice to managers.

However the CNE has its own characteristics that must be met in developing a management information system. Thus it is necessary to consider:

- The financial and administrative autonomy that each territorial level has.
- The fact that almost all the leading members are volunteers and therefore their availability is limited and uncertain.
- The diversity of academic and professional experiences of the managers who are responsible for the different organs and functions.
- The natural tendency, to be given primacy to what is the purpose of the organization-training and education of young people-as opposed to administrative and financial aspects.
- Breaks in the management of different organs as a result of the respective holders being elected for three years.
- The need to standardize the provision of information within the CNE as a whole and even within each territorial level
- The CNE activities are based on scouting projects and suppose the achievement of strategic objectives.

For the system's suitability to the CNE needs it must be: integrated; flexible; consolidated; understood by "nonaccountants"; continuous; useful and dynamic; integrated with the strategy.

The CNE management information system shall be constituted of a set of summary documents issued within the organization, serving the most responsible and interested in managing it. These documents should summarize, using indicators and summaries of reports, the direction of the organization. They should serve to inform, assess, and decide on the evolution of the organization, particularly on the resources, processes and outcomes.

The main documents should circulate among all regional departments. These can be important elements in support of internal communication, analysis and information processing, and ultimately to decision making. Furthermore they are also a source for information and communication with the outside of the CNE.

The management information system should make the bridge between different levels of decision making-strategic, tactical and operational-that take place at national, regional, and local core of the CNE. It should serve as a tool to:

- Transform strategy into action.
- Communicate the strategy to all the levels of the organization.
- Align from the top to the bottom of the organization the implementation of the strategy, linking it with the tactical decisions and operational decisions.

Performance Evaluation and Stakeholder's Accountability

In the second phase of fieldwork (in progress) we seek to identify criteria and indicators with which each stakeholder evaluates the performance of the CNE, in order to study the possibility of developing a model of performance evaluation that serves the information needs from stakeholders-internal and external to the CNE.

As a result of the fieldwork carried out, sixteen main categories of stakeholders were identified: parents, Scouts, former Scouts, Catholic Church, government agencies, municipalities, partners, sponsors, benefactors, suppliers, employees, officers of Unit, Executive Officers, Representatives, World Organization of Scouting and peers.

The current reporting of the CNE to its stakeholders on its performance is typically done through five parts:

- The management reports issued by executive bodies of the different territorial levels;
- Institutional data disclosed in various ways;
- The media owned by the association;
- Data transmitted through conversations and formal or informal meetings;
- The information requested by the stakeholder.

This reporting is not structured or standardized, to allow a direct assessment of the CNE performance for its stakeholders. The information provided, which appears to be more structured and standardized is the financial one and it is accomplished through the so called Scout Integrated Information System.

SUMMARY AND CONCLUSIONS

Some studies point to the difficulty of reconciling the information of interest to the various NPO stakeholders (Sowa et al., 2004). This is due to the multiple sectors and operational methods of NPOs and, to the existence of different interests of various stakeholders regarding a concrete NPO.

Yet, recent studies (Ebrahim, 2010; Molnár, 2008), point toward the provision of integrated and standardized information, arguing it is possible to create information systems that generally satisfy the information needs of NPO stakeholders. Although each stakeholder has its specific preferences, it seems to be possible and within the interest of most stakeholders that there is a set of indicators that provide them information about the performance of a particular NPO.

Regarding the traditional provision of information through financial statements, the heterogeneity of NPOs generates complexity when applying the comparability of accounts. In studies such as Jegers (2002) it is recognized that the issue of standardization is an open field research and in this context it is advantageous for stakeholders that NPO managers complement the financial statements with other quantitative and qualitative data.

In Molnár (2008) and Ebrahim (2010) it is said that it is relatively easy to find patterns for "accountability" for different activities and it is possible to set universal standards that serve NPOs. With this purpose they identify six key areas or factors that influence the organizational effectiveness of NPOs: the Mission and existing policies, programs and services; the activities of government, the financial management, activities of fundraising, management of human resources and public relations activities.

In the case study of the CNE it is concluded that although with different perspectives on the organization, stakeholders are generally interested in knowing how valuable the organization is and its performance over a given period. The assessment of organizational performance is made through the evaluation of the results of their action/activity and the variation of its value. The results of research to date allow us to conclude the importance of accountability to stakeholders through the complementarity between financial and nonfinancial indicators.

Five topics of information of interest to stakeholders about the value of the CNE were defined:

1 Mission, Purpose, Programs, Services Rendered;
2 Governance, Governing Bodies, Moral and Ethical Principles;
3 Human Capital (Knowledge, Skills ...);
4 Finance, Fundraising, Infrastructure;
5 Public Image, Corporate and Institutional Framework.

Based on these topics stakeholders can have a generic view of the NPO value, through the variety of indicators that give information. They are then allowed to evaluate the performance of the CNE in the medium and long term.

It also found four topics of interest to inform stakeholders about the results of their action/activity of CNE, taking into account the stages of the process:

1 *Inputs* – Resources, Economy;
2 *Processes* – Activities, Efficiency;
3 *Outputs* – Immediate Results, Effectiveness;
4 *Outcomes* – Future Results, Impact.

The literature concluded that measurement systems developed in the field of accounting and management control, that traditionally serve for the assessment of organizational performance in an internal perspective (the management), are extensible, adaptable and useful for providing information with the same purpose to stakeholders in general (Figge, Hahn, Schaltegger, & Wagner, 2002; Moullin et al., 2007; Perrini & Tencati, 2006).

Although not all the information collected by a performance measurement system with internal focus has an interest to stakeholders in general, using and make profitable the traditional systems of performance management, will have advantages. It seems therefore that any system of performance measurement can be adapted to provide information for assessing the performance of the NPO for its stakeholders.

As recognized by some authors (Kasperskaya, 2006; Sawhill & Williamson, 2001), every organization is a case, and may deserve different treatment. There are no "best", models or systems of performance measurement. All are protected, deepened, adapted and even successfully applied. However most of the current research recognizes the primacy of the *balanced scorecard* as the most valued system of performance measurement and more referred to in literature (Marr & Schiuma, 2003). The same is recognized in the case study of the CNE.

Bearing in mind that the work presented is an exploratory case study, it is limited by the characteristics of the method that was followed. As the research has been based upon a single case study, it is necessary to exercise caution in generalizing the observations. So, essentially some questions and hypotheses for future investigations are raised. In this sense, we highlight a few points worthy of further research, including the identification of the most relevant information for the various categories of stakeholders and the identification of the most appropriate indicators for assessing the performance of the organization.

NOTE

1. Baden-Powell (1857–1941) in 1907 was the founder of scouting worldwide.

ACKNOWLEDGMENTS

The authors would like to thank the Portuguese Foundation for Science and Technology and the Research Unit NECE/UBI for providing financial support to this research.

REFERENCES

ACCA (The Association of Chartered Certified Accountants). (2005). *Improving stakeholder engagement reporting*. London, UK: Certified Accountants Educational Trust.

Andreaus, M., Costa, E., & Ramus, T. (2009). Accountability as a managerial tool in non profit organizations: Evidence from Italian CSV's. *Proceedings of the 2nd EMES international conference on social enterprise*, July 1–4, Trento, Italy.

Atkinson, A., Waterhouse, J., & Wells, R. (1997). A stakeholder approach to strategic performance measurement. *Sloan Management Review*, *38*(3), 25–37.

Barman, E. (2007). What is the bottom line for nonprofit organizations? A history of measurement in the British voluntary sector. *Voluntas*, *18*, 101–115.

Barrett, M. (2001). A stakeholder approach to responsiveness and accountability in non-profit organisations. *Social Policy Journal of New Zealand*, *17*, 36–51.

Belal, A. R. (2002). Stakeholder accountability or stakeholder management: A review of UK firms' social and ethical accounting, auditing and reporting (SEAAR) practices. *Corporate Social Responsibility and Environmental Management*, *9*, 8–25.

Bourguignon, A., Malleret, V., & Nørreklit, H. (2004). The American balanced scorecard versus the French tableau de bord: The ideological dimension. *Management Accounting Research*, *15*, 107–134.

Carvalho, A. (2010). Quantifying the third sector in Portugal: An overview and evolution from 1997 to 2007. *Voluntas*, *21*, 588–610.

Charity Commission. (2005). *Accounting and reporting by charities-statement of recommended practice* (*the charities SORP*). London, UK: Charity Commission Publications.

CIM-The Centre for Innovation in Management. (2001). *Measuring the business value of stakeholder relationships – Part one*. Vancouver: CIM.

Clarkson, M. (1995). A stakeholder framework for analyzing and evaluating corporate social performance. *Academy of Management Review*, *20*(1), 92–117.

Connolly, C., & Hyndman, N. (2000). Charity accounting: An empirical analysis of the impact of recent changes. *British Accounting Review*, *32*, 77–100.

Connolly, C., & Hyndman, N. (2004). Performance reporting: A comparative study of British and Irish charities. *British Accounting Review*, *36*, 127–154.

Crittenden, W., Crittenden, V., Stone, M., & Robertson, C. (2004). An uneasy alliance: Planning and performance in nonprofit organizations. *International Journal of Organization Theory and Behavior*, *7*(1), 81–106.

Cutt, J. (1998). Performance measurement in non-profit organizations: Integration and focus within comprehensiveness. *Asian Journal of Public Administration*, *20*(1), 3–29.

Denscombe, M. (2003). *The good research guide, for small-scale social research projects* (2nd ed.). Philadelphia, PA: Open University Press, McGraw-Hill Education.

Donaldson, T., & Preston, L. (1995). The stakeholder theory of the corporation: Concepts, evidence and implications. *Academy of Management Review*, *20*(1), 65–91.

Drucker, P. (1989). What business can learn from nonprofits. *Harvard Business Review*, *July–August*, 88–93.

Dul, J., & Hak, T. (2008). *Case study methodology in business research*. Oxford, UK: Elsevier Ltd.

Ebrahim, A. (2003a). Accountability in practice: Mechanisms for NGOs. *World Development*, *31*(5), 813–829.

Ebrahim, A. (2003b). Making sense of accountability: Conceptual perspectives for northern and southern nonprofits. *Nonprofit Management & Leadership*, *14*(2), 191–212.

Ebrahim, A. (2005). Accountability myopia: Losing sight of organizational learning. *Nonprofit and Voluntary Sector Quarterly*, *34*(1), 56–87.

Ebrahim, A. (2010). *The many faces of nonprofit accountability*. Working Paper 10-069, Havard Business School, USA.

Eccles, R. (1991). The performance measurement manifesto. *Harvard Business Review*, *January–February*, 131–137.

Eccles, R., Herz, R., Keegan, E., & Philips, D. (2001). *The value reporting revolution*. New York, NY: John Wiley and Sons.

Farkas, F., & Molnár, M. (2005). Towards a universal standard of nonprofit accountability: 'Standard of standards' in NGO accountability? *Proceedings of the first ISTR-EMES international conference*, April, Paris, France.

Figge, F., Hahn, T., Schaltegger, S., & Wagner, M. (2002). The sustainability balanced scorecard – Linking sustainability management to business strategy. *Business Strategy and the Environment*, *11*, 269–284.

Fletcher, A., Guthrie, J., Steane, P., Roos, G., & Pike, S. (2003). Mapping stakeholder perceptions for the third sector organization. *Journal of Intellectual Capital*, *4*(4), 505–527.

Franco, R., Sokolowski, S., Hairel, E., & Salamon, L. (2005). *The Portuguese nonprofit sector in comparative perspective*. Porto, Portugal: Catholic University.

Freeman, R. (1984). *Strategic management: A stakeholder approach*. Boston, MA: Pitman.

Freeman, R., & McVea, J. (2001). *A stakeholder approach to strategic management*. Retrieved from http://papers.ssrn.com/paper.taf?abstract_id=263511

Freeman, R., Harrison, J., Wicks, A., Parmar, B., & De Colle, S. (2010). *Stakeholder theory: The state of the art*. Cambridge, UK: Cambridge University Press.

Friedman, A., & Miles, S. (2006). *Stakeholders, theory and practice*. New York, USA: Oxford University Press.

Fry, R. (1995). Accountability in organizational life: Problem or opportunity for nonprofits? *Nonprofit Management & Leadership*, *6*(2), 181–195.

Gillham, B. (2005). *Research interviewing – The range of techniques.* Berkshire, UK: Open University Press, McGraw-Hill Education.

Herman, R., & Renz, D. (1997). Multiple constituencies and the social construction of nonprofit organization effectiveness. *Nonprofit and Voluntary Sector Quarterly, 26*(2), 185–206.

Herman, R., & Renz, D. (1998). Nonprofit organizational effectiveness: Contrasts between especially effective and less effective organizations. *Nonprofit Management & Leadership, 9*(1), 23–38.

Herman, R., & Renz, D. (2004). Doing things right: Effectiveness in local nonprofit organizations, A panel study. *Public Administration Review, 64*(6), 694–704.

Hyndman, N. (1990). Charity accounting: An empirical study of the information needs of contributors to UK fundraising charities. *Financial Accountability and Management, 6*(4), 295–307.

Hyndman, N., & Anderson, R. (1995). The use of performance information in external reporting: An empirical study of UK executive agencies. *Financial Accountability & Management, 11*(1), 1–17.

IASB-International Accounting Standard Board. (2010). *Conceptual framework.* IASB, London.

Irvine, H., & Gaffikin, M. (2006). Methodological insights – Getting in, getting on and getting out: Reflections on a qualitative research project. *Accounting, Auditing & Accountability Journal, 19*(1), 115–145.

Ittner, C., & Larcker, D. (1998a). Innovations in performance measurement: Trends and research implications. *Journal of Management Accounting Research, 10*(Fall), 205–238.

Ittner, C., & Larcker, D. (1998b). Are nonfinancial measures leading indicators of financial performance? An analysis of customer satisfaction. *Journal of Accounting Research, 36*(Suppl.), 1–35.

Jegers, M. (2002). The economics of non profit accounting and auditing: Suggestions for a research agenda. *The International Journal of Not-for-Profit Law 5*(1). Retrieved from: http://www.icnl.org/research/journal/vol5iss1/art_1.htm. Accessed on February 10, 2012.

Jegers, M. (2010). The effects of board-manager agency conflicts on non-profit organizations' earnings and cost allocations manipulations. *Accounting and Business Research, 40*(5), 407–419.

Kaplan, R., & Norton, D. (1992). The balanced scorecard—Measures that drive performance. *Harvard Business Review, January–February*, 71–79.

Kaplan, R., & Norton, D. (1996). *The balanced scorecard: Translating strategy into action.* Boston, MA: Harvard Business School Press.

Kaplan, R. (2001). Strategic performance measurement and management in nonprofit organizations. *Nonprofit Management & Leadership, 11*(3), 353–370.

Kaplan, R., & Norton, D. (2001a). Transforming the balanced scorecard from performance measurement to strategic management: Part I. *Accounting Horizons, 15*(1), 87–104.

Kaplan, R., & Norton, D. (2001b). Transforming the balanced scorecard from performance measurement to strategic management: Part II. *Accounting Horizons, 15*(2), 147–160.

Kasperskaya, Y. (2006). *Essays on causal performance measurement models.* PhD theses, Universitat Pompeu Fabra, Barcelona, Spain.

Kearns, K. (1994). The strategic management of accountability in nonprofit organizations: An analytical framework. *Public Administration Review, 54*(2), 185–192.

Kearns, K. (1996). *Managing for accountability: Preserving the public trust in public and nonprofit organizations*. San Francisco, CA: Jossey-Bass Inc.

Lebas, M. (1994). Managerial accounting in France: Overview of past tradition and current practice. *The European Accounting Review*, *3*(3), 471–487.

Mainardes, E., Alves, H., & Raposo, M. (2011). Stakeholder theory: Issues to resolve. *Management Decision*, *49*(2), 226–252.

Marr, B., & Schiuma, G. (2003). Business performance measurement – Past, present and future. *Management Decision*, *41*(8), 680–681.

Martin, L., & Kettner, P. (1997). Performance measurement: The new accountability. *Administration in Social Work*, *21*(1), 17–29.

Mason, J. (2002). *Qualitative researching* (2nd ed.). London, UK: Sage Publications.

McAdam, R., Hazlett, S., & Casey, C. (2005). Performance management in the UK public sector: Addressing multiple stakeholder complexity. *The International Journal of Public Sector Management*, *18*(3), 256–273.

Mitchell, R., Agle, B., & Wood, D. (1997). Toward a theory of stakeholder identification and salience: Defining the principle of who and what really counts. *Academy of Management Review*, *22*(4), 853–886.

Molnár, M. (2008). The accountability paradigm: Standards of excellence. *Public Management Review*, *10*(1), 127–137.

Moullin, M., Soady, J., Skinner, J., Price, C., Cullen, J., & Gilligan, C. (2007). Using the public sector scorecard in public health. *International Journal of Health Care Quality Assurance*, *20*(4), 281–289.

Moxham, C., & Boaden, R. (2007). The impact of performance measurement in the voluntary sector: Identification of contextual and processual factors. *International Journal of Operations & Production Management*, *27*(8), 826–845.

Neely, A., Adams, C., & Kennerley, M. (2002). *The performance prism – The scorecard for measuring and managing business success*. London, UK: Financial Times Prentice Hall.

Niven, P. (2003). *Balanced scorecard, step-by-step, for government and not-for-profit agencies*. Hoboken, NJ: John Wiley & Sons, Inc.

Niven, P. (2005). Driving focus and alignment with the balanced scorecard. *Journal for Quality and Participation*, *28*(4), 21–25.

Perrini, F., & Tencati, A. (2006). Sustainability and stakeholder management: The need for new corporate performance evaluation and reporting systems. *Business Strategy and the Environment*, *15*, 296–308.

Poister, T. (2003). *Measuring performance in public and nonprofit organizations*. The Jossey-Bass nonprofit and public management series. San Francisco, CA: Wiley.

Ryan, B., Scapens, R., & Theobald, M. (2002). *Research method and methodology in finance and accounting* (2nd ed.). London, UK: Thomson.

Salamon, L., Geller, S., & Mengel, K. (2010). *Nonprofits, innovation, and performance measurement: Separating fact from fiction*. Communiqué No. 17, Listening Post Project, Centre for Civil Society Studies, Johns Hopkins University, MD.

Salamon, L., Sokolowski, S., & List, R. (2003). *Global civil society—An overview*. Baltimore, MD: Center for Civil Society Studies Institute for Policy Studies, The Johns Hopkins University.

Sawhill, J., & Williamson, D. (2001). Mission impossible? Measuring success in nonprofit organizations. *Nonprofit Management & Leadership*, *11*(3), 371–386.

Sowa, J., Selden, S., & Sandfort, J. (2004). No longer unmeasurable? A multidimensional integrated model of nonprofit organizational effectiveness. *Nonprofit and Voluntary Sector Quarterly*, *33*(4), 711–728.

Weisbrod, B. (1975). Toward a theory of the voluntary non-profit sector in a three-sector economy. In Edmund S. Phelps (Ed.), *Altruism, morality and economic theory* (pp. 171–195). New York, NY: Russel Sage Foundation.

Weisbrod, B. (1988). *The nonprofit economy*. Cambridge, USA: Harvard University Press.

Weisbrod, B. (1998). The nonprofit mission and its financing: Growing links between nonprofits and the rest of the economy. In B. A. Weisbrod (Ed.), *To profit or not to profit – The commercial transformation of the nonprofit sector* (pp. 1–22). New York, NY: Cambridge University Press.

Wisniewski, M., & Stewart, D. (2004). Performance measurement for stakeholders: The case of Scottish local authorities. *The International Journal of Public Sector Management*, *17*(3), 222–233.

Yin, R. (2003). *Case study research – Design and methods* (3rd ed.). Thousand Oaks, CA: Sage Publications.

Young, D. (1981). Entrepreneurship and the behavior of nonprofit organizations: Elements of a theory. In M. White (Ed.), *Nonprofit firms in a three-sector economy*. Washington, DC: The Urban Institute.

Zimmermann, J., & Stevens, B. (2006). The use of performance measurement in South Carolina nonprofits. *Nonprofit Management & Leadership*, *16*(3), 315–327.